HADAMARD EXPANSIONS AND HYPERASYMPTOTIC EVALUATION

An Extension of the Method of Steepest Descents

The author describes the recently developed theory of Hadamard expansions applied to the high-precision (hyperasymptotic) evaluation of Laplace and Laplace-type integrals. This new method builds on the well-known asymptotic method of steepest descents, of which the opening chapter gives a detailed account illustrated by a series of examples of increasing complexity. A discussion of uniformity problems associated with various coalescence phenomena, the Stokes phenomenon and hyperasymptotics of Laplace-type integrals follows. The remaining chapters deal with the Hadamard expansion of Laplace integrals, with and without saddle points. Problems of different types of saddle coalescence are also discussed. The text is illustrated with many numerical examples, which help the reader to understand the level of accuracy achievable. The author also considers applications to some important special functions.

This book is ideal for graduate students and researchers working in asymptotics.

R. B. PARIS is a Reader in Mathematics at the University of Abertay, Dundee.

Encyclopedia of Mathematics and Its Applications

This series is devoted to significant topics or themes that have wide application in mathematics or mathematical science and for which a detailed development of the abstract theory is less important than a thorough and concrete exploration of the implications and applications.

Books in the **Encyclopedia of Mathematics and Its Applications** cover their subjects comprehensively. Less important results may be summarized as exercises at the ends of chapters. For technicalities, readers can be referred to the bibliography, which is expected to be comprehensive. As a result, volumes are encyclopedic references or manageable guides to major subjects.

All the titles listed below can be obtained from good booksellers or from Cambridge University Press. For a complete series listing visit

http://www.cambridge.org/uk/series/sSeries.asp?code=EOM

Hadamard Expansions and Hyperasymptotic Evaluation

An Extension of the Method of Steepest Descents

R. B. PARIS

University of Abertay, Dundee

CAMBRIDGE UNIVERSITY PRESS

CAMBRIDGE UNIVERSITY PRESS
Cambridge, New York, Melbourne, Madrid, Cape Town,
Singapore, São Paulo, Delhi, Tokyo, Mexico City

Cambridge University Press
The Edinburgh Building, Cambridge CB2 8RU, UK

Published in the United States of America by Cambridge University Press, New York

www.cambridge.org
Information on this title: www.cambridge.org/9781107002586

© R. B. Paris 2011

First published 2011

Printed in the United Kingdom at the University Press, Cambridge

A catalogue record for the publication is available from the British Library

Library of Congress Cataloguing in Publication data
Paris, R. B. (Richard Bruce), 1946–
Hadamard Expansions and Hyperasymptotic Evaluation : An Extension of the Method
of Steepest Descents / R. B. Paris.
p.   cm. – (Encyclopedia of Mathematics and its Applications ; 141)
Includes bibliographical references and index.
ISBN 978-1-107-00258-6 (hardback)
1. Integral equations – Asymptotic theory.   2. Asymptotic expansions.   I. Title.
QA431.P287   2011
$515'.45$–dc22
2010051563

ISBN 978-1-107-00258-6 Hardback

Contents

Preface

The aims of this book are twofold. The first is to present a detailed account of the classical method of steepest descents applied to the asymptotic evaluation of Laplace-type integrals containing a large parameter, and the second is to give a coherent account of the theory of Hadamard expansions. This latter topic, which has been developed during the past decade, extends the method of steepest descents and effectively 'exactifies' the procedure since, in theory, the Hadamard expansion of a Laplace or Laplace-type integral can produce unlimited accuracy.

Many texts deal with the method of steepest descents, some in more detail than others. The well-known books by Copson *Asymptotic Expansions* (1965), Olver *Asymptotics and Special Functions* (1997), Bleistein and Handelsman *Asymptotic Expansion of Integrals* (1975), Wong *Asymptotic Approximations of Integrals* (1989) and Bender and Orszag *Advanced Mathematical Methods for Scientists and Engineers* (1978) are all good examples. It is our aim in the first chapter to give a comprehensive account of the method of steepest descents accompanied by a set of illustrative examples of increasing complexity. We also consider the common causes of non-uniformity in the asymptotic expansions of Laplace-type integrals and conclude the first chapter with a discussion of the Stokes phenomenon and hyperasymptotics.

The next two chapters present the Hadamard expansion theory of Laplace and of Laplace-type integrals possessing saddle points. A study of these chapters makes it apparent how this theory builds upon and extends the method of steepest descents. Considerable emphasis is devoted to explaining the problems associated with coalescence phenomena, such as a saddle point coalescing either with another saddle point or with an endpoint of the integration interval. Methods for dealing with these difficulties in the Hadamard expansion procedure are carefully described. The monograph closes with sophisticated applications of the ideas developed in the earlier chapters to four particular special functions: the Bessel function $J_\nu(\nu x)$ of large order and argument, the Pearcey integral (a two-variable generalisation of the classical Airy function), the parabolic cylinder function $U(a, z)$ of large order and argument, and the logarithm of the gamma function.

In keeping with the last-mentioned text above, many of the examples in the later chapters are illustrated with numerical studies to better display the calibre of the asymptotic approximations obtained, a strategy that gives the non-expert practitioner a good sense of the method being showcased. This book should be accessible to anyone with a solid undergraduate background in functions of a single complex variable.

The author acknowledges the support of his institution, the University of Abertay, Dundee, which facilitated the writing of this book. A considerable debt of gratitude is owed to several colleagues who generously undertook a careful inspection of various sections and for their critical comments that have helped to improve the presentation of this text. The whole of Chapter 1 was read by N. M. Temme, with the first half of this chapter and Chapter 2 being read by T. M. Dunster; Chapters 2 and 3 were read by D. Kaminski, and C. J. Howls inspected the section on hyperasymptotics of Laplace-type integrals in Chapter 1. Finally, the non-specialist comments on the first part of Chapter 1 by J. S. Dagpunar were helpful. It is almost inevitable, however, that in spite of this careful examination some errors or misprints will have remained undetected, and the author requests the reader's forebearance for those that prove to be vexatious.

1

Asymptotics of Laplace-type integrals

In this opening chapter we present a detailed account, together with a series of examples of increasing complexity, of the classical method of steepest descents applied to Laplace-type integrals. Consideration is also given to the common causes of non-uniformity in the asymptotic expansions so produced due to a variety of coalescence phenomena. The chapter concludes with a brief discussion of the Stokes phenomenon and hyperasymptotics, both of which have undergone intense development during the past two decades. Such a preliminary discussion, as well as hopefully being of general interest in its own right, is necessary for the remaining chapters, since the Hadamard expansion procedure can be viewed as an 'exactification' of the method of steepest descents yielding hyperasymptotic levels of accuracy. Considerable space in the later chapters is devoted to showing how the Hadamard expansion procedure can be modified to deal with various coalescence problems.

1.1 Historical introduction

One of the most important methods of asymptotic evaluation of certain types of integral is known as the method of steepest descents. This method has its origins in the observation made by Laplace in connection with the estimation of an integral arising in probability theory of the form (Laplace, 1820; Gillespie, 1997)

$$I_n = \int_a^b f(x)\{g(x)\}^n dx = \int_a^b f(x)e^{n\psi(x)}dx \qquad (n \to +\infty).$$

Here $f(x)$ and $g(x)$ are real continuous functions defined on the interval $[a, b]$ (which may be infinite), with $g(x) > 0$ and $\psi(x) = \log g(x)$. Laplace argued that the dominant contribution to this integral as $n \to +\infty$ should arise from a neighbourhood of the point where $g(x)$ (or $\psi(x)$) attains its maximum value. In the simplest situation where $\psi(x)$ possesses a single maximum at the point $x = x_0 \in (a, b)$, so that $\psi'(x_0) = 0$, $\psi''(x_0) < 0$ and $f(x_0) \neq 0$, then $\psi(x)$ and $f(x)$ may be replaced

1

by the leading terms in their Taylor series expansion. In a small neighbourhood of length δ either side of $x = x_0$, we then find

$$I_n \simeq \int_{x_0-\delta}^{x_0+\delta} f(x_0) e^{n\{\psi(x_0)+(x-x_0)^2\psi''(x_0)/2\}} \, dx$$

$$\simeq f(x_0) e^{n\psi(x_0)} \left(\frac{2}{-n\psi''(x_0)}\right)^{1/2} \int_{-u_*}^{u_*} e^{-u^2} \, du,$$

where $u_* = \delta(-n\psi''(x_0)/2)^{1/2}$. Assuming δ is chosen such that $n^{1/2}\delta \to \infty$ as $n \to +\infty$, we can replace the integration limits in the last integral over u by $\pm\infty$ and evaluate the integral, to obtain Laplace's result

$$I_n \simeq f(x_0) e^{n\psi(x_0)} \left(\frac{-2\pi}{n\psi''(x_0)}\right)^{1/2} \qquad (n \to +\infty).$$

This idea was subsequently employed by Cauchy (1829) in the estimation of the large-n behaviour of the coefficients a_n in certain series expansions, and in particular those in the Lagrange inversion series. This was motivated by the wish to determine the radius of convergence of such series expansions and to examine their behaviour on the circle of convergence. Let $g(z)$ denote a function that is analytic at the point $z = \alpha$ (with $g'(\alpha) \neq 0$) and $F(z)$ a function analytic at $z = 0$. Then the equation $z = wg(\alpha + z)$ has a unique solution $z = z(w)$ valid in a neighbourhood of $w = 0$ and $F(z(w)) = F(0) + \sum_{n=1}^{\infty} a_n w^n$. The coefficients a_n are given by

$$a_n = \frac{1}{2\pi i n} \oint_C F'(z) \frac{g^n(\alpha + z)}{z^n} \, dz = \frac{1}{2\pi i n} \oint_C F'(z) e^{n\psi(z)} \, dz,$$

where $\psi(z) = \log(g(\alpha + z)/z)$ and C is a circular contour surrounding $z = 0$. Cauchy then applied Laplace's argument in the complex plane (with $z = re^{i\theta}$): the circle C was expanded until it passed through the saddle point of $\psi(z)$ (given by the zeros of $\psi'(z) = 0$) closest to the origin. Cauchy never varied his choice of contour: he always took C to be a circle, thereby depriving his treatment of the necessary generality and incorrectly dealing with the case of multiple saddle points (Petrova and Solov'ev, 1997).

A quarter of a century later, Stokes (1850) investigated the asymptotics of the integral[1]

$$W(m) = \int_0^{\infty} \cos \tfrac{1}{2}\pi(w^3 - mw) \, dw \qquad (m \to +\infty)$$

in connection with the intensity of light in the neighbourhood of a caustic in the then new wave theory of light applied to the rainbow. The zeros of $W(m)$ corresponded to the location of darkbands in a system of supernumerary rainbows, of which up

---

[1] The integral $W(m)$ can be expressed in terms of the Airy function $\mathrm{Ai}(-\alpha m)$, where $\alpha = (\pi/2)^{2/3}3^{-1/3}$.

to 30 had been observed experimentally. By writing the cosine as the real part of its associated exponential and rotating the integration path through $-\pi/6$, Stokes reduced his integral to consideration of

$$\int_0^\infty \exp(-t^3 + 3q^2 t)\, dt, \quad q = (\pi/2)^{1/3} (m/3)^{1/2} e^{\pi i/6}.$$

He then proceeded to expand the integrand about $t = q$ (a saddle point) and took his integration path in the neighbourhood of this point in the direction in which the imaginary part of the exponent in the exponential factor remained constant (the path of steepest descent through $t = q$, see §1.2.1). By bounding the contribution from different parts of the deformed path, he was able to establish that the dominant contribution to this integral arose from a neighbourhood of $t = q$. Stokes thereby obtained the result

$$W(m) \sim (2/3)^{1/2} (m/3)^{-1/4} \cos\{\pi(m/3)^{3/2} - \pi/4\} \quad (m \to +\infty),$$

from which he was able to calculate approximations to the first 50 zeros of $W(m)$. Although Stokes did not mention the terms saddle point or path of steepest descent, he was, nevertheless, effectively employing the ideas of the saddle-point method in the complex plane. A more detailed account of this problem together with a discussion of Stokes' other mathematical contributions can be found in Paris (1996).

In 1863, Riemann investigated an asymptotic approximation for the Gauss hypergeometric function $_2F_1(n - c, n + a + 1; 2n + a + b + 2; x)$ when $n \to +\infty$ and the variable x has complex values. This function can be expressed in terms of the integral

$$I_n = \int_0^1 f(t) \left(\frac{t(1-t)}{1-xt}\right)^n dt = \int_0^1 f(t) e^{n\psi(t)} dt,$$

where $f(t) = t^a (1-t)^b (1-xt)^c$ and $\psi(t) = \log\{t(1-t)/(1-xt)\}$. This integral defines an analytic function in the complex x-plane cut along the real axis from $x = 1$ to $x = +\infty$. His paper (Riemann, 1863) on this calculation was never finished in his lifetime: the second part, containing the asymptotic calculation, contained only key expressions to guide him during the writing-up process. The text, together with some of the computations in this posthumous paper, were filled in by H. Schwarz. Riemann determined the two saddle points t_{s1} and t_{s2} of $\psi(t)$ (given by $\psi'(t) = 0$) as

$$t_{s1} = \frac{1}{1 + \sqrt{1-x}}, \quad t_{s2} = \frac{1}{1 - \sqrt{1-x}},$$

with the branch of the square root having positive real part in the cut x-plane. He then proceeded to show that the integration path could be deformed to pass through the saddle t_{s1} and argued that as $n \to +\infty$ the dominant contribution to I_n arises from a neighbourhood of t_{s1}. By integrating along the direction of steepest descent through the saddle, he obtained the result

$$I_n \sim \frac{(\pi/n)^{1/2}(1-x)^{(b+c)/2+\frac{1}{4}}}{(1+\sqrt{1-x})^{2n+a+b+1}} \qquad (n \to +\infty),$$

although no mention was made of the sector of validity (in the complex x-plane) of this approximation.

This paper is frequently cited as containing the germ of the idea of the method of steepest descents. If Riemann had used the whole path connecting $t = 0$ to $t = 1$, he would have obtained the full asymptotic expansion of I_n, instead of just the leading term. It is clear, however, that he must have been in possession of the steepest descent technique since he used this method with great skill in his famous expansion[2] of the function $Z(t)$, related to the Riemann zeta function $\zeta(s)$ on the critical line $s = \frac{1}{2}+it$, as $t \to +\infty$; for a detailed account, see Edwards (1974, Ch. 7).

An interesting survey paper by Petrova and Solov'ev (1997) discusses these developments (with the exception of those of Stokes) in greater detail. These authors also point out the work of the Russian mathematician Nekrosov who, about 20 years after Riemann, considered Cauchy's problem of the determination of the leading behaviour of the coefficients in the Langrange inversion series. He considered the problem in general and discussed the situation when there are several saddle points of arbitrary multiplicity. He showed the existence of a closed contour passing through the saddle points along the directions of steepest descent, but only obtained the dominant contribution from each saddle.

Finally, the first person to obtain a full asymptotic expansion in a specific case was the physicist Debye (Debye, 1909). He developed the method, after seeing Riemann's paper, for the Hankel functions defined in the form

$$H_\nu^{(1,2)}(x) = -\frac{1}{\pi} \int_{-\infty i}^{\infty i \mp \pi} e^{-ix\psi(t)}dt, \quad \psi(t) = \sin t - \alpha t, \quad \alpha = \frac{\nu}{x}$$

for large positive values of x and the order ν. The integrand has two saddle points at which $\psi'(t) = 0$ in the strip $-\pi < \mathrm{Re}(t) < \pi$, which are situated symmetrically about the origin on the real axis when $0 < \alpha < 1$ and on the imaginary axis when $\alpha > 1$. Debye deformed the contours into steepest descent paths passing through one or both saddles, and then converted the integrals leading up to and away from a saddle into a Laplace integral of the form $\int_0^\infty e^{-xu}\phi(u)\,du$ by an appropriate change of variable. Expansion of $\phi(u)$ about $u = 0$ into a series of fractional powers of u then enabled him to integrate term by term to obtain the asymptotic expansion of $H_\nu^{(1,2)}(x)$ for x and $\nu \to +\infty$.

Debye also considered the situation when the variable x and ν are large and nearly equal. In this case, the two saddles in the strip $-\pi < \mathrm{Re}(t) < \pi$ coalesce to form a double saddle point at the origin when $x = \nu$. A slight modification of his argument

---

[2] This is the Riemann–Siegel formula which was discovered by Siegel in Riemann's papers and published in 1932.

then enabled him to derive expansions for the Hankel functions as x and $\nu \to +\infty$ when $x \simeq \nu$.

1.2 The method of steepest descents

1.2.1 Preliminaries

The type of integral under consideration is

$$I(\lambda) = \int_C e^{\lambda \psi(t)} f(t) \, dt, \tag{1.2.1}$$

where λ is a large positive[3] parameter and C is a path of finite or infinite extent in the complex t-plane which is chosen such that $I(\lambda)$ converges. The amplitude function $f(t)$ and phase function $\psi(t)$ are assumed to be analytic on and near the path C and, for the purpose of this section, to be independent of λ.

Let $\psi(t) = U(x, y) + iV(x, y)$ where $t = x + iy$ and U, V, x, y are real. When λ is large a small displacement along the path C causing a small change in $V(x, y)$ will, in general, produce a rapid oscillation of the sinusoidal terms in $\exp(\lambda \psi(t))$. This has the consequence that the contribution to the integral will be subject to considerable cancellation between neighbouring parts of the path. An obvious remedy is to choose a path on which $V(x, y)$ is constant, thereby removing the rapid oscillations of the integrand. The most rapidly varying part of the integrand will then be $\exp(\lambda U)$ and the dominant contribution will arise from a neighbourhood of the point where $U(x, y)$ is greatest.

The choice of a path with $V(x, y) = $ constant has another major advantage. For, such paths are those on which $U(x, y)$ *changes the most rapidly*. To see this, let us consider a small displacement from the point t_0 given by $t = t_0 + se^{i\theta}$, where $s > 0$ and θ is a phase angle. Then the rate of change of $U(x, y)$ is given by

$$\frac{dU}{ds} = \frac{\partial U}{\partial x} \cos\theta + \frac{\partial U}{\partial y} \sin\theta = U_x \cos\theta + U_y \sin\theta.$$

Regarded as a function of θ, dU/ds will have a stationary point when its derivative D with respect to θ vanishes; that is, when $D := -U_x \sin\theta + U_y \cos\theta = 0$. Since $\psi(t)$ is an analytic function of the complex variable t in the neighbourhood of C, its derivatives are constrained to satisfy the Cauchy–Riemann equations

$$U_x = V_y, \qquad U_y = -V_x \tag{1.2.2}$$

in this neighbourhood. Substitution of these equations into the above stationary condition $D = 0$ yields

$$V_x \cos\theta + V_y \sin\theta = 0.$$

---

[3] If $\lambda = |\lambda| e^{i\phi}$ is complex, then we can take $|\lambda|$ as the large parameter and absorb the phase term $e^{i\phi}$ into $\psi(t)$.

But this last equation states that $dV/ds = 0$. Since

$$\left|\frac{dU}{ds}\right|^2 = |U_x \cos\theta + U_y \sin\theta|^2 = U_x^2 + U_y^2 - (U_x \sin\theta - U_y \cos\theta)^2$$

$$= U_x^2 + U_y^2 - D^2,$$

it is seen that the stationary direction ($D = 0$) corresponds to a maximum in the absolute value of the rate of change of $U(x, y)$ with respect to s. *Thus a path $V(x, y) =$ constant is one along which $U(x, y)$ changes the most rapidly.*

To obtain a geometrical insight into the nature of such paths we recall some well-known properties of functions of a complex variable. Since $|e^{\lambda\psi(t)}| = e^{\lambda U}$, we are interested in the modular surface S defined by $U = U(x, y)$, where the U-axis is perpendicular to the x, y-plane. A point (x_0, y_0) on this surface where $U_x(x_0, y_0) = U_y(x_0, y_0) = 0$ is a stationary point. From (1.2.2) we have $U_{xx} = V_{yx}$ and $U_{yy} = -V_{xy}$, so that

$$U_{xx} + U_{yy} = 0$$

and $U(x, y)$ is a harmonic function. Then, since the quantity

$$U_{xx}U_{yy} - U_{xy}^2 = -U_{xx}^2 - U_{xy}^2 < 0,$$

the stationary point (x_0, y_0) must be a saddle point. Hence all stationary points on S can only be saddle points (or cols); the surface S has no maxima[4] and no minima (except for isolated zeros of $\psi(t)$). Application of the Cauchy–Riemann equations again shows that $\psi'(t) = U_x(x, y) + iV_x(x, y) = U_x(x, y) - iU_y(x, y)$, so that the stationary point (x_0, y_0) must be a saddle point of the phase function $\psi(t)$.

The shape of the modular surface S can also be visualised on the x, y-plane by constructing the level curves on which $U(x, y) =$ constant. From (1.2.2) it follows that

$$U_x V_x + U_y V_y = \nabla U \cdot \nabla V = 0,$$

where $\nabla \equiv \mathbf{i}\partial/\partial x + \mathbf{j}\partial/\partial y$ is the two-dimensional gradient operator. Thus the families of curves corresponding to constant values of $U(x, y)$ and $V(x, y)$ are orthogonal at all their points of intersection. The regions where $U(x, y) > U(x_0, y_0)$ are called *hills* (or *ridges*) and those where $U(x, y) < U(x_0, y_0)$ are called *valleys*. The level curve through the saddle, $U(x, y) = U(x_0, y_0)$, separates the immediate neighbourhood of the saddle point (x_0, y_0) into a series of hills and valleys.

To see this topography, let us suppose that $t_s = x_0 + iy_0$ is a saddle point of order $m - 1$ (with $m \geq 2$); that is the first $m - 1$ derivatives of $\psi(t)$ at t_s all vanish

$$\psi'(t_s) = \psi''(t_s) = \cdots = \psi^{(m-1)}(t_s) = 0,$$

---

[4] This can also be seen by application of the maximum-modulus principle.

with $\psi^{(m)}(t_s) = Ae^{i\phi}$, $A > 0$. Then, if $t = t_s + re^{i\theta}$, $r > 0$, we have

$$\psi(t) - \psi(t_s) = (Ar^m/m!)e^{i(m\theta+\phi)} + \cdots$$

and hence

$$\begin{pmatrix} U(x, y) - U(x_0, y_0) \\ V(x, y) - V(x_0, y_0) \end{pmatrix} = \frac{Ar^m}{m!} \begin{pmatrix} \cos \\ \sin \end{pmatrix} (m\theta + \phi) + \cdots$$

in a neighbourhood of t_s. The directions of the paths of constant $U(x, y)$ and $V(x, y)$ are consequently determined by setting the right-hand side of the above expression equal to zero to find

$$\theta = (k + \tfrac{1}{2}\delta)\frac{\pi}{m} - \frac{\phi}{m}, \qquad \delta = \begin{pmatrix} 1 \\ 0 \end{pmatrix} \qquad (k = 0, 1, \ldots, 2m - 1), \qquad (1.2.3)$$

respectively. Therefore, there are $2m$ equally spaced steepest directions from t_s : m directions of steepest descent and m directions of steepest ascent. In the neighbourhood of t_s, the level curves $U = U(x_0, y_0)$ form the boundaries of m valleys surrounding the saddle point, in which $\cos(m\theta + \phi) < 0$, and m hills on which $\cos(m\theta + \phi) > 0$. The valleys and hills are situated respectively entirely below and above the saddle point, and each has angular width equal to π/m.

The steepest paths satisfy $\sin(m\theta + \phi) = 0$, and so have directions given by (1.2.3) with $\delta = 0$. The directions of steepest descent at the saddle point t_s are therefore

$$\theta = (2k + 1)\frac{\pi}{m} - \frac{\phi}{m} \qquad (k = 0, 1, \ldots, m - 1; \ m \geq 2). \qquad (1.2.4)$$

The topography of the surface S near the saddle point is shown in Fig. 1.1 when $\phi = \tfrac{1}{2}\pi$ for the cases $m = 2$ (which corresponds to the most commonly occurring situation of a first-order saddle) and $m = 3$. We remark that, away from the immediate neighbourhood of the saddle point, the projection of the steepest paths and the valley boundaries onto the t-plane will, in general, be curved paths.

A typical situation has the contour of integration C in (1.2.1) beginning and ending at infinity in the valleys (which is necessary for convergence). The contour is then deformed as far as possible into paths of steepest descent running along the bottoms of valleys and crossing over from one valley to the next over a saddle point; see Fig. 1.2. An interesting analogy has been given by De Bruijn (1958, p. 80) in the form of a person wishing to travel between two points in a mountainous region: if the two points are in different valleys, then the least effort should involve a passage via a col. A similar idea is presented in Greene and Knuth (1982, §4.3.3) in relation to a lazy hiker who will choose a path that crosses a ridge at its lowest point; but unlike the truly lazy hiker, who would probably choose a zig-zag path, the best path takes the steepest ascent to the col.

We remark that the determination of the paths of steepest descent in particular cases can be quite difficult. It is usually a simple matter to locate the saddle points of a given phase function $\psi(t)$ and the directions of steepest descent away from these

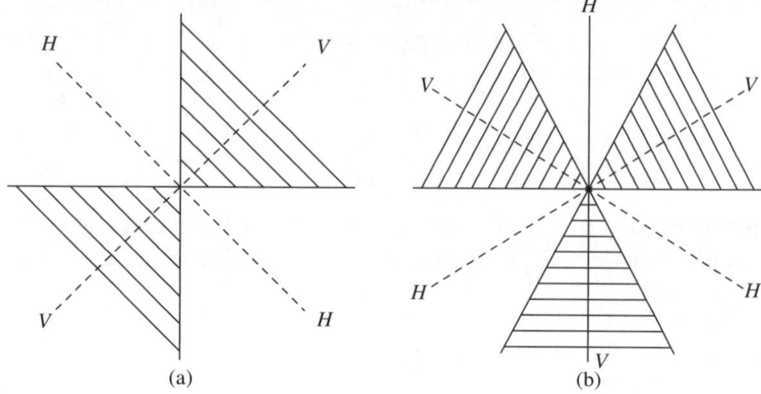

Figure 1.1 Paths of steepest descent and ascent (dashed lines) in the neighbourhood of the saddle point t_s when $\phi = \frac{1}{2}\pi$ and (a) $m = 2$ and (b) $m = 3$. The shaded regions denote the valleys (V) and the unshaded regions denote the hills (H).

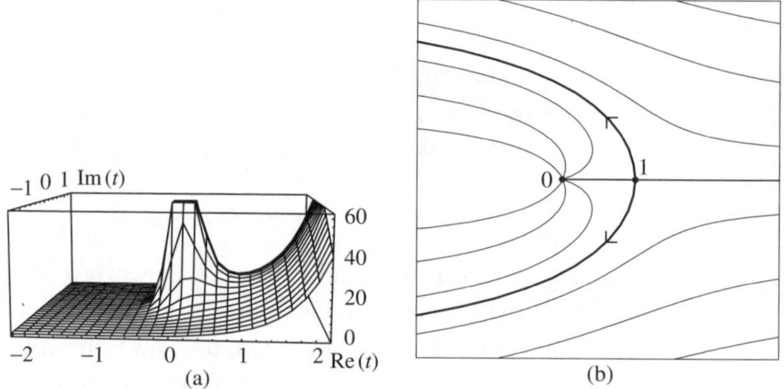

Figure 1.2 (a) The modular surface of $F = t^{-3}e^{3t}$ possessing a saddle point of order 1 ($m = 2$) at $t = 1$ and (b) the associated paths of constant $\mathrm{Im}(F)$. The heavy line denotes the steepest descent path through the saddle and the arrows denote the direction of descent.

saddles. It is also easy to determine the valleys at infinity. What is not so straightforward is how the various steepest descent paths connect up with the valleys. Often an intelligent guess is successful, especially when the variable λ is real. However, *Mathematica* can be employed to great advantage in the construction of the paths of steepest descent. This is the approach we adopt in all the examples in this book.

On a steepest path through a saddle point t_s we have

$$\psi(t) = \psi(t_s) - u,$$

where u is real. Unless this path[5] connects with another saddle or a singularity of $\psi(t)$, the variable u either increases monotonically to $+\infty$ along a steepest descent

---

[5] A path of steepest descent terminates only at infinity or at singular points of $\psi(t)$.

path or decreases monotonically to $-\infty$ along a steepest ascent path. In general, the contour will consist of a series of steepest descent paths, each running from a saddle point down a valley out to infinity or to a singularity of the phase function $\psi(t)$. This leads to finding the asymptotic expansion of integrals of the type

$$e^{\lambda\psi(t_s)} \int_{t_s}^{T} e^{-\lambda u} f(t)\,dt,$$

where T denotes some point on a steepest descent path, and adding the contributions from each relevant saddle point. The most common situation has $T = +\infty$, although T can be finite if the integration path encounters another saddle or, of course, if the original path C in (1.2.1) is finite.

1.2.2 Asymptotic expansion of $I(\lambda)$

In this section we determine the asymptotic expansion of the integral

$$I(\lambda) = \int_C e^{\lambda\psi(t)} f(t)\,dt \qquad (1.2.5)$$

for $\lambda \to +\infty$, where the path C is a steepest descent path that commences at a saddle point t_s of order $m - 1$. The derivation is formal but the expansion process is justified by a useful result known as Watson's lemma, which we state and prove in §1.2.4.

We commence by giving the definition of an asymptotic expansion.

Definition 1.1 Let $f(z)$ be a function of a real or complex variable z, $\sum c_k z^{-k}$ a formal power series (convergent or divergent) and $R_N(z)$ the difference between $f(z)$ and the Nth partial sum of the series; that is

$$f(z) = \sum_{k=0}^{N-1} c_k z^{-k} + R_N(z).$$

Then, if for each fixed value of N

$$R_N(z) = O(z^{-N})$$

as $z \to \infty$ in a certain unbounded region **R**, we say that the series $\sum c_k z^{-k}$ is an asymptotic expansion of $f(z)$ and write

$$f(z) \sim \sum_{k=0}^{\infty} c_k z^{-k} \qquad (z \to \infty \text{ in } \mathbf{R}). \qquad (1.2.6)$$

This definition is due to Poincaré (1886). The formal series so obtained is also referred to as an asymptotic expansion of *Poincaré type*, or an asymptotic expansion in the *Poincaré sense*, or more simply as a *Poincaré expansion*.

In the integral $I(\lambda)$ in (1.2.5) we put

$$\psi(t) = \psi(t_s) - u, \qquad (1.2.7)$$

where u is non-negative and monotonically increasing as one progresses down the steepest descent path. This produces

$$I(\lambda) = e^{\lambda\psi(t_s)} \int_{t_s}^{T} e^{-\lambda u} f(t)\, dt = e^{\lambda\psi(t_s)} \int_{0}^{T'} e^{-\lambda u} f(t) \frac{dt}{du}\, du, \qquad (1.2.8)$$

where $T' > 0$ is the map of T in the u-plane. For large positive λ, the exponential factor $e^{-\lambda u}$ in (1.2.8) decays rapidly so that the main contribution comes from the neighbourhood of $u = 0$. Accordingly, to determine the asymptotic expansion of $I(\lambda)$ for $\lambda \to +\infty$, we require the series expansion of $f(t)dt/du$ in ascending powers of u. This expansion is substituted into the integral (1.2.8) which is then integrated term by term.

If t_s is a saddle point of order $m - 1$, then

$$\psi(t) = \psi(t_s) - \sum_{k=0}^{\infty} a_k (t - t_s)^{m+k} \qquad (1.2.9)$$

valid in some disc surrounding t_s, where $a_k = -\psi^{(m+k)}(t_s)/(m+k)!$ and $a_0 \neq 0$. Comparison of this expansion with (1.2.7) shows that

$$u = \sum_{k=0}^{\infty} a_k (t - t_s)^{m+k}. \qquad (1.2.10)$$

If we let $u = \tau^m$, then for small $|t - t_s|$

$$\tau = a_0^{1/m}(t - t_s)\left\{ 1 + \frac{a_1}{ma_0}(t - t_s) + \frac{1}{m}\left(\frac{a_2}{a_0} - \frac{(m-1)a_1^2}{2ma_0^2} \right)(t - t_s)^2 + \cdots \right\},$$

where $a_0^{1/m}$ takes its principal value. It follows that τ is a single-valued analytic function of t in the neighbourhood of t_s and that $\tau'(t_s) \neq 0$. By the inverse function theorem (Copson, 1935, p. 121; Jeffreys and Jeffreys, 1972, p. 380) we then have

$$t - t_s = \sum_{k=1}^{\infty} \alpha_k \tau^k = \sum_{k=1}^{\infty} \alpha_k u^{k/m}, \qquad (1.2.11)$$

where

$$\alpha_1 = \frac{1}{a_0^{1/m}}, \qquad \alpha_2 = -\frac{a_1}{ma_0^{1+2/m}}, \qquad \alpha_3 = \frac{(m+3)a_1^2 - 2ma_0a_2}{2m^2 a_0^{2+3/m}}, \dots.$$

This gives one inversion of (1.2.10); the others are obtained by replacing u in (1.2.11) by $ue^{2\pi i n}$, with n an integer satisfying $1 \leq n \leq m - 1$.

In simple cases it is possible to determine the coefficients α_k in closed form by application of the Lagrange inversion theorem, but this is not practicable in more complicated cases. This important theorem can be stated as follows (Copson, 1935, p. 125; Whittaker and Watson, 1952, p. 133; Jeffreys and Jeffreys, 1972, p. 383).

Theorem 1.1 Lagrange's inversion theorem *If $f(z)$ is regular in a neighbourhood of z_0 and if $f(z_0) = w_0$, $f'(z_0) \neq 0$, then the equation*

$$f(z) = w$$

has a unique solution, regular in a neighbourhood of w_0, of the form

$$z = z_0 + \sum_{n=1}^{\infty} \frac{(w - w_0)^n}{n!} \left[\frac{d^{n-1}}{dz^{n-1}} \{\phi(z)\}^n \right]_{z=z_0}, \tag{1.2.12}$$

where

$$f(z) = w_0 + \frac{z - z_0}{\phi(z)}.$$

The particular case when $z_0 = 0$ and $w_0 = 0$, corresponding to the equation $w = z/\phi(z)$, is of special importance.

The use of *Mathematica* or *Maple* to generate the inversion coefficients is a great help, but if these coefficients are derived in symbolic form the limitations of machine capacity are soon reached. However, in specific cases (i.e., when the various parameters have numerical values), the use of the InverseSeries command in *Mathematica* generates numerical values of the inversion coefficients α_k for $k \gg 1$ with relative ease.

The amplitude function $f(t)$ in (1.2.5) can be expanded about the point t_s in the form

$$f(t) = \sum_{k=0}^{\infty} b_k (t - t_s)^k, \qquad b_k = \frac{f^{(k)}(t_s)}{k!}. \tag{1.2.13}$$

This expansion will converge in a disc centred at t_s with a radius determined by the nearest singularity of $f(t)$. Then, by differentiation of (1.2.11), we obtain dt/du and with this the expansion

$$f(t) \frac{dt}{du} = \sum_{k=0}^{\infty} c_k u^{(k+1-m)/m}, \tag{1.2.14}$$

where

$$c_0 = \frac{b_0}{m a_0^{1/m}}, \qquad c_1 = \frac{1}{m a_0^{2/m}} \left\{ b_1 - \frac{2 a_1 b_0}{a_0 m} \right\},$$

$$c_2 = \frac{1}{m a_0^{3/m}} \left\{ b_2 - \frac{3}{a_0 m}(a_1 b_1 + a_2 b_0) + \frac{3 a_1^2 b_0}{2 a_0^2 m^2}(m + 3) \right\},$$

$$c_3 = \frac{1}{ma_0^{4/m}}\left\{ b_3 - \frac{4}{a_0 m}(a_1 b_2 + a_2 b_1 + a_3 b_0) + \frac{2a_1}{a_0^2 m^2}(a_1 b_1 + 2a_2 b_0)(m+4)\right.$$

$$\left. - \frac{4a_1^3 b_0}{3a_0^3 m^3}(m+2)(m+4)\right\}, \dots . \tag{1.2.15}$$

The expansion (1.2.14) converges in a disc centred at t_s whose radius is controlled either by the closest singularities of $f(t)$ or by neighbouring saddle points of $\psi(t)$ (where dt/du is singular).

We are now in a position to determine the expansion of the integral (1.2.8) taken from the saddle t_s of order $m-1$ down one of the m valleys. These valleys are labelled by the integer n ($0 \le n \le m-1$) and which one is selected depends on the path C in (1.2.5). Formal substitution of (1.2.14), with u replaced by $ue^{2\pi in}$, into the integral (1.2.8) followed by term-by-term integration yields

$$I(\lambda) = e^{\lambda\psi(t_s)}\int_0^{T'} e^{-\lambda u} f(t)\frac{dt}{du}\,du$$

$$\sim e^{\lambda\psi(t_s)}\sum_{k=0}^{\infty} c_k \int_0^{T'} e^{-\lambda u}(ue^{2\pi in})^{(k+1-m)/m}\,du$$

$$\sim e^{\lambda\psi(t_s)}\sum_{k=0}^{\infty} \frac{c_k e^{2\pi in(k+1)/m}}{\lambda^{(k+1)/m}}\,\Gamma\left(\frac{k+1}{m}\right) \tag{1.2.16}$$

as $\lambda \to +\infty$. The upper limit of integration T' has been replaced by $+\infty$ (which introduces an exponentially small error of order $e^{-\lambda T'}$) and the resulting integrals have been evaluated in terms of the gamma function. This procedure requires justification since, in general, the circle of convergence of the series expansion of $f(t)dt/du$ in (1.2.14) will not completely include the domain of integration. This is given by Watson's lemma which we state and prove in the next section.

An integral representation for the coefficients c_k can be obtained by observing that with $u = \tau^m$, so that $dt/du = (\tau^{1-m}/m)dt/d\tau$, we find from (1.2.14)

$$f(t)\frac{dt}{d\tau} = m\sum_{k=0}^{\infty} c_k \tau^k.$$

Application of Cauchy's theorem, combined with (1.2.7), then shows that (Watson, 1952, p. 246; Copson, 1965, p. 69)

$$c_k = \frac{1}{2\pi im}\int^{(0+)} f(t)\frac{dt}{d\tau}\frac{d\tau}{\tau^{k+1}}$$

$$= \frac{1}{2\pi im}\int^{(t_s+)} \frac{f(t)\,dt}{\{\psi(t_s) - \psi(t)\}^{(k+1)/m}} \tag{1.2.17}$$

with an appropriate branch of $\{\psi(t_s) - \psi(t)\}^{1/m}$.

1.2.3 Quadratic and linear endpoint cases

The situation of most frequent occurrence is that of a simple saddle point corresponding to $m = 2$ (a quadratic endpoint). This case is so important that it is worthwhile giving a separate presentation of the expansion. There are two directions of steepest descent from the saddle: one leading away from t_s (corresponding to $n = 0$) and one leading up to t_s (corresponding to $n = 1$). The angular separation between these two paths is π; see Fig. 1.1(a). On these paths the variable u increases monotonically from 0. The expansion of (1.2.8) from the simple saddle t_s down the two steepest descent paths is, from (1.2.16),

$$I(\lambda) \sim \pm e^{\lambda \psi(t_s)} \sum_{k=0}^{\infty} \frac{(\pm)^k c_k}{\lambda^{(k+1)/2}} \Gamma(\tfrac{1}{2}k + \tfrac{1}{2}) \tag{1.2.18}$$

as $\lambda \to +\infty$, where the upper and lower signs correspond to $n = 0$ and $n = 1$, respectively.

When the integration path C has the simple saddle t_s as an interior point we label the two halves of the steepest descent path through t_s (corresponding to $n = 0$ and $n = 1$) by $t^{\pm}$, respectively. Then, from (1.2.18),

$$I(\lambda) = e^{\lambda \psi(t_s)} \int_0^{T'} e^{-\lambda u} \left\{ f(t^+) \frac{dt^+}{du} - f(t^-) \frac{dt^-}{du} \right\} du$$

$$\sim e^{\lambda \psi(t_s)} \sum_{k=0}^{\infty} \frac{c_k \Gamma(\tfrac{1}{2}k + \tfrac{1}{2})}{\lambda^{(k+1)/2}} \{1 + (-)^k\}$$

$$= 2 e^{\lambda \psi(t_s)} \sum_{k=0}^{\infty} \frac{c_{2k} \Gamma(k + \tfrac{1}{2})}{\lambda^{k + \tfrac{1}{2}}} \tag{1.2.19}$$

as $\lambda \to +\infty$, where, for convenience in presentation, we have taken the same value of T' on each steepest descent path. We observe that there is a cancellation of the terms with k odd in (1.2.19). The first few coefficients c_k can be obtained from (1.2.15) with $m = 2$ and, recalling that $a_k = -\psi^{(m+k)}/(m+k)!$ and $b_k = f^{(k)}/k!$, we find

$$c_0 = \frac{f}{(-2\psi'')^{1/2}}, \qquad c_1 = -\frac{1}{\psi''} \left(f' - \frac{\psi'''}{3\psi''} f \right),$$

$$c_2 = \frac{1}{(-2\psi'')^{3/2}} \left\{ 2f'' - 2\frac{\psi'''}{\psi''} f' + \left(\frac{5\psi'''^2}{6\psi''^2} - \frac{\psi^{iv}}{2\psi''} \right) f \right\},$$

$$c_3 = \frac{1}{3\psi''^2} \left\{ f''' - \frac{2\psi'''}{\psi''} f'' + \left(\frac{2\psi'''^2}{\psi''^2} - \frac{\psi^{iv}}{\psi''} \right) f' \right.$$
$$\left. - \left(\frac{8\psi'''^3}{9\psi''^3} - \frac{\psi''' \psi^{iv}}{\psi''^2} + \frac{\psi^{v}}{5\psi''} \right) f \right\},$$

$$c_4 = \frac{1}{(-2\psi'')^{5/2}}\left\{\frac{2}{3}f^{iv} - \frac{20\psi'''}{9\psi''}f''' + \frac{5}{3}\left(\frac{7\psi'''^2}{3\psi''^2} - \frac{\psi^{iv}}{\psi''}\right)f'' \right.$$

$$- \frac{35}{9}\left(\frac{\psi'''^3}{\psi''^3} - \frac{\psi'''\psi^{iv}}{\psi''^2} + \frac{6\psi^v}{35\psi''}\right)f'$$

$$\left. + \frac{35}{9}\left(\frac{11\psi'''^4}{24\psi''^4} - \frac{3}{4}\left(\frac{\psi'''^2}{\psi''^2} - \frac{\psi^{iv}}{6\psi''}\right)\frac{\psi^{iv}}{\psi''} + \frac{\psi'''\psi^v}{5\psi''^2} - \frac{\psi^{vi}}{35\psi''}\right)f\right\}, \ldots, \quad (1.2.20)$$

where ψ, f and their derivatives are evaluated at $t = t_s$. Expressions for the coefficients c_k for $k \leq 8$ when $m = 2$ are listed in Dingle (1973, p. 119). We remark that a closed-form representation for the c_k is given in Wojdylo (2006) in terms of combinatorial objects called partial ordinary Bell polynomials.

Finally, we consider the case of the integral (1.2.5) corresponding to a linear endpoint; that is, where the path commences at the *ordinary point* t_0 (where $\psi'(t_0) \neq 0$) and passes to infinity along the steepest descent path through t_0. The expansion of $\psi(t)$ about t_0 is given by (1.2.9), with $m = 1$ and t_s replaced by t_0. Similarly, the expansion of $f(t)$ about t_0 is given by (1.2.13) with t_s replaced by t_0. Then, from (1.2.16) with $m = 1$ and $n = 0$, we obtain

$$I(\lambda) = e^{\lambda\psi(t_0)}\int_{t_0}^{\infty} e^{-\lambda u}f(t)\,dt = e^{\lambda\psi(t_0)}\int_0^{\infty} e^{-\lambda u}f(t)\frac{dt}{du}\,du$$

$$\sim e^{\lambda\psi(t_0)}\sum_{k=0}^{\infty}\frac{c_k\Gamma(k+1)}{\lambda^{k+1}} \quad (1.2.21)$$

as $\lambda \to \infty$. The first four coefficients c_k are given by (1.2.15) with $m = 1$ and are found to be

$$c_0 = -\frac{f}{\psi'}, \quad c_1 = \frac{1}{\psi'^2}\left(f' - \frac{\psi''}{\psi'}f\right),$$

$$c_2 = -\frac{1}{2\psi'^3}\left\{f'' - \frac{3\psi''}{\psi'}f' + \left(\frac{3\psi''^2}{\psi'^2} - \frac{\psi'''}{\psi'}\right)f\right\},$$

$$c_3 = \frac{1}{6\psi'^4}\left\{f''' - \frac{6\psi''}{\psi'}f'' + 15\left(\frac{\psi''^2}{\psi'^2} - \frac{4\psi'''}{15\psi'}\right)f'\right.$$

$$\left. - 15\left(\frac{\psi''^3}{\psi'^3} - \frac{2\psi'''\psi''}{3\psi'^2} + \frac{\psi^{iv}}{15\psi'}\right)f\right\}, \quad (1.2.22)$$

where ψ, f and their derivatives are evaluated at $t = t_s$. Expressions for the coefficients c_k for $k \leq 7$ when $m = 1$ are listed in Dingle (1973, p. 114).

Explicit error bounds for the asymptotic expansion of integrals of the form (1.2.5) are discussed in Olver (1968, 1970). We also mention that a treatment of Laplace's method and the steepest descent method using multi-point Taylor series is given in Ferreira *et al.* (2007).

1.2.4 Watson's lemma

One of the most widely used results in the asymptotic expansion of Laplace integrals is Watson's lemma (Watson, 1918). It is well known that the Laplace transform of a piecewise continuous function $g(t)$ on the interval $[0, \infty)$ is $o(1)$ as the transform variable grows without bound. Knowledge of the small variable behaviour of $g(t)$ enables more information to be obtained about the growth at infinity of the transform.

Lemma 1.1 Watson's lemma *Let $g(t)$ be an analytic function of t when $|t| \le R'$ in the cut plane,*[6] *except at a branch point at $t = 0$, and let*

$$g(t) = \sum_{n=0}^{\infty} a_n t^{(n+\beta-\mu)/\mu} \tag{1.2.23}$$

when $|t| \le R$, where $0 < R < R'$, $\mu > 0$ and $\mathrm{Re}\,(\beta) > 0$. Let $|g(t)| < K e^{bt}$, where $K > 0$, $b > 0$ are independent of t, when $t \ge R$ on the positive real t-axis. Then, for $T > R'$, we have

$$\int_0^T e^{-zt} g(t)\, dt \sim \sum_{n=0}^{\infty} a_n \Gamma\left(\frac{n+\beta}{\mu}\right) z^{-(n+\beta)/\mu} \tag{1.2.24}$$

as $z \to \infty$ in the sector $|\arg z| \le \frac{1}{2}\pi - \delta$, $\delta > 0$, where $z^{(n+\beta)/\mu}$ takes its principal value.

Proof First observe that the condition $\mathrm{Re}(\beta) > 0$ is required for convergence of the integral at the origin and that $T > R'$ means that we are integrating beyond the circle of convergence of the series expansion for $g(t)$ in (1.2.23). Secondly, the integral over $[0, T]$ can be replaced by the integral taken over $[0, \infty)$, since

$$\left| \int_T^{\infty} e^{-zu} g(t)\, dt \right| < K \int_T^{\infty} e^{-\xi t} dt = K\xi^{-1} e^{-\xi T}, \qquad \xi = \mathrm{Re}(z) - b.$$

Since $\mathrm{Re}(z) \ge |z| \sin \delta$ this is $O(\exp(-|z| T \sin \delta))$ as $z \to \infty$ in $|\arg z| \le \frac{1}{2}\pi - \delta$. This contribution is therefore exponentially small.

For fixed positive integer N we can choose a constant C such that

$$\left| g(t) - \sum_{n=0}^{N-1} a_n t^{(n+\beta-\mu)/\mu} \right| < C e^{bt} t^{(N+\beta-\mu)/\mu}$$

when $t \ge 0$. Hence

$$\int_0^{\infty} e^{-zt} g(t)\, dt = \sum_{n=0}^{N-1} a_n \Gamma\left(\frac{n+\beta}{\mu}\right) z^{-(n+\beta)/\mu} + R_N(z),$$

---

[6] If $\mu = 1$, $g(t)$ does not have a branch point at $t = 0$ and the cut plane is not necessary.

where

$$R_N(z) = \int_0^\infty e^{-zt} \left\{ g(t) - \sum_{n=0}^{N-1} a_n t^{(n+\beta-\mu)/\mu} \right\} dt.$$

From the above bound we find

$$|R_N(z)| < C \int_0^\infty e^{-\xi t} t^{(N+\mathrm{Re}(\beta)-\mu)/\mu} dt$$

$$= C\,\Gamma\left(\frac{N+\mathrm{Re}(\beta)}{\mu}\right) \xi^{-(N+\mathrm{Re}(\beta))/\mu} = O(z^{-(N+\beta)/\mu})$$

as $z \to \infty$ in $|\arg z| \le \frac{1}{2}\pi - \delta$. Consequently we have, for $N = 1, 2, \ldots$,

$$\int_0^T e^{-zt} g(t)\,dt = \sum_{n=0}^{N-1} a_n \Gamma\left(\frac{n+\beta}{\mu}\right) z^{-(n+\beta)/\mu} + O(z^{-(N+\beta)/\mu}) \qquad (1.2.25)$$

as $z \to \infty$ in the sector $|\arg z| \le \frac{1}{2}\pi - \delta < \frac{1}{2}\pi$. From the definition of an asymptotic expansion in (1.2.6) this completes the proof. $\qquad\square$

It has been pointed out in Olver (1997, p. 114), using a path rotation argument, that if $g(t)$ is holomorphic within the sector $-\alpha_1 < \arg t < \alpha_2$, where $\alpha_1, \alpha_2 > 0$, then the range of validity of the expansion in Watson's lemma can be extended beyond the right half-plane. For each $\delta \in (0, (\alpha_1 + \alpha_2)/2)$, if the expansion (1.2.23) (with $\mu > 0$ and $\mathrm{Re}(\beta) > 0$) holds as $t \to 0$ in the sector $S_\delta : -\alpha_1 + \delta \le \arg t \le \alpha_2 - \delta$, and $g(t) = O(e^{bt})$ as $t \to \infty$ in S_δ, then the integral $\int_0^\infty e^{-zt} g(t)\,dt$, or its analytic continuation, has the expansion (1.2.24) as $z \to \infty$ in the wider sector

$$-\tfrac{1}{2}\pi - \alpha_2 + \delta \le \arg z \le \tfrac{1}{2}\pi + \alpha_1 - \delta. \qquad (1.2.26)$$

A generalisation of (1.2.24), known as Watson's lemma for loop integrals, deals with an integral of the form

$$I(z) = \frac{1}{2\pi i} \int_{-\infty}^{(0+)} e^{zt} g(t)\,dt, \qquad (1.2.27)$$

where the path is a loop which starts at $-\infty$, encircles $t = 0$ in the positive sense and returns to $-\infty$. Let us take the integration path to be the lower and upper sides of the negative real axis to the left of the point $-R$, together with the circle $|t| = R$. We assume that $g(t)$ is holomorphic, but not necessarily single valued, in an annulus $0 < |t| < R'$, where $R' > R$, and also that $g(t)$ is continuous on the path of integration. Then we have the following.

Lemma 1.2 Watson's lemma for loop integrals *With the above conditions on* $g(t)$, *let*

$$g(t) \sim \sum_{n=0}^\infty a_n t^{(n+\beta-\mu)/\mu}$$

as $t \to 0$ in $|\arg t| \leq \pi$, where $\mu > 0$ but β is an unrestricted (real or complex) constant. Then

$$I(z) \sim \sum_{n=0}^{\infty} \frac{a_n z^{-(n+\beta)/\mu}}{\Gamma((\mu - \beta - n)/\mu)} \tag{1.2.28}$$

as $z \to \infty$ in the sector $|\arg z| \leq \frac{1}{2}\pi - \delta$, $\delta > 0$, where $z^{(n+\beta)/\mu}$ takes its principal value.

The proof of this theorem is given in Olver (1997, p. 120). The expansion for this integral was first given by Barnes (1906); see also Wyman and Wong (1969). We remark that, in contrast to the result in (1.2.24), the parameter β is no longer restricted to the right half-plane. Extension of the expansion (1.2.28) for $I(z)$ (or its analytic continuation) to the wider sector (1.2.26) is also possible; see Olver (1997, p. 121).

An alternative form of expansion for $x > 0$ of integrals of the type

$$I(x) = \int_0^{\infty} e^{-xt} t^{\alpha - 1} f(t)\, dt,$$

where $f(t)$ is an analytic function and $\mathrm{Re}(\alpha) > 0$, has been given in Franklin and Friedman (1957). The leading behaviour of $I(x)$ from Watson's lemma is $\Gamma(\alpha) x^{-\alpha} f(0)\{1 + O(x^{-1})\}$; these authors showed that a more accurate result could be obtained by changing the argument of the amplitude function $f(t)$ to α/x, to find

$$I(x) = \Gamma(\alpha) x^{-\alpha} f(\alpha/x)\{1 + O(x^{-2})\}.$$

In this way, they were able to derive the expansion for $I(x)$ in the form

$$I(x) \sim \Gamma(\alpha) x^{-\alpha} \sum_{k=0}^{\infty} (\alpha)_k x^{-2k} f_k(\xi_k), \qquad \xi_k = \frac{\alpha + k}{x}$$

for $x > 0$, where

$$f_0(t) = f(t), \qquad f_{k+1}(t) = \frac{d}{dt}\left\{ \frac{f_k(t) - f_k(\xi_k)}{t - \xi_k} \right\} \qquad (k \geq 0).$$

This expansion can be shown to be asymptotic as $x \to +\infty$ when the successive derivatives of $f(t)$ satisfy certain bounds, and, in addition, to be convergent for a certain class of $f(t)$. Although this expansion is more accurate than that obtained from Watson's lemma, it suffers from the disadvantage that the coefficients are much more complicated.

1.2.5 Approximate methods

There are two well-known methods for obtaining the dominant large-λ behaviour of integrals of the type (1.2.1), both of which are related to Laplace's approximation. These are the saddle-point method and the method of stationary phase. The former method is a localised version of the method of steepest descents whereas the latter method applies to Fourier-type integrals. For completeness, we give a brief description of both these methods.

The saddle-point method

In many cases the construction of steepest descent paths and the determination of the inversion coefficients can be a daunting process.[7] A method that replaces such paths with simpler ones and avoids the inversion problem has been referred to as the *saddle-point method* by Copson (1965, p. 91) and De Bruijn (1958, p. 85). For the integral

$$I(\lambda) = \int_C e^{\lambda \psi(t)} f(t)\, dt,$$

let us suppose that the path C passes over a *simple* saddle point at t_s and lies in the valleys below this saddle point. Then the saddle-point method consists of *estimating* the integral from a suitably chosen path in a neighbourhood of t_s. The path of integration used in this neighbourhood can be a straight line segment tangential to the steepest descent path through t_s, but there is considerable latitude in the choice of direction of the line segment.[8]

The justification of this approximation – without knowledge of the steepest descents expansion in (1.2.19) – relies on a subtle argument for choosing the neighbourhood of t_s: this must be small enough to enable the approximation $\psi(t) \simeq \psi(t_s) + \frac{1}{2}\psi''(t_s)(t - t_s)^2$ to be valid but large enough to capture the dominant contribution of the integral. The proof we describe is based on that given by Copson (1965, p. 92).

Near t_s, we can write

$$\psi(t) = \psi(t_s) + \tfrac{1}{2}\psi''(t_s)(t - t_s)^2 + \Psi(t),$$

where

$$\Psi(t) = \sum_{n=3}^{\infty} b_n (t - t_s)^n, \qquad b_n = \frac{\psi^{(n)}(t)}{n!}$$

valid in some disc $|t - t_s| \le R$, say. If M denotes the maximum of $|\psi(t)|$ on $|t - t_s| = R$, then $|b_n| \le M/R^n$ by Cauchy's inequality. It then follows that

$$|\Psi(t)| \le M \sum_{n=3}^{\infty} \frac{|t - t_s|^n}{R^n} = \frac{M|t - t_s|^3}{R^2\{R - |t - t_s|\}}.$$

Let ϵ be a positive parameter in the interval $\frac{1}{3} < \epsilon < \frac{1}{2}$ and set $\delta = \lambda^{-\epsilon}$. For λ sufficiently large it is possible to choose $\delta \le R/2$, so that when $|t - t_s| \le \delta$, we have $|\Psi(t)| \le 2M\delta^3/R^3 = 2M\lambda^{-3\epsilon}/R^3$. We therefore obtain in $|t - t_s| \le \delta$

---

[7] This was certainly true before the advent of computer packages such as *Mathematica* or *Maple*.
[8] The path does not have to be a line segment but can be some arc.

$$e^{\lambda \psi(t)} = e^{\lambda\{\psi(t_s)+\frac{1}{2}\psi''(t_s)(t-t_s)^2\}}\{1 + O(\lambda^{1-3\epsilon})\}$$

and

$$f(t) = f(t_s) + O(\lambda^{-\epsilon}) = f(t_s)\{1 + o(\lambda^{1-3\epsilon})\},$$

the little-o estimate following since $\epsilon < \frac{1}{2}$.

The direction of integration is such that $\psi''(t_s)(t - t_s)^2$ is real and negative. With $\psi''(t_s) = Ae^{i\phi}$, where $A > 0$, this gives $\arg(t - t_s) = \pm\frac{1}{2}\pi - \frac{1}{2}\phi$; these values correspond to the two directions of steepest descent at t_s. Let us put $t - t_s = re^{\frac{1}{2}(\pi-\phi)i}$ with $-\delta \leq r \leq \delta$. Then the contribution to $I(\lambda)$ from the neighbourhood of the saddle point is

$$f(t_s)e^{\lambda\psi(t_s)+\frac{1}{2}(\pi-\phi)i} \int_{-\delta}^{\delta} e^{-\frac{1}{2}\lambda Ar^2}\{1 + o(1)\}\,dr$$

$$= f(t_s)e^{\lambda\psi(t_s)+\frac{1}{2}(\pi-\phi)i} \left(\frac{2}{\lambda A}\right)^{1/2} \int_{-\eta}^{\eta} e^{-\tau^2}\{1 + o(1)\}\,d\tau, \quad (1.2.29)$$

where we have made the change of variable $\tau = (\frac{1}{2}\lambda A)^{1/2} r$ and

$$\eta = (\tfrac{1}{2}\lambda A)^{1/2}\delta = O(\lambda^{\frac{1}{2}-\epsilon}).$$

Since $\epsilon < \frac{1}{2}$, η tends to $+\infty$ with λ. The final step in the argument consists of replacing the limits of integration in (1.2.29) by $\pm\infty$, since the contribution $\int_{\eta}^{\infty} e^{-\tau^2}\,d\tau = O(\eta^{-1}e^{-\eta^2})$ is exponentially small, and similarly for the integral over $(-\infty, -\eta]$. We then find upon straightforward evaluation of the integral that the contribution to $I(\lambda)$ from the neighbourhood of the saddle point t_s is

$$f(t_s)e^{\lambda\psi(t_s)} \left(\frac{2\pi}{-\lambda\psi''(t_s)}\right)^{1/2} \qquad (\lambda \to \infty). \qquad (1.2.30)$$

This result, of course, agrees with the leading term in the expansion (1.2.19) in the case $m = 2$.

We remark that it is not necessary to take the path tangential to the steepest descent path at the saddle point, but that it may be inclined at any angle $< \frac{1}{4}\pi$ with this latter direction; that is, the integration path can be any line segment contained in the two valleys. This follows from the fact that $\psi''(t_s)(t - t_s)^2$ is then complex with negative real part and the integral (1.2.29) is unchanged in value by Cauchy's theorem. In the case of saddle points of higher order $m \geq 3$, the procedure is similar and the path can be taken at any angle $< \pi/(2m)$ to the direction of the steepest descent path at the saddle in a particular valley.

The method of stationary phase

The method of stationary phase can be employed to deal with Fourier-type integrals of the form

$$\int_a^b f(t)e^{i\lambda g(t)}\,dt, \qquad (1.2.31)$$

where a, b, λ and the functions $f(t)$, $g(t)$ are real. For large λ, the real and imaginary parts of the integrand oscillate rapidly, thereby producing a large degree of cancellation. Such cancellation does not occur at the endpoints (when finite) or in the neighbourhood of stationary points where $g'(t) = 0$. The reason for a significant contribution to the integral in these two cases is a result of lack of symmetry at the endpoints and the fact that $g(t)$ varies slowly near stationary points.

If $g(t)$ has no stationary point in $[a, b]$ a straightforward integration by parts shows that

$$\int_a^b f(t)e^{i\lambda g(t)}\,dt = \frac{f(b)e^{i\lambda g(b)}}{i\lambda g'(b)} - \frac{f(a)e^{i\lambda g(a)}}{i\lambda g'(a)} + O(\lambda^{-2}). \tag{1.2.32}$$

In the case of a single stationary point t_s in the interior of the interval $[a, b]$, we have

$$\int_a^b f(t)e^{i\lambda g(t)}\,dt \sim f(t_s)e^{i(\lambda g(t_s)\pm\pi/4)}\left|\frac{2\pi}{\lambda g''(t_s)}\right|^{1/2} \tag{1.2.33}$$

as $\lambda \to \pm\infty$, where the upper or lower sign is chosen according as $\lambda g''(t_s)$ is positive or negative. It can be seen that the contribution from a stationary point is $O(\lambda^{-1/2})$ whereas that from an endpoint is $O(\lambda^{-1})$. The argument used to obtain (1.2.33) is very similar to that used for the saddle-point method and we do not repeat this here; see, for example, Copson (1965, p. 31) and Olver (1997, pp. 96–102). It should be stressed that the stationary phase procedure can be extended to yield an asymptotic expansion; see Wong (1989, p. 77).

Referring to the integral (1.2.1), we can see that the method of stationary phase corresponds to choosing paths of constant $\mathrm{Re}\{\psi(t)\}$, instead of paths of constant $\mathrm{Im}\{\psi(t)\}$ in the method of steepest descents. An alternative procedure, which avoids stationary phase altogether but which requires analyticity of $\psi(t)$ and $f(t)$, is to treat the integral (1.2.31) as a Laplace-type integral and employ the method of steepest descents. This is the approach we shall adopt in this book.

1.3 Examples

In this section we present a series of examples to illustrate the use of the method of steepest descents in the asymptotic evaluation of Laplace-type integrals. The examples selected are basically in order of increasing difficulty. The first two examples are particularly simple since the original integration paths coincide with a steepest descent path through a saddle point and so require no path deformation. Some further, more difficult examples are considered in §1.4.

Example 1.1 One of the simplest examples illustrating the ideas of the method of steepest descents is the evaluation of the integral

$$I(\lambda) = \frac{1}{\sqrt{\pi}}\int_0^\infty e^{-\lambda(t^2-2t)}t^\nu\,dt \qquad (\mathrm{Re}(\nu) > -1) \tag{1.3.1}$$

for $\lambda \to +\infty$. With $\psi(t) = 2t - t^2$, the exponential factor has a single saddle point where $\psi'(t) = 0$, namely at $t_s = 1$. The steepest descent paths through the saddle are defined by

$$\psi(t) - \psi(1) = -(t-1)^2 = -u \qquad (u \geq 0);$$

since $\psi''(1) = -2$, we find from (1.2.4) (with $\phi = \pi$) that the directions of these paths through $t_s = 1$ are 0 and π. The steepest paths through $t_s = 1$ are described by

$$\text{Im}\{\psi(t)\} = 2(1-x)y = 0, \qquad t = x + iy.$$

Consequently, the line $y = 0$ on either side of $t_s = 1$ corresponds to the steepest descent paths and the vertical line $x = 1$ corresponds to the steepest ascent paths; see Fig. 1.3(a). The integration path $[0, \infty)$ is then seen to coincide with the steepest descent path through $t_s = 1$ and so, for $\lambda > 0$, no path deformation is required.

The inversion of the above quadratic change of variable $t \mapsto u$ is particularly simple and takes the form

$$t \equiv t^{\pm}(u) = 1 \pm u^{1/2}, \ u \geq 0 \qquad \begin{cases} t \geq 1 \\ t \leq 1, \end{cases}$$

where $t^{\pm}(u)$ refer to the two halves of the steepest descent path through $t_s = 1$. Accordingly, with $f(t) = t^{\nu}$, we have for $\lambda \to +\infty$

$$I(\lambda) \sim \frac{e^{\lambda}}{\sqrt{\pi}} \int_0^{\infty} e^{-\lambda u} \left(f(t^+)\frac{dt^+}{du} - f(t^-)\frac{dt^-}{du} \right) du,$$

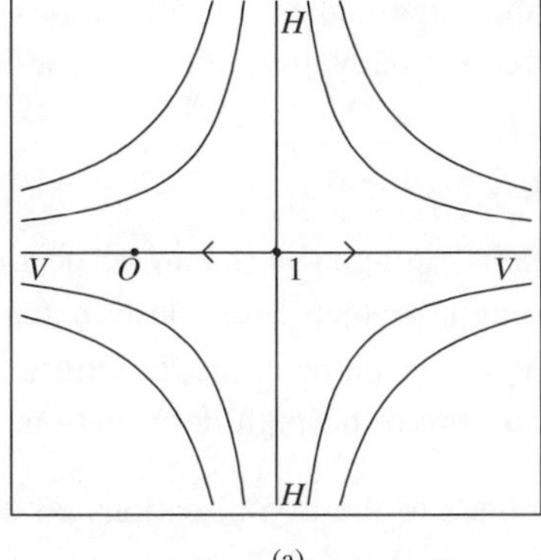

(a)

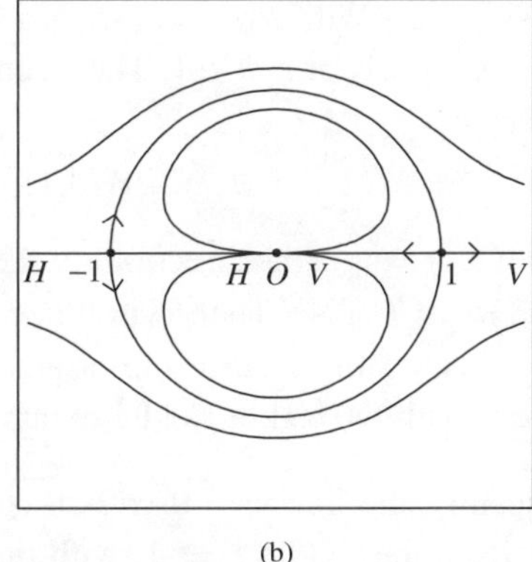

(b)

Figure 1.3 The paths of steepest descent and ascent for (a) $\psi(t) = 2t - t^2, t_s = 1$ and (b) $\psi(t) = -(t + 1/t), t_s = \pm 1$. The valleys (V) and the hills (H) at infinity are indicated and the arrows denote directions of steepest descent from the saddles.

where

$$f(t^+)\frac{dt^+}{du} - f(t^-)\frac{dt^-}{du} = \frac{1}{2}u^{-1/2}(1 + u^{1/2})^\nu + \frac{1}{2}u^{-1/2}(1 - u^{1/2})^\nu$$

$$= \frac{1}{2}\sum_{k=0}^{\infty}\binom{\nu}{k}u^{(k-1)/2}\{1 + (-)^k\}$$

$$= \sum_{k=0}^{\infty}\binom{\nu}{2k}u^{k-\frac{1}{2}} \qquad (|u| < 1).$$

Then, from (1.2.19), we obtain the expansion

$$I(\lambda) \sim \frac{e^\lambda}{\sqrt{\pi}}\sum_{k=0}^{\infty}\binom{\nu}{2k}\int_0^\infty u^{k-\frac{1}{2}}e^{-\lambda u}du = \frac{e^\lambda}{\sqrt{\pi}}\sum_{k=0}^{\infty}\binom{\nu}{2k}\frac{\Gamma(k + \frac{1}{2})}{\lambda^{k+\frac{1}{2}}}.$$

Use of the duplication formula for the gamma function and the result

$$\frac{\nu!}{(\nu - 2k)!} = (-\nu)_{2k} = (-\tfrac{1}{2}\nu)_k(\tfrac{1}{2} - \tfrac{1}{2}\nu)_k 2^{2k}$$

enables us to express the above expansion in the alternative form

$$I(\lambda) \sim e^\lambda \sum_{k=0}^{\infty}\frac{(-\frac{1}{2}\nu)_k(\frac{1}{2} - \frac{1}{2}\nu)_k}{k!\,\lambda^{k+\frac{1}{2}}} \qquad (\lambda \to +\infty). \tag{1.3.2}$$

This example is considered further in Example 3.4 where the Hadamard expansion of $I(\lambda)$ is derived.

Example 1.2 Consider the integral

$$I(\lambda) = \int_0^\infty e^{-\lambda(t+t^{-1})}dt \tag{1.3.3}$$

for $\lambda \to +\infty$. With $\psi(t) = -(t + t^{-1})$, the integrand has saddle points where $\psi'(t) = 0$, namely at $t = \pm 1$. The steepest descent paths through these saddles are defined by

$$\psi(t) - \psi(\pm 1) = -u \qquad (u \geq 0);$$

since $\psi''(\pm 1) = \mp 2$, the directions of these paths through $t = \pm 1$ are respectively 0, π and $\pm\frac{1}{2}\pi$ by (1.2.4). The steepest paths through the saddles are described by

$$\mathrm{Im}\{\psi(t)\} = -y\left(1 - \frac{1}{x^2 + y^2}\right) = 0, \qquad t = x + iy.$$

Consequently, the line $y = 0$, $x > 0$ corresponds to the two directions of steepest descent from the saddle $t = 1$, with the line $y = 0$, $x < 0$ corresponding to the two directions of steepest ascent from the saddle at $t = -1$. The steepest ascent path through $t = 1$ is the unit circle $x^2 + y^2 = 1$. This circular path connects with the saddle at $t = -1$, as shown in Fig. 1.3(b), where it is locally the steepest descent path. Inside this circle, the paths of constant $\text{Im}\{\psi(t)\}$ emanate from the origin in $\text{Re}(t) > 0$ and return to the origin in $\text{Re}(t) < 0$; on these closed paths $\text{Re}\{\psi(t)\}$ increases steadily from $-\infty$ to $+\infty$. For $\lambda > 0$, the integration path in (1.3.3) passes over the saddle point at $t = 1$ and coincides with the steepest descent paths. Thus no path deformation is required.

Since $\psi^{(n)}(1) = (-)^{n-1}n!\ (n = 2, 3, \ldots)$, we obtain from (1.2.20) the coefficients

$$c_0 = c_1 = \tfrac{1}{2}, \quad c_2 = \tfrac{3}{16}, \quad c_3 = 0, \quad c_4 = -\tfrac{5}{256}.$$

Then, from (1.2.19), it follows that

$$I(\lambda) \sim \sqrt{\frac{\pi}{\lambda}}\, e^{-2\lambda} \left(1 + \frac{3}{16\lambda} - \frac{15}{512\lambda^2} + \cdots \right) \qquad (\lambda \to +\infty). \tag{1.3.4}$$

Higher terms in this expansion can be obtained by inversion of the equation defining the steepest descent paths through $t = 1$ given by $\psi(t) - \psi(1) = -u$, or

$$\frac{(t - 1)}{t^{1/2}} = u^{1/2}.$$

By means of the Lagrange inversion theorem in (1.2.12) with $\phi(t) = t^{1/2}$, or with the InverseSeries command in *Mathematica*, we find

$$t - 1 = \sum_{n=1}^{\infty} \frac{u^{n/2}}{n!} \left[\frac{d^{n-1}}{dt^{n-1}} t^{n/2} \right]_{t=1} = \frac{1}{2}u + \sum_{j=0}^{\infty} \frac{u^{j+\frac{1}{2}}}{(2j+1)!} \left[\frac{d^{2j}}{dt^{2j}} t^{j+\frac{1}{2}} \right]_{t=1}, \tag{1.3.5}$$

since the terms corresponding to even $n \geq 4$ all vanish. By means of the duplication formula for the gamma function it is then straightforward to show that

$$\frac{1}{(2j+1)!} \left[\frac{d^{2j}}{dt^{2j}} t^{j+\frac{1}{2}} \right]_{t=1} = \frac{(j+\frac{1}{2})\ldots(-j+\frac{3}{2})}{(2j+1)!} = \frac{(-)^{j-1}\Gamma(j-\frac{1}{2})}{\sqrt{\pi}\, 2^{2j+1}\, j!}.$$

The ratio test shows that the series in (1.3.5) converges [9] in the disc $|u| < 4$. It then follows that

$$\frac{dt}{du} = \sum_{k=0}^{\infty} c_k u^{(k-1)/2} \qquad (|u| < 4),$$

where $c_1 = \frac{1}{2}, c_3 = c_5 = c_7 = \ldots = 0$ and

$$c_{2k} = \frac{(-)^{k-1}\Gamma(k-\frac{1}{2})(k+\frac{1}{2})}{\sqrt{\pi}\, 2^{2k+1}\, k!} \qquad (k = 0, 1, 2, \ldots).$$

---

[9] This can also be seen from the fact that the nearest singularity in the mapping $t \mapsto u$ corresponds to the saddle at $t = -1$, where $u = -4$. The singularity at $t = 0$ maps to ∞ in the u-plane.

Then, from (1.2.19), we obtain the expansion

$$
I(\lambda) \sim \frac{e^{-2\lambda}}{\sqrt{\pi}} \sum_{k=0}^{\infty} \frac{(-)^{k-1}\Gamma(k-\frac{1}{2})\Gamma(k+\frac{3}{2})}{2^{2k}\,k!\,\lambda^{k+\frac{1}{2}}}
$$

$$
= \sqrt{\frac{\pi}{\lambda}}\, e^{-2\lambda}\left(1 + \frac{3}{16\lambda} - \frac{15}{512\lambda^2} + \frac{105}{8192\lambda^3} - \frac{4725}{524288\lambda^4} + \cdots \right) \qquad (1.3.6)
$$

as $\lambda \to +\infty$.

The result in (1.3.6) is easily verified since $I(x)$ can be expressed in terms of a modified Bessel function. From Watson (1952, p. 183), we find

$$
K_\nu(2\lambda) = \frac{1}{2}\int_0^\infty e^{-\lambda(t+t^{-1})} t^{-\nu-1} dt,
$$

so that $I(\lambda) = 2K_{-1}(2\lambda)$. The expansion (1.3.6) then follows from the well-known asymptotic expansion of $K_\nu(z)$ given in Abramowitz and Stegun (1965, p. 378).

It is worth noting that for the integral

$$
\hat{I}(\lambda) = \int_1^\infty e^{-\lambda(t+t^{-1})} dt,
$$

with $\lambda > 0$, the integration path coincides with the steepest descent path emanating from the saddle at $t = 1$ to the right. Then, by (1.2.18), we obtain the expansion

$$
\hat{I}(\lambda) \sim e^{-2\lambda} \sum_{k=0}^{\infty} \frac{c_k\,\Gamma(\frac{1}{2}k + \frac{1}{2})}{\lambda^{(k+1)/2}}
$$

$$
= \frac{e^{-2\lambda}}{2\lambda} + \frac{1}{2}\sqrt{\frac{\pi}{\lambda}}\, e^{-2\lambda}\left(1 + \frac{3}{16\lambda} - \frac{15}{512\lambda^2} + \frac{105}{8192\lambda^3} - \cdots \right) \qquad (1.3.7)
$$

as $\lambda \to +\infty$.

Example 1.3 We consider the derivation of the asymptotic expansion of the gamma function. We commence with Hankel's integral [10] for the reciprocal of $\Gamma(z)$

$$
\frac{1}{\Gamma(z)} = \frac{1}{2\pi i}\int_C e^\tau \tau^{-z} d\tau,
$$

which holds for all real and complex z when the path C is a loop that starts at $\infty e^{-\pi i}$, encircles the origin in the positive sense and ends at $\infty e^{\pi i}$. Let us first take z to be real and positive. Then we can put $\tau = zt$ to find

$$
\frac{1}{\Gamma(z)} = \frac{e^z z^{1-z}}{2\pi i}\int_C e^{z\psi(t)} dt, \qquad \psi(t) = t - 1 - \log t, \qquad (1.3.8)
$$

where the t-plane is cut along the negative real axis.

---

[10] The quantity τ^{-z} denotes $\exp(-z\log\tau)$, where $\log\tau$ has its principal value.

The integrand has a saddle point where $\psi'(t) = 0$, namely at $t_s = 1$, and $\psi(t_s) = 0$. The steepest descent path is therefore given by

$$\psi(t) - \psi(t_s) = -u, \tag{1.3.9}$$

where $u \geq 0$; since $\psi''(t_s) = 1$, it follows from (1.2.4) that the directions of these paths at the saddle point are $\pm\frac{1}{2}\pi$. The steepest paths through t_s are described by

$$\mathrm{Im}\{\psi(t)\} = y - \arctan(y/x) = 0, \qquad t = x + iy.$$

The line $y = 0$, $x > 0$ corresponds to the two directions of the steepest ascent from t_s, whereas the curve specified by $x = y/\tan y$ gives the two halves of the steepest descent path. These latter paths are symmetrical about the real axis and it is easily seen that $y \to \pm\pi$ as $x \to -\infty$; see Fig. 1.4(a). By Cauchy's theorem, the loop C can then be made to coincide with the steepest descent path through the saddle point $t_s = 1$.

If, in (1.3.9), we make use of the intermediate variable $u = -\frac{1}{2}w^2$, we obtain

$$\psi(t) = \tfrac{1}{2}(t-1)^2 \left\{ 1 - \tfrac{2}{3}(t-1) + \tfrac{2}{4}(t-1)^2 + \cdots \right\} = \tfrac{1}{2}w^2.$$

Inversion of this last expression then yields

$$t^{\pm}(w) - 1 = \sum_{k=1}^{\infty} \alpha_k(\pm w)^k = \sum_{k=1}^{\infty} \alpha_k(\pm i)^k (2u)^{k/2} \tag{1.3.10}$$

valid in a neighourhood of $w = 0$ ($u = 0$), where the upper or lower sign signifies the upper or lower half of the steepest descent path, respectively. The first few coefficients α_k are

$$\alpha_1 = 1, \quad \alpha_2 = \tfrac{1}{3}, \quad \alpha_3 = \tfrac{1}{36}, \quad \alpha_4 = -\tfrac{1}{270}, \quad \alpha_5 = \tfrac{1}{4320}, \quad \alpha_6 = \tfrac{1}{17010}, \cdots .$$

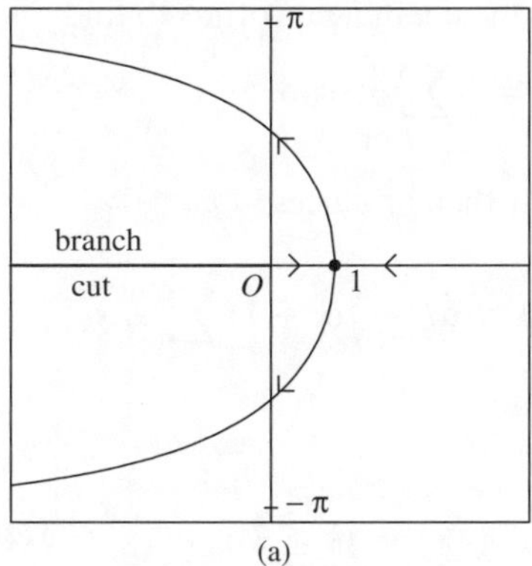

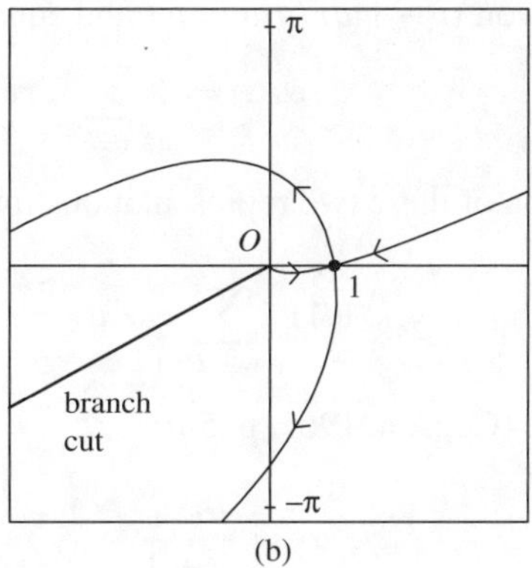

Figure 1.4 The paths of steepest descent and ascent through the saddle $t_s = 1$ when (a) $\arg z = 0$ and (b) $\arg z = -\pi/8$. The arrows denote directions of descent and the heavy line denotes the branch cut.

The mapping $t \mapsto w$ has singular points where $dt/dw = wt/(t-1)$ is singular; that is, at $t = 1$. Taking into account the multi-valuedness of the logarithm, we find that there are branch points of $t(w)$ at $w = \pm 2\sqrt{(\pi n)}e^{\pm \pi i/4}$ for non-negative integer n. The expansion (1.3.10) for $t(w)$ therefore converges in a disc about $w = 0$ controlled by the nearest singularities; that is, in the disc $|w| < 2\sqrt{\pi}$, or in the disc $|u| < 2\pi$ in the u-plane. It follows that

$$\frac{dt^{\pm}}{du} = \sum_{k=1}^{\infty} k\alpha_k (\pm i)^k (2u)^{k/2-1} = \sum_{k=0}^{\infty} (\pm i)^{k+1} c_k u^{(k-1)/2},$$

in $|u| < 2\pi$, where $c_k = 2^{(k-1)/2}(k+1)\alpha_{k+1}$.

The contribution from the saddle t_s from both halves of the steepest descent path is, from (1.2.19),

$$\frac{1}{\Gamma(z)} = \frac{e^z z^{1-z}}{2\pi i} \int_0^{\infty} e^{-zu} \left(\frac{dt^+}{du} - \frac{dt^-}{du}\right) du$$

$$\sim \frac{e^z z^{1-z}}{\pi} \sum_{k=0}^{\infty} (-)^k c_{2k} \frac{\Gamma(k + \frac{1}{2})}{z^{k+\frac{1}{2}}}. \tag{1.3.11}$$

This then yields the expansion

$$\frac{1}{\Gamma(z)} = \frac{e^z z^{\frac{1}{2}-z}}{\sqrt{2\pi}} \sum_{k=0}^{\infty} \gamma_k z^{-k}, \tag{1.3.12}$$

where

$$\gamma_k = (-)^k \sqrt{\frac{2}{\pi}} c_{2k} \Gamma(k + \tfrac{1}{2}) = (-)^k 2^{k+1} \alpha_{2k+1} \frac{\Gamma(k + \frac{3}{2})}{\sqrt{\pi}}. \tag{1.3.13}$$

The values of the so-called Stirling coefficients γ_k can be generated by means of the recurrence relation obtained by substitution of the expansion (1.3.10) into the expression $(t-1)dt/dw = wt$ and equating the coefficients of w^k to find

$$\alpha_{k-1} = \sum_{r+s=k+1} r\alpha_r \alpha_s = \sum_{r+s=k+1} s\alpha_s \alpha_r.$$

Addition of these two representations for α_{k-1} then produces

$$\alpha_{k-1} = \tfrac{1}{2}(k+1) \sum_{r+s=k+1} \alpha_r \alpha_s = (k+1)\alpha_1 \alpha_k + \tfrac{1}{2}(k+1) \sum_{r=2}^{k-1} \alpha_r \alpha_{k+1-r},$$

whence (Copson, 1965, p. 56)

$$\alpha_k = \frac{\alpha_{k-1}}{k+1} - \frac{1}{2} \sum_{r=2}^{k-1} \alpha_r \alpha_{k+1-r} \qquad (k \geq 2), \tag{1.3.14}$$

where $\alpha_1 = 1$ and an empty sum is interpreted as zero. The values of γ_k for $1 \leq k \leq 10$ (with $\gamma_0 = 1$) are presented in Table 1.1; see also Wrench (1968). We note

Table 1.1 *The Stirling coefficients γ_k for $1 \le k \le 10$* (with $\gamma_0 = 1$)

k	γ_k	k	γ_k
1	$-\dfrac{1}{12}$	2	$\dfrac{1}{288}$
3	$\dfrac{139}{51\,840}$	4	$-\dfrac{571}{2\,488\,320}$
5	$-\dfrac{163\,879}{209\,018\,880}$	6	$\dfrac{5\,246\,819}{75\,246\,796\,800}$
7	$\dfrac{534\,703\,531}{902\,961\,561\,600}$	8	$-\dfrac{4\,483\,131\,259}{86\,684\,309\,913\,600}$
9	$-\dfrac{432\,261\,921\,612\,371}{514\,904\,800\,886\,784\,000}$	10	$\dfrac{6\,232\,523\,202\,521\,089}{86\,504\,006\,548\,979\,712\,000}$

that a representation for the Stirling coefficients has been obtained in López, Pagola and Pérez Sinusía (2009) in the form

$$\gamma_k = \sum_{m=0}^{2k} \sum_{r=0}^{m} \frac{(\frac{1}{2})_{k+m}}{r!\,2^{r-k-m}} \sum_{j=0}^{m-r} \frac{(-)^{j+k} S_{2k+2m-2r-j}^{(m-r-j)}}{j!\,(2k+2m-2r-j)!},$$

where $S_n^{(m)}$ denotes the Sirling number of the first kind (Abramowitz and Stegun, 1965, p. 824).

When z is complex the integral in (1.3.8) defines $1/\Gamma(z)$ in the sector $|\arg z| \le \frac{1}{2}\pi - \delta$, $\delta > 0$. If we put $z = |z|e^{i\theta}$, the phase factor $e^{i\theta}$ can be absorbed into $\psi(t)$ and we have[11]

$$\psi(t) = e^{i\theta}(t - 1 - \log t).$$

The directions of the steepest descent path at $t_s = 1$ are now $\pm \frac{1}{2}\pi - \frac{1}{2}\theta$ and the associated steepest descent paths are described by

$$\mathrm{Im}\{\psi(t)\} = \cos\theta\,(y - \arg t) + \sin\theta\,(x - 1 - \log|t|) = 0.$$

This shows that the paths of steepest descent and ascent at large distance from the origin are parallel to the line $y = -x\tan\theta$, and so are similar to those when $\arg z = 0$, but rotated through an angle $-\theta$; see Fig. 1.4(b). The analysis is then essentially unchanged with the result that the expansion (1.3.12) is valid for $|z| \to \infty$ in the sector $|\arg z| \le \frac{1}{2}\pi - \delta$. This result can also be deduced from Watson's lemma in §1.2.4 applied to (1.3.11). Since dt/du has branch points at $\pm 2\pi ni$ on the imaginary u-axis, it follows that dt/du is holomorphic within the sector $|\arg u| < \frac{1}{2}\pi$. So, by (1.2.26), the expansion (1.3.12) is in fact valid for $|z| \to \infty$ in the wider sector $|\arg z| \le \pi - \delta$.

---

[11] A different transformation $\tau = |z|t$ is employed in Bleistein and Handelsman (1975, p. 286) to obtain the integral in (1.3.8) with $\psi(t) = e^{i\theta}\log t - t$ valid for all θ. In this case, the principal saddle point is $t_s = e^{i\theta}$.

A similar treatment can be brought to bear on the integral

$$\Gamma(z) = z^{-1} \int_0^\infty e^{-\tau} \tau^z d\tau = e^{-z} z^{z-1} \int_0^\infty e^{-z\psi(t)} dt,$$

where again $\psi(t) = t - 1 - \log t$ and initially we suppose $z > 0$. The steepest descent path through the saddle point $t_s = 1$ is now the positive real axis. Then, with $\psi(t) - \psi(t_s) = u$, $u \geq 0$, we have

$$\frac{dt^\pm}{du} = \sum_{k=0}^\infty (\pm)^{k+1} c_k u^{(k-1)/2} \qquad (|u| < 2\pi)$$

and

$$\Gamma(z) \sim \sqrt{2\pi}\, e^{-z} z^{z-\frac{1}{2}} \sum_{k=0}^\infty (-)^k \gamma_k z^{-k} \tag{1.3.15}$$

as $z \to +\infty$, where the coefficients γ_k are those appearing in (1.3.12); see also Temme (1996, p. 71). By Watson's lemma and (1.2.26), this expansion also holds for $|z| \to \infty$ in the sector $|\arg z| \leq \pi - \delta$. Error bounds for the asymptotic expansion of $\Gamma(z)$ are given in Olver (1968, 1970).

Example 1.4 The Airy integral is defined by

$$\mathrm{Ai}(z) = \frac{1}{2\pi i} \int_C e^{\frac{1}{3}\tau^3 - z\tau} d\tau,$$

where C is any path that starts at infinity in the sector $-\frac{1}{2}\pi < \arg \tau < -\frac{1}{6}\pi$ and ends at infinity in the sector $\frac{1}{6}\pi < \arg \tau < \frac{1}{2}\pi$. To put the integral in its standard form, we initially suppose that $z > 0$ and make the substitution $\tau = z^{1/2} t$ to find

$$\mathrm{Ai}(z) = \frac{z^{1/2}}{2\pi i} \int_C e^{\lambda \psi(t)} dt, \qquad \psi(t) = \tfrac{1}{3} t^3 - t, \tag{1.3.16}$$

where $\lambda = z^{3/2}$. By analytic continuation, we note that the integral (1.3.16) defines $\mathrm{Ai}(z)$ in the sector $|\arg z| < \frac{1}{3}\pi$.

The integrand has saddle points where $\psi'(t) = 0$, namely at the points $t_s = \pm 1$, where $\psi(\pm 1) = \mp\frac{2}{3}$. The steepest descent paths are given by

$$\psi(t) - \psi(t_s) = -u, \tag{1.3.17}$$

where $u \geq 0$; since $\psi''(\pm 1) = \pm 2$, it follows from (1.2.4) that the directions of these paths at the saddle $t_s = 1$ are $\pm\frac{1}{2}\pi$, whereas those at the saddle $t_s = -1$ have the directions 0 and π. The bottoms of the valleys at infinity are along the rays $\arg t = \pm\frac{1}{3}\pi$ and $\arg t = \pi$, whereas the ridges of the hills are along the rays $\arg t = 0$ and $\arg t = \pm\frac{2}{3}\pi$. The steepest paths through $t_s = \pm 1$ (when $z > 0$) are described by

$$\mathrm{Im}\{\psi(t)\} = y(x^2 - \tfrac{1}{3} y^2 - 1) = 0, \qquad t = x + iy.$$

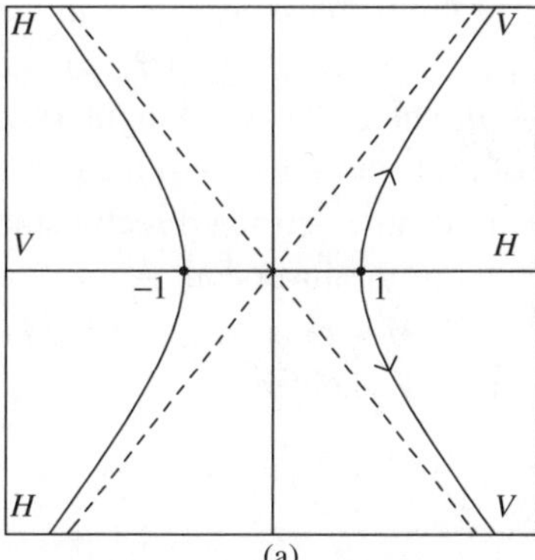 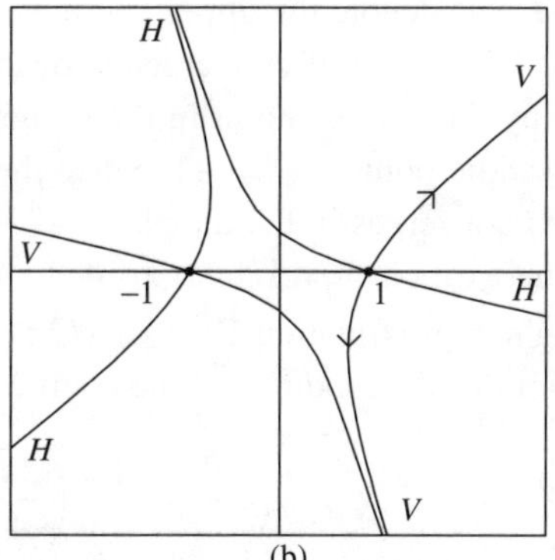

Figure 1.5 The steepest paths for the Airy integral (1.3.16) when (a) arg $z = 0$ and (b) arg $z = \pi/4$. The dashed lines are the asymptotes arg $t = \pm\pi/3$. The valleys (V) and the hills (H) at infinity are indicated and the arrows denote directions of descent.

The part of the line $y = 0$, $x > -1$ is a steepest ascent path for the saddle at $t_s = 1$, whereas the part $x < 1$ is a steepest descent path for the saddle at $t_s = -1$. The other paths of $\mathrm{Im}\{\psi(t)\} = 0$ are the hyperbola $x^2 - \frac{1}{3}y^2 = 1$, with the branches through $t_s = \pm 1$ being steepest descent and ascent paths, respectively. The asymptotes of the hyperbola are $y = \pm\sqrt{3}x$, so that the steepest descent path through $t_s = 1$ approaches the bottoms of the valleys along the rays arg $t = \pm\frac{1}{3}\pi$ as $u \to +\infty$. The steepest paths through $t_s = \pm 1$ in the case arg $z = 0$ are illustrated in Fig. 1.5(a). By Cauchy's theorem, the original path C can be made to coincide with the steepest descent path through the saddle point at $t_s = 1$.

From (1.3.17) wth $t_s = 1$, we find

$$\psi(t) - \psi(1) = \tfrac{1}{3}t^3 - t + \tfrac{2}{3} = \tfrac{1}{3}(t-1)^3 + (t-1)^2 = -u,$$

or

$$(t-1)\,(\tfrac{1}{3}t + \tfrac{2}{3})^{1/2} = \pm iu^{1/2},$$

where the square root $(\tfrac{1}{3}t + \tfrac{2}{3})^{1/2}$ takes the value 1 at $t = 1$. In this example we are in the fortunate position of being able to determine the inversion coefficients in the mapping $t \mapsto u$ explicitly. By Lagrange's inversion theorem in (1.2.12) we obtain

$$t^{\pm}(u) - 1 = \sum_{k=1}^{\infty} \alpha_k \frac{(\pm iu^{\frac{1}{2}})^k}{k!}, \qquad \alpha_k = \frac{d^{k-1}}{dt^{k-1}}\,(\tfrac{1}{3}t + \tfrac{2}{3})^{-k/2}\Big|_{t=1},$$

$$t^{\pm}(u) - 1 = \sum_{k=1}^{\infty} \frac{(-)^{k-1}(\pm iu^{\frac{1}{2}})^k}{3^{k-1}\,k!}\,\frac{\Gamma(\tfrac{3}{2}k - 1)}{\Gamma(\tfrac{1}{2}k)}, \tag{1.3.18}$$

and hence that

$$\frac{dt^{\pm}}{du} = \pm\frac{1}{2}i \sum_{k=0}^{\infty} \frac{(\mp i)^k \Gamma(\tfrac{3}{2}k + \tfrac{1}{2})}{3^k\,k!\,\Gamma(\tfrac{1}{2}k + \tfrac{1}{2})}\,u^{(k-1)/2}, \tag{1.3.19}$$

where $t^{\pm}(u)$ denote the upper and lower halves of the steepest descent path through $t_s = 1$. The circle of convergence of the expansions (1.3.18) and (1.3.19) is determined by the nearest point in the mapping $t \mapsto u$ where dt/du is singular; that is, at the saddle point at $t = -1$. Since the value of u at $t = -1$ is $-\frac{4}{3}$, the expansion (1.3.18) converges in the disc $|u| < \frac{4}{3}$. This result can be verified directly since we have the explicit representation of the coefficients in the inversion. Application of Stirling's approximation $\Gamma(z) \simeq \sqrt{2\pi}\, e^{-z} z^{z-\frac{1}{2}}$ as $z \to +\infty$ shows that the large-k behaviour of the modulus of the coefficients in (1.3.18) is given by

$$\sqrt{\frac{2}{3\pi}}\, k^{-3/2} \left(\frac{3u}{4}\right)^{k/2} \qquad (k \to \infty). \tag{1.3.20}$$

Use of the ratio test then confirms that (1.3.18) and (1.3.19) converge in the disc $|u| < \frac{4}{3}$.

The contribution from the saddle point $t_s = 1$ from both halves of the steepest descent path is, from (1.2.19), (1.3.17) and (1.3.19),

$$\begin{aligned}
\mathrm{Ai}(z) &= \frac{z^{1/2}}{2\pi i} e^{-2\lambda/3} \int_0^\infty e^{-\lambda u} \left(\frac{dt^+}{du} - \frac{dt^-}{du}\right) du \\
&\sim \frac{z^{1/2}}{2\pi} e^{-2\lambda/3} \sum_{k=0}^\infty \frac{(-)^k \Gamma(3k + \frac{1}{2})}{3^{2k}(2k)!\,\lambda^{k+1/2}}.
\end{aligned}$$

Upon substituting the value $\lambda = z^{3/2}$, we finally obtain the desired asymptotic expansion of $\mathrm{Ai}(z)$ given by

$$\mathrm{Ai}(z) \sim \frac{z^{-1/4}}{2\pi} \exp(-\tfrac{2}{3}z^{3/2}) \sum_{k=0}^\infty \frac{(-)^k \Gamma(3k + \frac{1}{2})}{3^{2k}(2k)!\, z^{3k/2}} \tag{1.3.21}$$

as $z \to +\infty$.

When z is complex, we let $\theta = \arg z$ and write the exponential factor in (1.3.16) as $\exp\{|\lambda|\psi(t)\}$, where now $\psi(t) = e^{3i\theta/2}(\frac{1}{3}t^3 - t)$. Then it is easily seen that the directions of the steepest descent paths at the saddle point $t_s = 1$ are $\pm\frac{1}{2}\pi - \frac{3}{4}\theta$, whereas those at the saddle point $t_s = -1$ are $-\frac{3}{4}\theta$ and $\pi - \frac{3}{4}\theta$. The bottoms of the valleys at infinity are along the rays $\arg t = \pm\frac{1}{3}\pi - \frac{1}{2}\theta$ and $\pi - \frac{1}{2}\theta$. The steepest paths through $t_s = \pm 1$ are described by $\mathrm{Im}\{\psi(t)\} = \mp\frac{2}{3}\sin\frac{3}{2}\theta$, respectively; that is

$$\tfrac{1}{3}\rho^3 \sin(3\phi + \tfrac{3}{2}\theta) - \rho\sin(\phi + \tfrac{3}{2}\theta) = \mp\tfrac{2}{3}\sin\tfrac{3}{2}\theta,$$

where $t = \rho e^{i\phi}$, $\rho > 0$. As $\rho \to \infty$, these paths must be asymptotic to the rays defined by $\sin(3\phi + \frac{3}{2}\theta) = 0$; that is, the steepest descent paths also approach the rays $\arg t = \pm\frac{1}{3}\pi - \frac{1}{2}\theta$ and $\arg t = \pi - \frac{1}{2}\theta$. An illustration of the steepest paths is shown in Fig. 1.5(b) in the particular case $\theta = \frac{1}{4}\pi$. It is evident that when $\theta \neq 0$ the saddle points become disconnected in the sense that they no longer share a common path of constant $\mathrm{Im}\{\psi(t)\}$ as they do in the case $\theta = 0$. Provided that $|\theta| < \frac{1}{3}\pi$, the steepest descent path through $t_s = 1$ passes to infinity in the sectors $\frac{1}{2}\pi < |\arg t| < \frac{1}{6}\pi$. This latter path is therefore an acceptable integration path in (1.3.16) and, consequently, the expansion (1.3.21) is valid in the sector $|\arg z| \leq \frac{1}{3}\pi - \delta$, $\delta > 0$. This last result can also be deduced from Watson's lemma (1.2.24).

Extension of the expansion of $\mathrm{Ai}(z)$ to a wider sector that includes the rays $\arg z = \pm\pi$ relies on the analytic connection between Airy functions of rotated argument given in (3.3.28) and is not discussed here; see, for example, Copson (1965, p. 103) and Olver (1997, p. 117).

Example 1.5 Let

$$I(x) = \frac{1}{\Gamma(1 + 1/n)} \int_0^1 e^{ixt^n} dt,$$

where x is a large positive variable and n denotes an integer satisfying $n \geq 2$. The integrand has a saddle point of order $n - 1$ at $t = 0$. If we set $\psi(t) = it^n$ and $t = \rho e^{i\phi}$ (where ρ and ϕ are real), the steepest paths from the origin are given by $\mathrm{Im}\{\psi(t)\} = \rho^n \cos n\phi = 0$; that is, the rays

$$\arg t = \pm\frac{(2k - 1)\pi}{2n} \qquad (1 \leq k \leq n).$$

The steepest path through $t = 1$ is described by

$$\psi(t) - \psi(1) = i(t^n - 1) = -u \qquad (-\infty < u < \infty)$$

or $t = (1 + iu)^{1/n}$, where the principal value of the nth root is taken. This describes the steepest path in parametric form; in terms of the real and imaginary parts of t this path is a polynomial curve of degree n. In Fig. 1.6 we show the steepest paths from the origin closest to the positive real axis together with the steepest path through the point $t = 1$.

The integration path $[0, 1]$ is not a steepest path. However, it can be decomposed into such paths by taking the ray $\arg t = \pi/(2n)$ from 0 to ∞ and then from ∞ to 1 along the upper half of the steepest path through $t = 1$. Accordingly, we have

$$I(x) = \frac{1}{\Gamma(1 + 1/n)} \left\{ \int_0^{\infty e^{\pi i/(2n)}} - \int_1^{\infty e^{\pi i/(2n)}} \right\} e^{x\psi(t)} dt \equiv \frac{I_1 - I_2}{\Gamma(1 + 1/n)}. \qquad (1.3.22)$$

The integral I_1 along the ray $\arg t = \pi/(2n)$ can be evaluated in terms of the gamma function as follows

$$I_1 = \int_0^{\infty e^{\pi i/(2n)}} e^{x\psi(t)} dt = e^{\pi i/(2n)} \int_0^\infty e^{-x\tau^n} d\tau$$

$$= \frac{e^{\pi i/(2n)}}{n} \int_0^\infty e^{-xw} w^{(1-n)/n} dw$$

$$= e^{\pi i/(2n)} \Gamma(1 + 1/n) x^{-1/n}. \qquad (1.3.23)$$

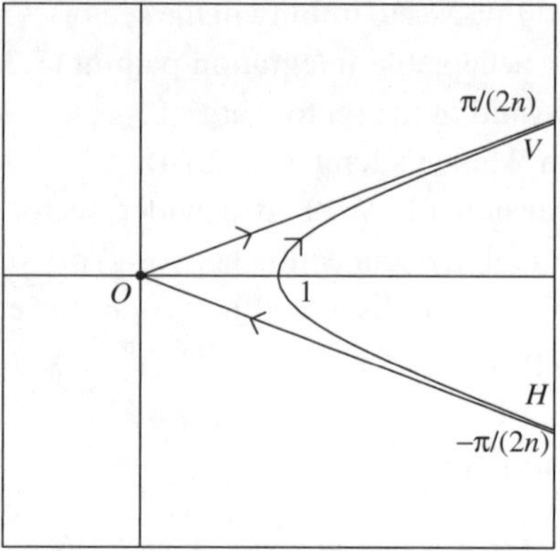

Figure 1.6 The steepest paths for the integral $I(x)$. The valley (V) and hill (H) along the rays $\arg t = \pm \pi/(2n)$, respectively, are indicated and the arrows denote the directions of descent.

In the integral I_2, we make the change of variable $t = (1 + iu)^{1/n}$ to find

$$I_2 = \int_1^{\infty e^{\pi i/(2n)}} e^{x\psi(t)} dt = e^{ix} \int_0^\infty e^{-xu} \frac{dt}{du} \, du$$

$$= \frac{ie^{ix}}{n} \int_0^\infty e^{-xu} (1 + iu)^{(1-n)/n} du.$$

Expansion of $(1 + iu)^{-\alpha}$ by the binomial theorem, where $\alpha = (n - 1)/n$, produces

$$(1 + iu)^{-\alpha} = \frac{1}{\Gamma(\alpha)} \sum_{k=0}^\infty \frac{(-iu)^k}{k!} \Gamma(k + \alpha) \qquad (|u| < 1).$$

By Watson's lemma (1.2.24) we then find the expansion

$$I_2 \sim \frac{e^{ix}}{n\Gamma(1 - 1/n)} \sum_{k=0}^\infty \frac{\Gamma(k + 1 - 1/n)}{(ix)^{k+1}} \tag{1.3.24}$$

as $x \to +\infty$.

Substituting the results (1.3.23) and (1.3.24) into (1.3.22), we obtain the asymptotic expansion of $I(x)$ as

$$I(x) \sim e^{\pi i/(2n)} x^{-1/n} + \frac{e^{ix}}{\pi} \sin(\pi/n) \sum_{k=0}^\infty \frac{\Gamma(k + 1 - 1/n)}{(ix)^{k+1}} \tag{1.3.25}$$

as $x \to +\infty$. From the conditions in Watson's lemma it is easy to see that the expansion (1.3.25) also holds for large complex values of x in the sector $|\arg x| \leq \frac{1}{2}\pi - \delta$, $\delta > 0$. Furthermore, it is not difficult to establish that (1.3.25) is valid when n is an arbitrary parameter satisfying $n > 1$. The trivial case $n = 1$ can be verified by using a limiting process to (1.3.25).

Examples of this integral in the cases $n = 2$ and $n = 3$ have been given in Bender and Orszag (1978, p. 283) and Erdélyi (1956, p. 45), respectively.

Example 1.6 Consider the Fourier integral [12]

$$I(x) = \int_0^\pi e^{ix(t-\sin t)} dt \qquad (1.3.26)$$

for large positive values of x. With $\psi(t) = i(t - \sin t)$, the saddle points of the integrand are given by $\psi'(t) = i(1 - \cos t) = 0$; that is, at the points $t = 0, \pm 2\pi, \pm 4\pi, \dots$. At these points, it is seen that $\psi''(t) = 0$ (with $\psi'''(t) \neq 0$) and the saddles are all double. In the neighbourhood of $t = 0$, we have $\psi(t) \simeq it^3/3!$ so that the directions of steepest descent at $t = 0$ are $\frac{1}{6}\pi$, $\frac{5}{6}\pi$ and $-\frac{1}{2}\pi$, with the directions of steepest ascent arranged so as to bisect the angular spacing between these directions. The paths of constant $\mathrm{Im}\{\psi(t)\}$ emanating from the origin are described by

$$\mathrm{Im}\{\psi(t)\} = \sigma - \sin\sigma \cosh\tau = 0, \qquad t = \sigma + i\tau.$$

The axis $\sigma = 0$ is consequently a steepest path, with the others defined by the curves

$$\tau = \pm\mathrm{arccosh}\left(\frac{\sigma}{\sin\sigma}\right) \qquad (-\pi < \sigma < \pi).$$

These latter paths are consequently asymptotic to the lines $\sigma = \pm\pi$ and are illustrated in Fig. 1.7. The steepest path through the point $t = \pi$ is readily shown to be the line $\sigma = \pi$.

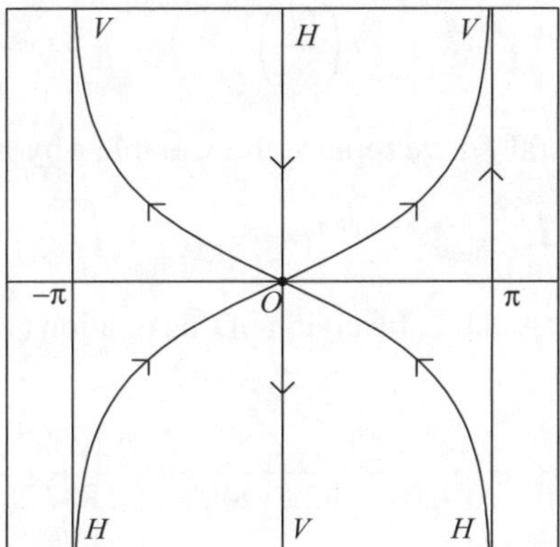

Figure 1.7 The steepest paths for the integral $I(x)$ in (1.3.26) in the interval $-\pi \leq \mathrm{Re}(t) \leq \pi$. The valleys ($V$) and hills ($H$) at infinity are indicated and the arrows denote the directions of descent.

---

[12] The integral $I(x)$ can be expressed in terms of the Anger function $\mathbf{J}_x(x)$ and the Weber function $\mathbf{E}_x(x)$; see Olver (1997, p. 103).

As in Example 1.5, the integration path $[0, \pi]$ is not a steepest path; the endpoints can be connected by steepest paths if we choose the steepest descent path from the origin that approaches the point $\pi + \infty i$ and then the upper half of the path $\sigma = \pi$ from $\pi + \infty i$ to π. In this way we have

$$I(x) = \left\{ \int_0^{\pi+\infty i} - \int_{\pi}^{\pi+\infty i} \right\} e^{x\psi(t)} dt \equiv I_1 - I_2.$$

For the integral I_1, we introduce the new variable u by $\psi(t) = -u = it^3/3! - it^5/5! + \cdots$. Upon inversion (using *Mathematica*), we obtain

$$t = (6iu)^{1/3} + \frac{6iu}{60} + \frac{(6iu)^{5/3}}{1400} + \frac{(6iu)^{7/3}}{25200} + \frac{129(6iu)^3}{51\,744\,000} + \cdots$$

and hence the expansion

$$\frac{dt}{du} = \sum_{k=0}^{\infty} c_k^{(1)} (6i)^{(2k+1)/3} u^{(2k-2)/3} \qquad (|u| < 2\pi),$$

where the first few coefficients $c_k^{(1)}$ are

$$c_0^{(1)} = \tfrac{1}{3}, \quad c_1^{(1)} = \tfrac{1}{60}, \quad c_2^{(1)} = \tfrac{1}{840}, \quad c_3^{(1)} = \tfrac{1}{10800}, \quad c_4^{(1)} = \tfrac{129}{17\,248\,000}, \cdots.$$

The convergence of the above expansions for t and dt/du is controlled by the presence of the adjacent saddles at $t = \pm 2\pi$, where $u = \pm 2\pi i$; hence these series converge in the disc $|u| < 2\pi$. Then, by (1.2.16), we obtain

$$I_1 = \int_0^{\pi+\infty i} e^{x\psi(t)} dt = \int_0^{\infty} e^{-xu} \frac{dt}{du}\, du$$

$$\sim \sum_{k=0}^{\infty} c_k^{(1)} \Gamma(\tfrac{2}{3}k + \tfrac{1}{3}) \left(\frac{6i}{x} \right)^{(2k+1)/3} \qquad (x \to +\infty). \qquad (1.3.27)$$

In the case of the integral I_2, we replace the variable t by $\pi + i\tau$ to find

$$I_2 = ie^{\pi i x} \int_0^{\infty} e^{-x(\tau+\sinh \tau)} d\tau = ie^{\pi i x} \int_0^{\infty} e^{-xu} \frac{d\tau}{du}\, du,$$

where we have put $u = \tau + \sinh \tau$. Inversion of this relation (when $\sinh \tau$ is expanded in powers of τ) yields

$$\tau = \frac{u}{2} - \frac{u^3}{96} + \frac{u^5}{1920} - \frac{u^7}{1\,290\,240} + \frac{223u^9}{92\,897\,280} - \cdots,$$

and hence

$$\frac{d\tau}{du} = \sum_{k=0}^{\infty} (-)^k c_k^{(2)} u^{2k} \qquad (|u| < \pi),$$

where

$$c_0^{(2)} = \tfrac{1}{2}, \quad c_1^{(2)} = \tfrac{1}{12}, \quad c_2^{(2)} = \tfrac{1}{384}, \quad c_3^{(2)} = \tfrac{43}{184\,320}, \quad c_4^{(2)} = \tfrac{223}{10\,321\,920}, \cdots.$$

The convergence of the above expansions for τ and $d\tau/du$ is controlled by the presence of the neighbouring saddles at $t = 0$ and $t = 2\pi$ ($\tau = \pm\pi i$), where $u = \pm\pi i$; hence these series converge in the disc $|u| < \pi$. Then, since the point $t = \pi$ is a linear endpoint, we obtain from (1.2.21)

$$I_2 \sim ie^{\pi ix} \sum_{k=0}^{\infty} (-)^k c_k^{(2)} \frac{(2k)!}{x^{2k+1}} \qquad (x \to +\infty). \tag{1.3.28}$$

Substitution of (1.3.27) and (1.3.28) into (1.3.26) then yields the compound expansion

$$I(x) \sim \sum_{k=0}^{\infty} c_k^{(1)} \Gamma(\tfrac{2}{3}k + \tfrac{1}{3}) \left(\frac{6i}{x}\right)^{(2k+1)/3} - ie^{\pi ix} \sum_{k=0}^{\infty} (-)^k c_k^{(2)} \frac{(2k)!}{x^{2k+1}} \tag{1.3.29}$$

as $x \to +\infty$. This expansion has been obtained using the method of stationary phase in Wong (1989, p. 82).

Example 1.7 Our final example in this section is the integral

$$I(x) = \int_{-\infty+ic}^{\infty+ic} e^{-x\psi(t)} dt, \qquad \psi(t) = t^2 + \frac{2}{t} - 3e^{-2\pi i/3}, \tag{1.3.30}$$

where, with $c > 0$, the integration path is taken parallel [13] to the real axis in the upper half-plane and x is a positive variable (Lauwerier, 1966, p. 52). The phase function $\psi(t)$ has a pole at $t = 0$ and three saddles where $\psi'(t) = 0$ situated on the unit circle at $t_s = 1$ and $t_s = e^{\pm 2\pi i/3}$. The saddles at $t_s = e^{\pm 2\pi i/3}$ are of unit height (that is, $e^{-x\psi(t_s)} = 1$), whereas the saddle at $t_s = 1$ has height $\exp\{-3x(1 + i\sqrt{3})/2\}$ and so is subdominant for $x > 0$.

Since $\psi(t) = \psi(t_s) + 3(t - t_s)^2 + \cdots$ in the neighbourhood of each saddle point t_s, the steepest descent paths are all horizontal at these points. The steepest paths through t_s are determined by $\text{Im}\{\psi(t) - \psi(t_s)\} = 0$ or, with $t = \sigma + i\tau$, by

$$\sigma\tau - \frac{\tau}{\sigma^2 + \tau^2} + \frac{3\sqrt{3}}{4} \begin{Bmatrix} 0 \\ 1 \end{Bmatrix} = 0, \qquad t_s = \begin{cases} 1 \\ e^{\pm 2\pi i/3}. \end{cases}$$

These paths are therefore given by

$$t_s = 1 : \qquad \tau = 0, \ \sigma(\sigma^2 + \tau^2) = 1$$
$$t_s = e^{\pm 2\pi i/3} : \quad (\sigma\tau + \tfrac{3}{4}\sqrt{3})(\sigma^2 + \tau^2) = \tau.$$

The steepest paths through the saddle points are shown in Fig. 1.8. The positive real axis is a path of steepest descent for the subdominant saddle point at $t_s = 1$, whereas those through the saddles at $t_s = e^{\pm 2\pi i/3}$ start at $-\infty$ and end at the pole at $t = 0$. The presence of the pole results in the steepest descent paths through the saddles

---

[13] The integration path does not need to be parallel to the real axis and can be taken as any path in the upper half-plane starting at $-\infty + ic$ and ending at $\infty + ic'$, where $c, c' > 0$.

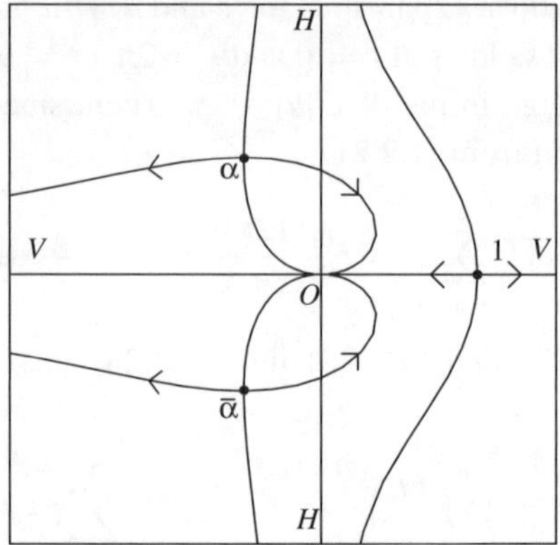

Figure 1.8 The steepest descent and ascent paths for the phase function in (1.3.30). The saddle points are denoted by heavy dots with $\alpha = \exp(2\pi i/3)$. The valleys (V) and hills (H) at infinity are indicated and the arrows denote the directions of descent.

connecting at a common point. When $c > 0$, the integration path in (1.3.30) can be deformed to pass over the saddle points at $t_s = e^{2\pi i/3}$ and $t_s = 1$, approaching the pole at the origin in directions situated in $\mathrm{Re}(t) > 0$.

Let us consider the saddle at $t_s = e^{2\pi i/3}$. If we introduce the new variable $u = \psi(t) - \psi(t_s)$, we find upon inversion (using *Mathematica*) that

$$t - t_s = \frac{u^{1/2}}{\sqrt{3}} - \frac{e^{\pi i/3} u}{9} - \frac{e^{2\pi i/3} u^{3/2}}{54\sqrt{3}} - \frac{u^2}{243} + \frac{5 e^{4\pi i/3} u^{5/2}}{1944\sqrt{3}} + \cdots .$$

This yields the expansion

$$\frac{dt}{du} = \sum_{k=0}^{\infty} c_k e^{2\pi i k/3} u^{(k-1)/2}, \tag{1.3.31}$$

where the even-order coefficients c_{2k} (which are the only ones we require here) are given by

$$c_0 = \frac{1}{2\sqrt{3}}, \quad c_2 = -\frac{1}{36\sqrt{3}}, \quad c_4 = \frac{25}{3888\sqrt{3}}, \quad c_6 = -\frac{245}{209\,952\sqrt{3}},$$

$$c_8 = \frac{2555}{15\,116\,544\sqrt{3}}, \quad c_{10} = -\frac{41503}{2\,448\,880\,128\sqrt{3}}, \quad \cdots .$$

The radius of convergence [14] of the expansions for $t - t_s$ and dt/du is $|u| < 3\sqrt{3}$, since in the u-plane the nearest singular points are the saddles at $t = e^{-2\pi i/3}$ and $t = 1$. Although the precise value of this quantity is not essential to the application of the method of steepest descents, it is given here because this is of great significance in the Hadamard expansion procedure; see Example 3.5.

---

[14] Based on arguments using Cauchy's integral formula similar to those given in Olver (1997, p. 314), it can be shown that $|c_k| \sim \{3^{1/4}/(2\sqrt{\pi})\}(3\sqrt{3})^{-k/2}(k/2)^{-3/2} f_k$ as $k \to +\infty$, where $f_k = |\sin\{(7k + 5)\pi/12\}|$. An analogous expression holds for the coefficients about $t = 1$.

Then, from (1.2.19), the expansion that results from the dominant saddle point $t_s = e^{2\pi i/3}$ as $x \to +\infty$ is

$$2 \sum_{k=0}^{\infty} c_{2k} e^{2\pi i k/3} \frac{\Gamma(k + \frac{1}{2})}{x^{k+\frac{1}{2}}}.$$

The same procedure applied to the subdominant saddle point $t_s = 1$ produces the same expansion as above multiplied by the exponential factor $e^{-x\psi(1)}$, but without the factors $\exp(2\pi i k/3)$. Hence the expansion of the integral $I(x)$ as $x \to +\infty$ is given by

$$I(x) \sim \sqrt{\frac{\pi}{3x}} \left\{ 1 - \frac{e^{2\pi i/3}}{36x} + \frac{25 e^{4\pi i/3}}{2592x^2} - \frac{1225}{279\,936x^3} + \cdots \right\}$$
$$+ \sqrt{\frac{\pi}{3x}} \exp(-3^{3/2} x e^{\pi i/6}) \left\{ 1 - \frac{1}{36x} + \frac{25}{2592x^2} - \frac{1225}{279\,936x^3} + \cdots \right\}.$$

$$(1.3.32)$$

We remark that we have included the exponentially small contribution. This has become standard practice in precision asymptotics, although a correct interpretation of (1.3.32) would require optimal truncation (see §1.6.1) of the dominant expansion at its least term. An example of the importance of retaining exponentially small terms is discussed in Olver (1997, p. 76). Although such terms are negligible in the Poincaré sense, their inclusion can significantly improve the accuracy of a numerical computation.

1.4 Further examples

In this section we present three additional examples that require a more detailed and involved application of the method of steepest descents. The examples chosen deal with the asymptotics of the modified Bessel function of the third kind of imaginary order, Gordeyev's integral arising in plasma physics and, finally, an exponential integral which contains in its asymptotic structure subdominant expansions of different levels of subdominance.

Example 1.8 The modified Bessel function of the third kind of purely imaginary order is defined by (Watson, 1952, p. 181)

$$K_{i\nu}(x) = \int_0^\infty e^{-x \cosh t} \cos \nu t \, dt,$$

where we shall assume throughout that $x > 0$, $\nu \geq 0$ so that $K_{i\nu}(x)$ is real. When $x < \nu$, $K_{i\nu}(x)$ is oscillatory – and so is difficult to compute when the variables are large – whereas when $x > \nu$ the function goes over into a monotonic decay; see Fig. 1.9. The presentation we give here is based on that in Temme (1994) which

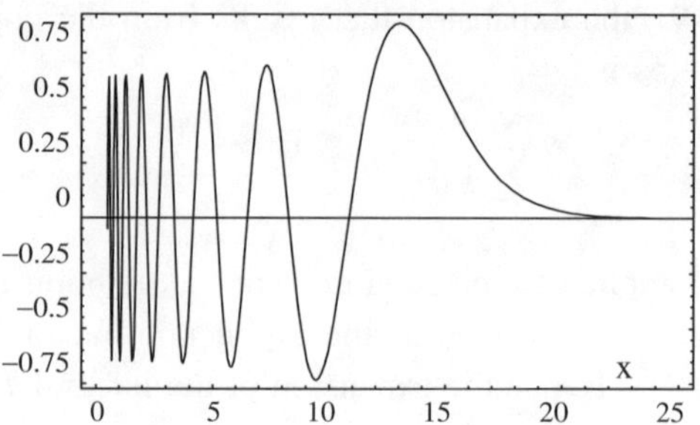

Figure 1.9 The graph of $e^{\pi\nu/2}K_{i\nu}(x)$ when $\nu = 15$.

was concerned with the representation of $K_{i\nu}(x)$ in terms of non-oscillatory integrals suitable for numerical quadrature.

We write

$$K_{i\nu}(x) = \tfrac{1}{2}\int_{-\infty}^{\infty} e^{-x\psi(t)}dt, \qquad \psi(t) = \cosh t - i\nu t/x. \qquad (1.4.1)$$

The saddle points are given by $\psi'(t) = 0$, that is the solutions of the equation $\sinh t = i\nu/x$. We define the new variable $u = \psi(t) - \psi(t_s)$, so that in the case of a simple saddle point t_s

$$u = \cosh t_s\left\{\frac{(t-t_s)^2}{2!} + \tanh t_s\frac{(t-t_s)^3}{3!} + \frac{(t-t_s)^4}{4!} + \cdots\right\}. \qquad (1.4.2)$$

Upon inversion (using *Mathematica*) we find

$$t - t_s = (2\hat{u})^{1/2} - \frac{1}{3}\hat{u}\tanh t_s + \frac{(-3 + 5\tanh^2 t_s)}{18\sqrt{2}}\hat{u}^{3/2}$$

$$+ \frac{2\hat{u}^2}{135}(9 - 10\tanh^2 t_s)\tanh t_s + \frac{(81 - 462\tanh^2 t_s + 385\tanh^4 t_s)}{2160\sqrt{2}}\hat{u}^{5/2} + \cdots,$$

where $\hat{u} = u\,\text{sech}\,t_s$. The disc of convergence of this expansion has a radius u_0 in the u-plane controlled by the neighbouring saddle points (see below). Differentiation of the above expansion then yields

$$\frac{dt}{du} = \sum_{k=0}^{\infty} c_k u^{(k-1)/2} \qquad (|u| < u_0), \qquad (1.4.3)$$

where

$$c_0 = (2\cosh t_s)^{-1/2}, \qquad c_1 = -\frac{\tanh t_s}{3\cosh t_s}, \qquad c_2 = \frac{(-3 + 5\tanh^2 t_s)}{12\cosh t_s}c_0,$$

$$c_3 = -\frac{4(9 - 10\tanh^2 t_s)}{45\cosh t_s}c_1, \qquad c_4 = \frac{(81 - 462\tanh^2 t_s + 385\tanh^4 t_s)}{864\cosh t_s}c_0, \ldots.$$

$$(1.4.4)$$

Let us consider the monotonic region $x > \nu$ first and put

$$\nu = x \cos\theta \qquad (0 \le \theta \le \tfrac{1}{2}\pi).$$

The relevant saddle points are $t_s = i\theta$ and its neighbour at $i(\pi - \theta)$. Since $\mathrm{Im}\{\psi(i\theta)\} = 0$, the steepest descent paths through t_s are defined by $\mathrm{Im}\{\psi(t)\} = 0$. If we write $t = \sigma + i\tau$, the steepest descent path through t_s is given by

$$\tau(\sigma) = \arcsin\left(\frac{\sigma \sin\theta}{\sinh\sigma}\right) \qquad (-\infty < \sigma < \infty),$$

whereas the steepest ascent path through the saddle at $i(\pi - \theta)$ is $\pi - \tau(\sigma)$. When $\theta < \tfrac{1}{2}\pi$, the steepest descent paths are as shown in Fig. 1.10(a); in this case the integration path $(-\infty, \infty)$ can be deformed to coincide with the steepest descent path through $t_s = i\theta$; when $\theta = 0$, the real axis is the path of steepest descent through the saddle at $t = 0$. When $\theta = \tfrac{1}{2}\pi$, the saddles at $i\theta$ and $i(\pi - \theta)$ coalesce to form a double saddle point; see Fig. 1.10(b).

When $0 \le \theta < \tfrac{1}{2}\pi$, the disc of convergence of (1.4.3) has finite radius given by $u_0 = |\psi(i\theta) - \psi(i(\pi - \theta))| = 2\sin\theta(\theta - \tfrac{1}{2}\pi + \cot\theta)$. Then, from (1.2.19) and (1.4.3), we obtain the expansion

$$K_{i\nu}(x) \sim e^{-x\psi(i\theta)} \sum_{k=0}^{\infty} c_{2k} \frac{\Gamma(k + \tfrac{1}{2})}{x^{k+\frac{1}{2}}},$$

where the coefficients c_{2k} are given by (1.4.4) with $t_s = i\theta$. If, for convenience in making a comparison with standard results, we employ the complementary angle β, given by $\theta + \beta = \tfrac{1}{2}\pi$, then

$$e^{-x\psi(i\theta)} = e^{-\nu(\cot\theta + \theta)} = e^{-\nu(\tan\beta - \beta)} e^{-\frac{1}{2}\pi\nu}$$

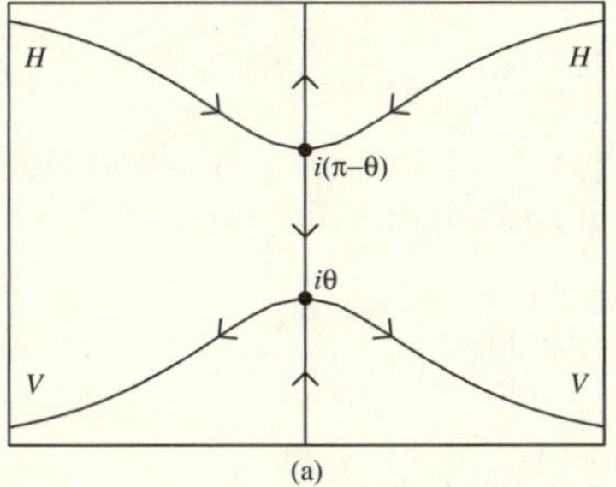

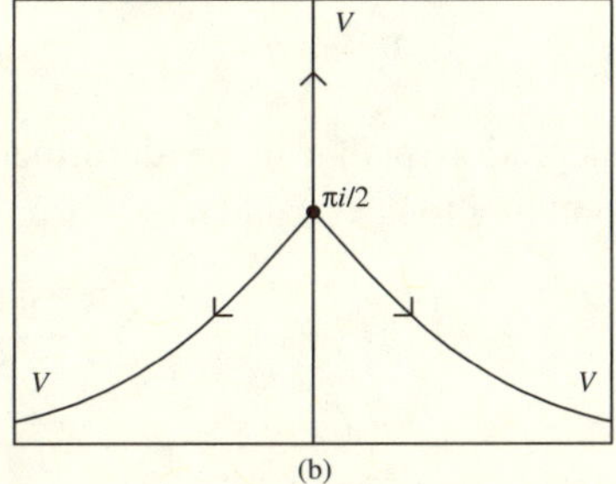

Figure 1.10 (a) The paths of steepest descent and ascent through the saddles $t = i\theta$ and $i(\pi - \theta)$ when $0 < \theta < \tfrac{1}{2}\pi$ and (b) the paths of steepest descent through the double saddle $t_s = \tfrac{1}{2}\pi i$ when $\theta = \tfrac{1}{2}\pi$. The arrows denote the directions of descent and the valleys (V) and hills (H) at infinity are indicated.

and the above expansion can then be expressed in the form (see Watson, 1952, p. 244)

$$e^{\frac{1}{2}\pi\nu}K_{i\nu}(\nu\sec\beta)\sim\frac{\sqrt{\pi}\,e^{-\nu(\tan\beta-\beta)}}{\sqrt{2\nu\tan\beta}}\sum_{k=0}^{\infty}\frac{(-)^{k}(\frac{1}{2})_{k}A_{k}}{(\frac{1}{2}\nu\tan\beta)^{k}} \tag{1.4.5}$$

valid for large positive ν when β is a fixed angle in the interval $[\delta,\frac{1}{2}\pi]$, $\delta>0$. The first few coefficients A_{k} are

$$A_{0}=1,\quad A_{1}=\tfrac{1}{8}+\tfrac{5}{24}\cot^{2}\beta,$$

$$A_{2}=\tfrac{3}{128}+\tfrac{77}{576}\cot^{2}\beta+\tfrac{385}{3456}\cot^{4}\beta,\ \ldots.$$

Observe that (1.4.5) does not hold uniformly as $\theta\to\frac{1}{2}\pi$, since the radius of the convergence disc $u_{0}\to0$ in this limit.

When $\theta=\frac{1}{2}\pi$ ($\beta=0$), the saddle point at $t_{s}=\frac{1}{2}\pi i$ is double and $m=3$ in (1.2.16). In this case, (1.4.2) becomes

$$u=i\left\{\frac{(t-t_{s})^{3}}{3!}+\frac{(t-t_{s})^{5}}{5!}+\frac{(t-t_{s})^{7}}{7!}+\cdots\right\}\qquad(t_{s}=\tfrac{1}{2}\pi i),$$

so that in the neighbourhood of the saddle $t-\frac{1}{2}\pi i\simeq(6u)^{1/3}e^{-\pi i/6}$. This gives one direction of steepest descent from the saddle as $-\pi/6$; the other directions are found by replacing u by $ue^{2\pi in}$ with $n=\pm1$, that is $\frac{1}{2}\pi$ and $-5\pi/6$; see Fig. 1.10(b). Inversion leads to the expansion [15] (with $\hat{u}=-iu$)

$$t-\tfrac{1}{2}\pi i=(6\hat{u})^{1/3}-\frac{\hat{u}}{10}+\frac{(6\hat{u})^{5/3}}{1400}-\frac{(6\hat{u})^{7/3}}{25200}+\frac{1161\hat{u}^{2}}{2\,156\,000}-\frac{1213(6\hat{u})^{11/3}}{7\,207\,200\,000}+\cdots$$

and hence

$$\frac{dt}{du}=\sum_{k=0}^{\infty}c_{2k}u^{2(k-1)/3}\qquad(|u|<2\pi),$$

where

$$c_{0}=\tfrac{1}{3}6^{1/3}e^{-\pi i/6},\quad c_{2}=\tfrac{1}{10}i,\quad c_{4}=-\tfrac{1}{140}6^{2/3}e^{\pi i/6},$$

$$c_{6}=\tfrac{1}{300}6^{1/3}e^{-\pi i/6},\quad c_{8}=\tfrac{3483}{2\,156\,000}i,\quad c_{10}=-\tfrac{1213}{9\,100\,000}6^{2/3}e^{\pi i/6},\ \ldots.$$

The integration path can be deformed to pass over the steepest descent curves corresponding to $n=0$ and $n=-1$ in (1.2.16) and we therefore find

$$e^{\frac{1}{2}\pi\nu}K_{i\nu}(\nu)\sim\frac{1}{2}\sum_{k=0}^{\infty}\frac{c_{2k}\Gamma(\frac{2}{3}k+\frac{1}{3})}{\nu^{(2k+1)/3}}\{1-e^{-2\pi i(2k+1)/3}\}$$

$$=\frac{1}{3}\sum_{k=0}^{\infty}\frac{B_{k}\Gamma(\frac{2}{3}k+\frac{1}{3})}{(\frac{1}{6}\nu)^{(2k+1)/3}}\sin(\tfrac{2}{3}k+\tfrac{1}{3})\pi \tag{1.4.6}$$

---

[15]   The radius of convergence is determined by the adjacent double saddles at $\frac{5}{2}\pi i$ and $-\frac{3}{2}\pi i$ and is given by $u_{0}=|\psi(\frac{1}{2}\pi i)-\psi(\frac{5}{2}\pi i)|=2\pi$.

as $\nu \to +\infty$, where

$$B_0 = 1, \quad B_2 = \tfrac{1}{280}, \quad B_3 = \tfrac{1}{3600}, \quad B_5 = \tfrac{1213}{65\,520\,000}, \quad \cdots$$

We note the omission of the coefficients with index $k = 1, 4, 7, \ldots$; these terms do not contribute to (1.4.6) on account of the vanishing of the sine factor.

In the oscillatory region $x < \nu$, we put

$$\nu = x \cosh \mu \quad (\mu > 0).$$

The saddle points, given by $\sinh t = i \cosh \mu$, are

$$t_k^{\pm} = \pm \mu + (2k + \tfrac{1}{2})\pi i \qquad (k = 0, \pm 1, \pm 2, \ldots).$$

The paths of steepest descent and ascent through the saddles are defined by $\mathrm{Im}\{\psi(t)\} = \mathrm{Im}\{\psi(t_k^{\pm})\} = \pm(\sinh \mu - \mu \cosh \mu)$ and have equation (with $t = \sigma + i\tau$)

$$\sin \tau = \frac{\cosh \mu}{\sinh \sigma}\{\sigma \pm (\tanh \mu - \mu)\}$$

which is *independent* of k. These paths cannot intersect the imaginary axis along which $\mathrm{Im}\{\psi(t)\} = 0$; consequently, the imaginary axis must separate the groups of paths through t_k^+ (situated in $\mathrm{Re}(t) > 0$) and t_k^- (situated in $\mathrm{Re}(t) < 0$). The complex t-plane is divided into horizontal strips of width π: the bottoms of the valleys V_j lie on the lines $\mathrm{Im}(t) = 2\pi j$ and the ridges of the hills H_j lie on the lines $\mathrm{Im}(t) = (2j + 1)\pi$ $(j = 0, \pm 1, \pm 2, \ldots)$. The steepest curves are shown in Fig. 1.11. On each parabolic-like curve are situated two saddle points: as a consequence, the part of each path after the second saddle must be a path of steepest ascent. We label the first saddles t_k^- in the upper half-plane by A, B, C, D and the first saddles t_k^+ by H, G, F, E. The heights of the saddles decrease montonically with k, since

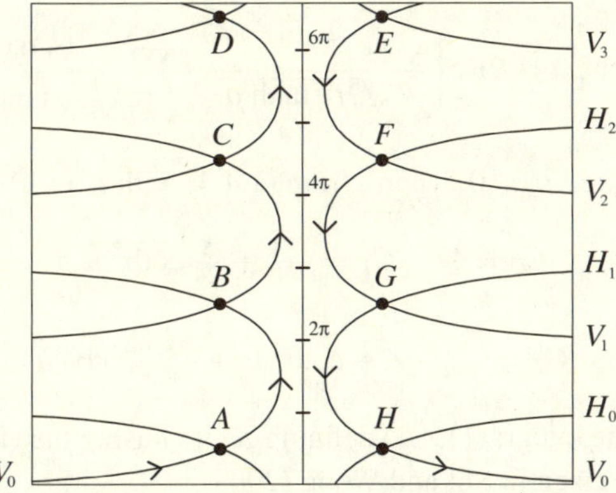

Figure 1.11 The steepest paths through the saddle points $t_k^{\pm}$ (denoted by heavy dots) when $x < \nu$. The arrows denote the direction of integration (Temme, 1994).

$$|e^{-x\psi(t_k^{\pm})}| = e^{-(2k+\frac{1}{2})\pi\nu}. \tag{1.4.7}$$

Then the integration path in (1.4.1) can be split into two parts [16] $\mathcal{L}^- \cup \mathcal{L}^+$, where $\mathcal{L}^-$ runs $-\infty$ to 0 and from 0 to $+\infty i$ and $\mathcal{L}^+$ runs from $+\infty i$ to 0 and from 0 to $+\infty$. The rectilinear paths $\mathcal{L}^{\mp}$ can now be deformed to coincide with the paths $V_0(-\infty)ABCD\ldots$ and $\ldots EFGHV_0(+\infty)$, respectively, both of which comprise an infinite string of saddles. Due to the symmetry, it is sufficient to consider $\mathcal{L}^+$ only and take twice the real part.

Hence

$$K_{i\nu}(x) = \mathrm{Re}\left\{\int_{t_0^+}^{\infty} - \int_{t_0^+}^{\infty i}\right\}e^{-x\psi(t)}dt,$$

where the integration is along the steepest descent paths indicated in Fig. 1.11. We remark that the contribution from the inter-saddle paths HG, GF, FE, … can be simplified, since these paths are similar and, by (1.4.7), the saddle heights form a geometrically decreasing progression, so that

$$\int_{t_0^+}^{\infty i} e^{-x\psi(t)}dt = \frac{1}{1 - e^{-2\pi\nu}} \int_{t_0^+}^{t_1^+} e^{-x\psi(t)}dt.$$

Here, we shall content ourselves with evaluating only the contribution from the dominant saddles $t_0^{\pm}$ of order $\exp(-\frac{1}{2}\pi\nu)$ and neglect the subdominant saddles. The expansion (1.4.3) about the saddle t_0^+ converges in a disc of radius u_0 determined by the proximity of the two neighbouring saddles in the u-plane, where

$$u_0 = \min\{|\psi(t_0^+) - \psi(t_1^+)|, |\psi(t_0^+) - \psi(t_0^-)|\}$$
$$= 2\cosh\mu \, \min\{\pi, \mu - \tanh\mu\}.$$

Then, from (1.2.19), we find

$$e^{\frac{1}{2}\pi\nu} K_{i\nu}(x\,\mathrm{sech}\,\mu) \sim 2\mathrm{Re}\left\{\frac{\sqrt{\pi}\,e^{-i\nu(\tanh\mu-\mu)}}{\sqrt{2i\nu\tanh\mu}} \sum_{k=0}^{\infty} \frac{(\frac{1}{2})_k C_k}{(\frac{1}{2}i\nu\tanh\mu)^k}\right\}, \tag{1.4.8}$$

for $\nu \to +\infty$ and fixed $\mu > 0$, where, from (1.4.4) with $t_s = \mu + \frac{1}{2}\pi i$,

$$C_0 = 1, \quad C_1 = -\tfrac{1}{8} + \tfrac{5}{24}\coth^2\mu,$$

$$C_2 = \tfrac{3}{128} - \tfrac{77}{576}\coth^2\mu + \tfrac{385}{3456}\coth^4\mu, \ldots.$$

A discussion of the integral (1.4.1) defining $K_{i\nu}(x)$ using the Hadamard expansion procedure has been given in Shi and Wong (2009).

---

[16] The integrals converge as $t \to +\infty i$ since $\nu > 0$.

Example 1.9 Consider the integral

$$G_v(\omega, \lambda) = \int_0^\infty \exp\{i\omega t - \lambda(1 - \cos t) - \tfrac{1}{2}vt^2\}\, dt \qquad (1.4.9)$$

for positive values of the parameters ω, λ and v (Paris, 1997). This function, known as Gordeyev's integral, arises in the propagation of electrostatic waves in a hot magnetised Maxwellian plasma. The parameter ω is a normalised wave frequency and λ, v are respectively the squares of the perpendicular and parallel components (with respect to the magnetic field) of the wave vector normalised to the ion Larmor gyration radius. We shall be interested here in estimating the real part of $G_v(\omega, \lambda)$ for large ω and λ, such that $\omega/\lambda = O(1)$, and $v \to 0+$.

Let the parameter α be specified by $\sinh\alpha = \omega/\lambda$ and define

$$\psi(t) = 1 - \cos t - it \sinh\alpha, \qquad f(t) = e^{-vt^2/2}.$$

The integral in (1.4.9) can be expressed over a finite interval by dividing the range of integration into intervals of length 2π to find

$$G_v(\omega, \lambda) = \int_0^{2\pi} e^{-\lambda\psi(t)} f(t)\{1 + \Theta(t)\}\, dt, \qquad (1.4.10)$$

where

$$\Theta(t) = \sum_{n=1}^\infty q^{n^2} \exp\{2\pi in(\Delta\omega + ivt)\}, \qquad q = e^{-2\pi v^2} \qquad (1.4.11)$$

and $\omega = N + \Delta\omega$, with N a non-negative integer chosen such that $|\Delta\omega| \leq \tfrac{1}{2}$. The function $\Theta(t)$ is easily shown to satisfy the quasi-periodicity relation

$$\Theta(t) = e^{2\pi i\chi(t)}\{1 + \Theta(t + 2\pi)\}, \qquad \chi(t) = \Delta\omega + iv(\pi + t). \qquad (1.4.12)$$

We now divide the interval in (1.4.10) into $[0, \pi]$ and $[\pi, 2\pi]$, and in the second subinterval make the change of variable $t \mapsto t + 2\pi$, to find

$$G_v(\omega, \lambda) = \left(\int_0^\pi + \int_\pi^{2\pi}\right) e^{-\lambda\psi(t)} f(t)\{1 + \Theta(t)\}\, dt$$

$$= \int_0^\pi e^{-\lambda\psi(t)} f(t)\{1 + \Theta(t)\}\, dt$$

$$+ \int_{-\pi}^0 e^{-\lambda\psi(t)} f(t) e^{2\pi i\chi(t)}\{1 + \Theta(t + 2\pi)\}\, dt$$

upon noting that $e^{-\lambda\psi(t+2\pi)} f(t + 2\pi) = e^{-\lambda\psi(t)} f(t) e^{2\pi i\chi(t)}$. It then follows from (1.4.12) that

$$G_v(\omega, \lambda) = \int_0^\pi e^{-\lambda\psi(t)} f(t)\, dt + \int_{-\pi}^\pi e^{-\lambda\psi(t)} f(t)\, \Theta(t)\, dt. \qquad (1.4.13)$$

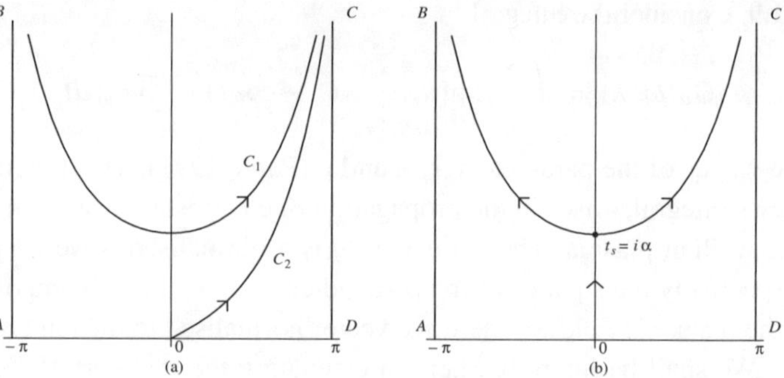

Figure 1.12 (a) The paths of integration C_1 and C_2 and (b) the paths of steepest descent and ascent through the saddle $t_s = i\alpha$. The arrows in (b) denote directions of descent.

Consider the path C_1 with endpoints at $\pm\pi + \infty i$ in the t-plane and a second path C_2 starting from the origin and passing to infinity at $\pi + \infty i$, as shown in Fig. 1.12(a). Then, since $\lambda > 0$, the first integral in (1.4.13) over $[0, \pi]$ can be deformed to pass over the path C_2 together with the semi-infinite path CD parallel to the imaginary axis, whereas the second integral over $[-\pi, \pi]$ can be deformed along the path C_1 together with the semi-infinite paths AB and CD. We then find that

$$G_\nu(\omega, \lambda) = I_1 + I_2, \tag{1.4.14}$$

where

$$I_1 = \int_{C_1} e^{-\lambda\psi(t)} f(t)\Theta(t)\, dt, \qquad I_2 = \int_{C_2} e^{-\lambda\psi(t)} f(t)\, dt, \tag{1.4.15}$$

since the contributions from the paths AB and CD combine to give

$$\int_{AB} e^{-\lambda\psi(t)} f(t)\Theta(t)\, dt + \int_{CD} e^{-\lambda\psi(t)} f(t)\{1 + \Theta(t)\}\, dt$$
$$= \int_{-\pi}^{-\pi+\infty i} e^{-\lambda\psi(t)} f(t)\{\Theta(t) - e^{2\pi i \chi(t)}(1 + \Theta(t + 2\pi))\}\, dt = 0$$

by (1.4.12).

Since $\Theta(t)$ does not depend on the large variables N and λ, it is a slowly varying function of t. We also note that, since $\nu \to 0+$, we have included the factor $\exp(-\frac{1}{2}\nu t^2)$ in the slowly varying function $f(t)$. The integrands in (1.4.15) possess saddle points at the zeros of $\psi'(t) = 0$, that is, at the points given by $\sin t = i \sinh\alpha$. The saddle point in the domain of interest in the strip $-\pi \leq \mathrm{Re}(t) \leq \pi$ is therefore situated at $t_s = i\alpha$ on the imaginary axis. The path of steepest descent is parallel to the real t-axis at t_s and passes to infinity at $\pm\pi + \infty i$, whereas the steepest ascent path through t_s is the imaginary t-axis; see Fig. 1.12(b). By Cauchy's theorem, the path C_1 can be deformed to coincide with the steepest descent path through t_s, whereas C_2 can be deformed into the path with $\mathrm{Im}\{\psi(t)\} = 0$, which corresponds to the part of the imaginary axis between O and t_s, and the half of the steepest descent path situated in $\mathrm{Re}(t) > 0$.

We are now in a position to apply the method of steepest descents to the integrals I_1 and I_2 when $\omega, \lambda \to +\infty$, with ω/λ finite, and $\nu \to 0+$. For the integral I_1, the coefficients a_k and b_k in (1.2.9) and (1.2.13) are given by

$$a_k = \frac{\psi^{(k+2)}(t_s)}{(k+2)!}, \qquad b_k = \frac{\{f(t_s)\Theta(t_s)\}^{(k)}}{k!}.$$

In particular, we obtain $a_0 = \frac{1}{2}\cosh\alpha$, $a_1 = -\frac{1}{6}i\sinh\alpha$, $a_2 = -\frac{1}{24}\cosh\alpha$ and $b_0 = e^{\nu\alpha^2/2}\Theta(i\alpha)$, where, from (1.4.11),

$$\Theta(i\alpha) = \sum_{n=1}^{\infty} q^{n^2} e^{2\pi i n\Omega}, \qquad \Omega = \Delta\omega - \alpha\nu. \tag{1.4.16}$$

From (1.2.15), the first few coefficients $c_k^{(1)}$ are

$$c_0^{(1)} = \frac{b_0}{(2\cosh\alpha)^{1/2}}, \qquad c_1^{(1)} = \frac{1}{\cosh\alpha}(\tfrac{1}{3}ib_0\tanh\alpha + b_1),$$

$$c_2^{(1)} = \frac{2}{(2\cosh\alpha)^{3/2}}(\tfrac{1}{12}b_0(3 - 5\tanh^2\alpha) + ib_1\tanh\alpha + 2b_2), \ldots.$$

For the integral I_2, the corresponding coefficients $c_k^{(2)}$ about the saddle point are obtained from the $c_k^{(1)}$ by replacing $\Theta(t)$ by unity in the coefficients b_k. Then, from (1.2.18) and (1.2.19), we find

$$I_1 \sim 2e^{-\lambda\psi(i\alpha)} \sum_{k=0}^{\infty} c_{2k}^{(1)} \frac{\Gamma(k + \frac{1}{2})}{\lambda^{k+\frac{1}{2}}}$$

and

$$I_2 \sim J_\alpha + e^{-\lambda\psi(i\alpha)} \sum_{k=0}^{\infty} c_k^{(2)} \frac{\Gamma(\frac{1}{2}k + \frac{1}{2})}{\lambda^{(k+1)/2}},$$

where J_α is the contribution to I_2 from the path between O and t_s on the imaginary axis given by

$$J_\alpha = i \int_0^\alpha e^{-\lambda\psi(i\tau)} f(i\tau)\, d\tau.$$

The expansion of this last integral is of the type (1.2.21), since the origin is a linear endpoint, and has the leading behaviour $J_\alpha \sim i/\omega$.

For real parameters, J_α is purely imaginary and it is readily seen that the coefficients $c_k^{(2)}$ of even and odd order are respectively real and purely imaginary. Hence, the real part of $G_\nu(\omega, \lambda)$ has the expansion

$$\mathrm{Re}\, G_\nu(\omega, \lambda) \sim e^{-\lambda\psi(i\alpha)} \sum_{k=0}^{\infty} C_{2k} \frac{\Gamma(k + \frac{1}{2})}{\lambda^{k+\frac{1}{2}}}, \qquad C_{2k} = 2\mathrm{Re}\,(c_{2k}^{(1)}) + c_{2k}^{(2)} \tag{1.4.17}$$

for $\lambda \to +\infty$ with $\omega/\lambda = O(1)$. Since $C_0 = (2\cosh\alpha)^{-1/2} f(i\alpha)\{1 + 2\mathrm{Re}\,\Theta(i\alpha)\}$, the leading behaviour is therefore given by

$$\mathrm{Re}\,G_\nu(\omega,\lambda) \sim \left(\frac{\pi}{2\lambda\cosh\alpha}\right)^{1/2} e^{-\lambda\psi(i\alpha)+\frac{1}{2}\nu\alpha^2}\,\vartheta(\Omega,q), \qquad (1.4.18)$$

where

$$\vartheta(\Omega,q) = 1 + 2\sum_{n=1}^{\infty} q^{n^2} \cos 2\pi n\Omega$$

is the Jacobian theta function of the third kind (Whittaker and Watson, 1952, p. 464). The real part of $G_\nu(\omega,\lambda)$ is consequently exponentially small for $\alpha > 0$, whereas the imaginary part is dominated by the algebraic expansion J_α. In the limit $\nu \to 0+$ ($q \to 1-$), the sum $\vartheta(\Omega,q)$ is slowly convergent and a more convenient form can be obtained by means of the Poisson–Jacobi transformation (Whittaker and Watson, 1952, p. 475) which shows that

$$\vartheta(\Omega,q) = \frac{e^{-\Omega^2/(2\nu)}}{\sqrt{2\pi\nu}}\left\{1 + 2\sum_{n=1}^{\infty} e^{-n^2/(2\nu)} \cosh\left(n\Omega/\nu\right)\right\}.$$

It should be noted that (1.4.17) and (1.4.18) hold *uniformly* in $\Delta\omega$ through a harmonic $\Omega = 0$, where $\vartheta(\Omega,q)$ is strongly peaked as $\nu \to 0+$; see Fig. 1.13. It is this latter factor that contains the fine structure of $\mathrm{Re}\,G_\nu(\omega,\lambda)$ in ω. We present in Table 1.2 a comparison between the computed values of $\mathrm{Re}\,G_\nu(\omega,\lambda)$ and the first two terms of the expansion (1.4.17) for different values of ω in the neighbourhood of a harmonic. The computed values were obtained by expanding the factor $\exp(\lambda\cos t)$ in (1.4.9) in terms of modified Bessel functions $I_n(\lambda)$ to find (Stix, 1962, p. 176)

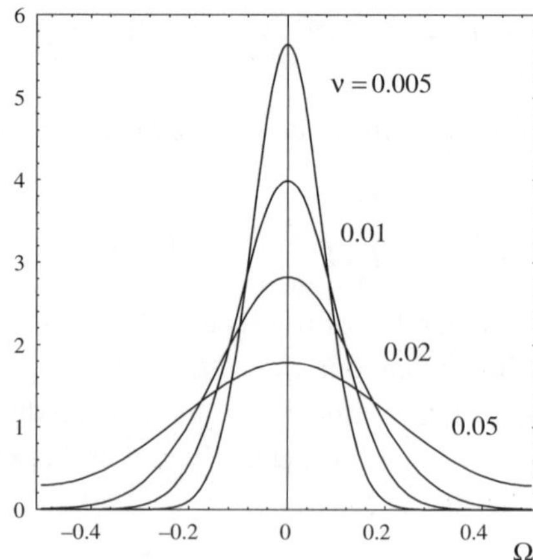

Figure 1.13 The variation of $\vartheta(\Omega,q)$ against Ω for small values of ν when $q = e^{-2\pi^2\nu}$.

Table 1.2 *The computed and asymptotic values of*
$\mathrm{Re}\, G_\nu(\omega, \lambda)$ *when* $\omega = N + \Delta\omega$, $N = 50$, $\lambda = 100$
and $\nu = 0.01$

$\Delta\omega$	$\mathrm{Re}\, G_\nu(\omega, \lambda)$	Asymptotic
0.0	2.2482012×10^{-6}	2.2482006×10^{-6}
0.1	1.3636030×10^{-6}	1.3636006×10^{-6}
0.2	3.0426096×10^{-7}	3.0426008×10^{-7}
0.3	2.4975260×10^{-8}	2.4975169×10^{-8}
0.4	$7.5420851 \times 10^{-10}$	$7.5420552 \times 10^{-10}$
0.5	$1.3522807 \times 10^{-11}$	$1.3522820 \times 10^{-11}$

$$G_\nu(\omega, \lambda) = \sqrt{\frac{\pi}{2\nu}}\, e^{-\lambda} \sum_{n=-\infty}^{\infty} I_n(\lambda)\, e^{-X_n^2}\, \mathrm{erfc}(-i X_n), \quad X_n = \frac{\omega - n}{\sqrt{(2\nu)}}.$$

Although this form eliminates the problems associated with a highly oscillatory integrand, a large number of terms of the expansion must be employed when λ is large.

Finally, we remark that the asymptotics of $G_\nu(\omega, \lambda)$ in (1.4.17) and (1.4.18) remain valid for $\omega/\lambda = o(1)$, that is when $\alpha \to 0$ as $\lambda \to +\infty$. In this case, although the saddle t_s moves down the imaginary axis towards the origin, the form of the contribution from the saddle point remains unchanged, so that (1.4.17) and (1.4.18) hold uniformly as $\alpha \to 0$. The imaginary part of $G_\nu(\omega, \lambda)$, however, will not remain valid since the expansion for J_α is not uniformly valid in this limit. Further details of these expansions for different scalings of ω with λ and for complex parameters can be found in Paris (1997).

Example 1.10 Our final example concerns the integral

$$I_p(x) = \int_0^\infty e^{-\tau^p + ix\tau}\, d\tau \tag{1.4.19}$$

for $x \to +\infty$ when $p > 1$. This integral, for complex x and integer p, has found application in number theory in the discussion of the major arcs in Waring's problem in Hardy and Littlewood (1920). Let us introduce the notation

$$X = \kappa(hx)^{1/\kappa}, \quad A_0 = (2\pi)^{\frac{1}{2}} \kappa^{-1/p} p^{-\vartheta},$$

$$\kappa = (p - 1)/p, \quad h = (1/p)^{1/p}, \quad \vartheta = (2 - p)/2p \tag{1.4.20}$$

and make the change of variable $\tau \mapsto \lambda h t$, $\lambda = (hx)^{1/(p-1)}$. We then obtain the integral in the form

$$I_p(x) = (X/\kappa p)^{1/p} \int_0^\infty e^{-X\psi(t)}\, dt, \quad \psi(t) = (t^p/p - it)/\kappa. \tag{1.4.21}$$

The saddle points of $\psi(t)$ are given by $t_r^{p-1} = i$, that is

$$t_r = \exp\left\{(2r + \tfrac{1}{2})\pi i/(p-1)\right\} \qquad (r = 0, \pm 1, \pm 2, \dots).$$

$$(1.4.22)$$

For integer values of p, there are exactly $p-1$ saddles situated on the unit circle in the t-plane. When p is non-integer, however, the t-plane must be cut along the negative real axis: in this case there will, in general, be saddles situated on the principal sheet $-\pi < \arg t \leq \pi$, with other saddles situated on adjacent Riemann sheets.[17] When $1 < p < \frac{3}{2}$, there are no saddles situated on the principal sheet. As p increases beyond this interval, the saddle t_0 crosses the branch cut $\arg t = \pi$ from the upper adjacent sheet onto the principal sheet and moves steadily round the unit circle, passing through the point $t = i$ when $p = 2$. As p continues to increase, a second saddle t_{-1} moves off the lower adjacent sheet, crossing the branch cut $\arg t = -\pi$ when $p = \frac{5}{2}$, and moves round the unit circle in the opposite sense. This pattern of oppositely directed rotation of the saddles towards the point $t = 1$ continues as p increases further, with a new saddle crossing over (alternately from the upper and lower adjacent sheets) onto the principal sheet each time p equals a half-integer value.

The distribution of the saddles and the topology of the associated paths of steepest descent and ascent for $I_p(x)$ are illustrated in Fig. 1.14 for $1 < p < 3$ and in Fig. 1.15 for integer p. These paths are described by $\mathrm{Im}\{\psi(t) - \psi(t_r)\} = 0$ in the case of the saddles t_r and $\mathrm{Im}\{\psi(t)\} = 0$ in the case of the path emanating from the origin, and were constructed with *Mathematica*. The valleys V_k (resp. hills H_k) of $\psi(t)$, corresponding to the directions along which $\psi(t) \to +\infty$ (resp. $-\infty$) as $|t| \to \infty$, are given by the rays $\arg t = 2\pi k/p$ (resp. $\arg t = (2k + 1)\pi/p$), where $k = 0, \pm 1, \pm 2, \dots$.

For $1 < p < 2$, the path of steepest descent from the origin on the principal sheet does not pass over a saddle, with the result that the path of integration in (1.4.21) can be reconciled with this contour. As $p \to 2-$, the path $\mathrm{Im}\{\psi(t)\} = 0$ becomes progressively deformed in the neighbourhood of the saddle t_0 until, when $p = 2$, it connects with this saddle. For $p > 2$, the path of steepest descent from the origin now passes to infinity down the valley V_1. In this case, the path of integration in (1.4.21) can be deformed to coincide with this path together with the steepest descent path that connects V_1 to the adjacent valley V_0 passing over the saddle t_0. In Fig. 1.15 we show only the path from $t = 0$ that eventually connects with infinity along the valley V_0. As p increases, more saddles move round the unit circle into the right half-plane, with a new saddle crossing the positive imaginary axis whenever $p = 4N + 2$, $N = 0, 1, 2, \dots$.

---

[17] An alternative approach is to use the change of variable $t = e^u$. This transformation causes all the Riemann sheets in the t-plane to appear as horizontal strips of width 2π in the u-plane. The saddle points are then situated on the imaginary u-axis.

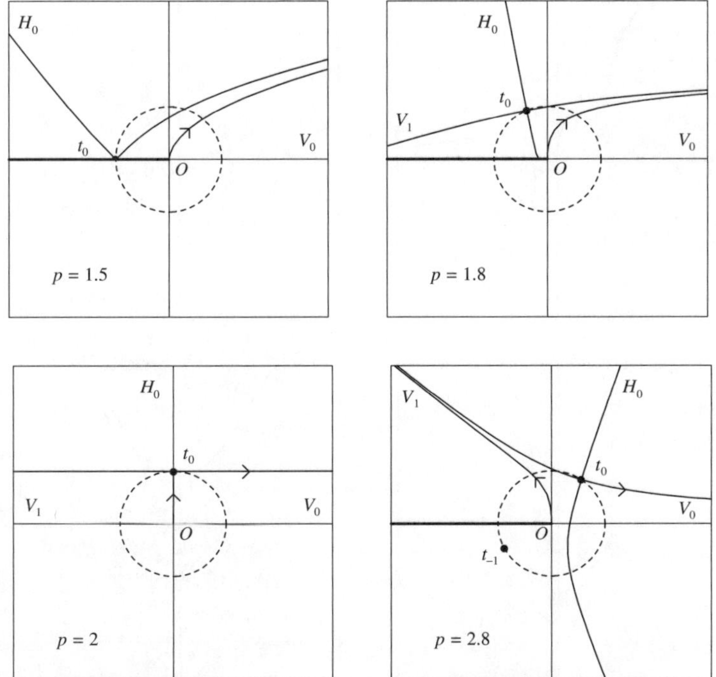

Figure 1.14 The paths of steepest descent and ascent when $1 < p < 3$. The branch cut along the negative real axis in the case of non-integer p is indicated by the heavy line. Arrows indicate the integration path and the heavy dots denote the saddle points on the unit circle. In the first two figures the steepest descent path from $t = 0$ is asymptotically parallel to the $\mathrm{Re}(t)$-axis.

In the case of integer p, the expansion that results from the path emanating from the origin can be determined from (1.2.21), where the coefficients c_k are given by (1.2.15) with $m = 1$. This is inapplicable, however, for non-integer values of p since higher derivatives of $\psi(t)$ at the origin become singular. To obtain an expansion valid for arbitrary $p > 1$, we take the path in (1.4.19) to be along the positive imaginary axis (the steepest descent direction at $\tau = 0$), expand $\exp(-\tau^p)$ as a Maclaurin series and integrate term by term. Then, with $\tau = iu$, we find the asymptotic contribution

$$i \sum_{k=0}^{\infty} \frac{(-)^k i^{kp}}{k!} \int_0^{\infty} u^{kp} e^{-xu}\, du = i \sum_{k=0}^{\infty} \frac{(-)^k i^{kp}}{k!} \frac{\Gamma(kp+1)}{x^{kp+1}}.$$

This gives the algebraic expansion

$$H_p(x) := i \sum_{k=0}^{\infty} \frac{(-)^k}{k!} e^{\frac{1}{2}\pi i k p} \Gamma(kp+1) x^{-kp-1}. \tag{1.4.23}$$

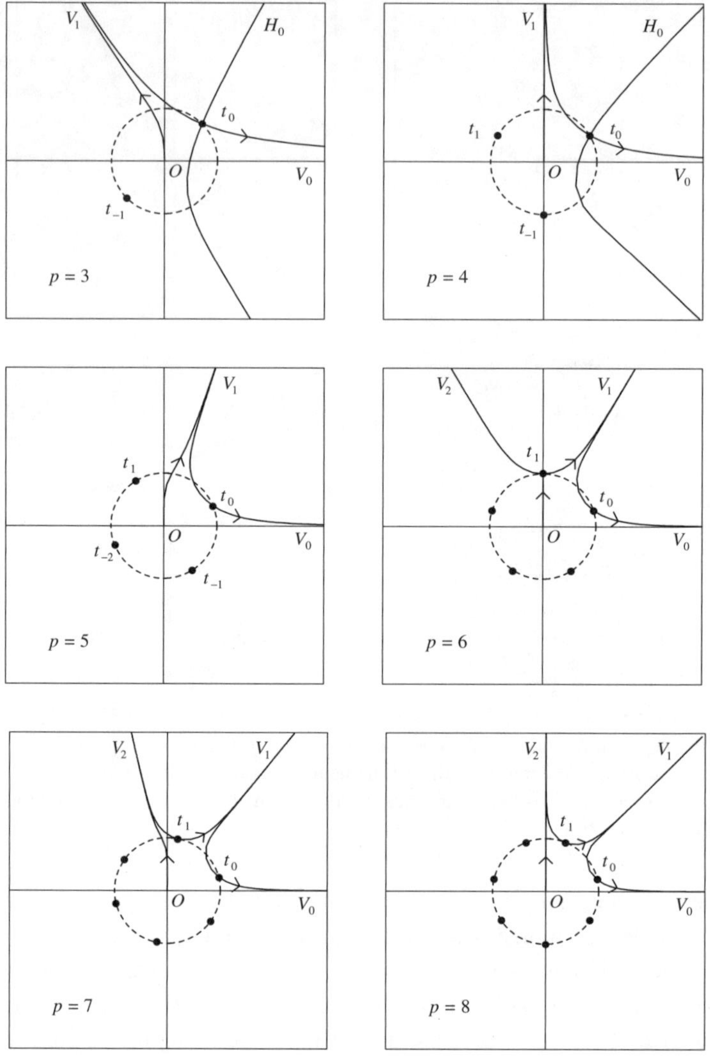

Figure 1.15 The paths of steepest descent for integer values of p. (The steepest ascent paths have been omitted in the lower figures for clarity.) Arrows indicate the integration path and the heavy dots denote the saddle points on the unit circle.

In the case $1 < p < 2$, where the steepest descent path does not pass over a saddle, the expansion of $I_p(x)$ is then described by $I_p(x) \sim H_p(x)$ as $x \to +\infty$.

The contributory saddle points are located in the first quadrant in the t-plane; that is, from (1.4.22), those points t_r corresponding to $0 \leq r \leq r'$, where r' is the greatest integer satisfying $4r' + 2 \leq p$. It is easily verified that each of these saddles results in an exponentially small contribution to $I_p(x)$, since on the unit circle $\text{Re}\{\psi(e^{i\theta})\} = \kappa^{-1}\{\sin\theta + \cos(p\theta)/p\} > 0$ for $0 \leq \theta \leq \pi/2$ and $p \geq 2$. To determine the directions of the steepest descent paths at t_r, we have $\psi''(t_r) = pt_r^{p-2}$ and

$$\arg \psi''(t_r) = \arg t_r^{p-2} = \frac{(4r+1)(p-2)\pi}{2(p-1)}$$

$$= 2\omega_r + 2\pi r, \quad \omega_r = \frac{(p-4r-2)\pi}{4(p-1)}. \tag{1.4.24}$$

Then, in the neighbourhood of t_r we have

$$\psi(t) = \psi(t_r) + \tfrac{1}{2}p(t-t_r)^2 e^{2i\omega_r} + \dots ,$$

so that the directions of the paths of steepest descent from the saddle t_r are given by $\arg(t-t_r) = -\omega_r$ and $\pi - \omega_r$. For $0 \le r \le r'$, it is easily verified that $0 \le \omega_r < \pi/4$.

For arbitrary $p \ge 2$, let us consider the contribution from a contributory saddle t_r passing along one half of the steepest descent path down the valley V_r. Upon noting that

$$\psi^{(k+2)}(t) = (k+2)! \frac{a_k' \psi''(t)}{p t^k}, \qquad a_k' = \frac{p}{(k+2)!} \prod_{\ell=2}^{k+1} (p-\ell) \quad (k \ge 0),$$

we make the change of variable $u = \psi(t) - \psi(t_r)$ to find

$$U \equiv \frac{u}{(t_r e^{i\omega_r})^2} = \sum_{k=0}^{\infty} a_k' \left(\frac{t-t_r}{t_r} \right)^{k+2}.$$

Inversion of this series then yields

$$\frac{dt}{dU} = t_r \sum_{k=0}^{\infty} c_k U^{(k-1)/2},$$

where, from (1.2.15) with $m = 2$, the coefficients c_k are *independent* of r and

$$c_0 = \frac{1}{\sqrt{2p}}, \quad c_1 = \frac{2-p}{3p}, \quad c_2 = \frac{(2p-1)(p-2)}{12p\sqrt{2p}},$$

$$c_3 = \frac{2}{135 p^2}(2-p)(2p-1)(p+1), \quad c_4 = \frac{(2p-1)(p-2)}{864 p^2 \sqrt{2p}}(2p^2+19p+2), \dots .$$

We therefore obtain

$$\frac{dt}{du} = e^{-i\omega_r} \sum_{k=0}^{\infty} c_k \frac{u^{(k-1)/2}}{(t_r e^{i\omega_r})^k} = e^{-i\omega_r} \sum_{k=0}^{\infty} (-)^{rk} c_k \frac{u^{(k-1)/2}}{t_r^{kp/2}},$$

since, from (1.4.22) and (1.4.24),

$$(t_r e^{i\omega_r})^{-k} = e^{-ik\omega_r} t_r^{-kp/2} t_r^{k(p-2)/2} = (-)^{rk} t_r^{-kp/2}.$$

We now introduce the variable X_r defined by

$$X_r := X e^{(2r+\frac{1}{2})\pi i/\kappa} = X t_r^p \tag{1.4.25}$$

and make use of the results

$$\psi(t_r) = -t_r^p, \qquad (X/\kappa p)^{1/p}\{\pi/(2pX)\}^{1/2}\,e^{-i\omega_r} = \frac{(-)^r}{2p}A_0 X_r^\vartheta$$

obtained from (1.4.20), (1.4.24) and the fact that $t_r^{p-1} = i$. Then, as $x \to +\infty$, the contribution to the asymptotic expansion of $I_p(x)$ associated with a path from the saddle t_r along the steepest descent path down the valley V_r is given by

$$(X/\kappa p)^{1/p}e^{-X\psi(t_r)}\int_0^\infty e^{-Xu}\frac{dt}{du}du$$

$$\sim (X/\kappa p)^{1/p}e^{-X\psi(t_r)-i\omega_r}\sum_{k=0}^\infty (-)^{rk}\frac{c_k\,\Gamma(\tfrac{1}{2}k + \tfrac{1}{2})}{X^{(k+1)/2}\,t_r^{kp/2}}$$

$$= \frac{(-)^r}{2p}X_r^\vartheta e^{X_r}\sum_{k=0}^\infty (-)^{rk}A_k X_r^{-k/2}; \tag{1.4.26}$$

compare (1.2.18). The coefficients $A_k := A_0(c_k/c_0)\Gamma(\tfrac{1}{2}k + \tfrac{1}{2})/\Gamma(\tfrac{1}{2})$ depend only on p and are given by

$$\frac{A_{2k}}{A_0} = \frac{(2p-1)(p-2)}{k(24p)^k}d_{2k}, \qquad \frac{A_{2k+1}}{A_1} = \frac{(2p-1)(p+1)\pi^{\frac{1}{2}}}{2\Gamma(k+\tfrac{5}{2})(6p)^k}d_{2k+1} \tag{1.4.27}$$

for $k \geq 1$, where

$$A_1 = \frac{(2-p)}{3p}\sqrt{\frac{2p}{\pi}}\,A_0 = \tfrac{2}{3}(2-p)(\kappa p)^{-1/p}$$

and the first few values of d_k are

$$d_2 = d_3 = 1, \quad d_4 = 2p^2 + 19 + 2, \quad d_5 = -(2p^2 - 23p + 2),$$

$$d_6 = -\tfrac{1}{10}(556p^4 - 1628p^3 - 9093p^2 - 1628p + 556),$$

$$d_7 = -18(2p^4 + 5p^3 - 66p^2 + 5p + 2),$$

$$d_8 = -\tfrac{1}{30}(4568p^6 + 226668p^5 - 465702p^4 - 2013479p^3 - 465702p^2$$
$$+ 226668p + 4568),$$

$$d_9 = \tfrac{2}{5}(1124p^6 - 21192p^5 - 8133p^4 + 271366p^3 - 8133p^2 - 21192p + 1124), \ldots.$$

The values of the even-order coefficients d_{2k} for $0 \leq k \leq 8$ are given in Paris and Kaminski (2001, §8.1.6). We remark that $A_k = 0$ ($k \geq 1$) when $p = 2$.

From (1.4.26), the contribution to $I_p(x)$ from both halves of the steepest descent path through the saddle t_r is accordingly

$$e_r(x; p) := \frac{(-)^r}{p}X_r^\vartheta e^{X_r}\sum_{k=0}^\infty A_{2k}X_r^{-k}, \tag{1.4.28}$$

since the terms with odd k cancel; compare (1.2.19). When $p = 4N + 2$ ($N = 1, 2, \ldots$), the integration path from the origin runs along the positive imaginary axis to the saddle point $t_N = i$ and subsequently turns through a right angle to follow the steepest descent path down the valley V_N; see Fig. 1.15. The contribution from this path is therefore

$$H_p(x) + \tfrac{1}{2}e_N(x; p) \qquad (p = 4N + 2),$$

where $H_p(x)$ is defined in (1.4.23) and, since $X_N = Xe^{(2N+1)\pi i}$,

$$e_N(x; p) := \frac{X^\vartheta}{p} e^{-X} \sum_{k=0}^{\infty} (-i)^k A_k X^{-k/2}.$$

The asymptotic expansion of $I_p(x)$ may now be stated as follows.

Theorem 1.2 *With the notation in (1.4.20), we have the asymptotic expansion*

$$I_p(x) \sim E_p(x) + H_p(x) \qquad (p > 1) \tag{1.4.29}$$

as $x \to +\infty$, where $E_p(x)$ and $H_p(x)$ are the exponentially small and algebraic contributions given by

$$E_p(x) := \sum_{r=0}^{N-1} e_r(x; p) + \tfrac{1}{2}e_N(x; p)\,\delta_{pp'} \qquad (p' = 4N + 2) \tag{1.4.30}$$

and

$$H_p(x) := i \sum_{k=0}^{\infty} \frac{(-)^k}{k!} e^{\frac{1}{2}\pi ikp} \Gamma(kp + 1) x^{-kp-1}, \tag{1.4.31}$$

with N being the nearest integer part[18] of $\tfrac{1}{4}p$ and δ_{mn} the Kronecker symbol. The expansions $e_r(x; p)$ $(0 \le r \le N - 1)$ are defined by

$$e_r(x; p) := \frac{(-)^r}{p} X_r^\vartheta e^{X_r} \sum_{k=0}^{\infty} A_{2k} X_r^{-k},$$

where X_r is defined in (1.4.25), and

$$e_N(x; p) := \frac{X^\vartheta}{p} e^{-X} \sum_{k=0}^{\infty} (-i)^k A_k X^{-k/2}.$$

The coefficients A_k are given in (1.4.27) and $A_0 = (2\pi)^{1/2}\kappa^{-1/p} p^{-\vartheta}$.

The sum of exponential expansions $e_r(x; p)$ corresponds to the contribution obtained from all the contributory saddles in (1.4.22). The presence of the expansion $e_N(x; p)$ results from the appearance of an additional contributory saddle on

---

[18] The nearest integer part of x is the integer N when x is in the interval $(N - \tfrac{1}{2}, N + \tfrac{1}{2}]$.

the positive imaginary t-axis, when the path of steepest descent from the origin connects with this saddle to produce a Stokes phenomenon – see §1.7. This arises when $p = 4N + 2$, for non-negative integer N; inspection of (1.4.23) reveals that the terms in the algebraic expansion then all have the same phase. The treatment of the case when p is allowed to increase through these critical values would require taking into account the smooth appearance of the additional exponentially small expansion $e_N(x; p)$ in the neighbourhood of $p = 4N + 2$. An example of this type of problem in a related context involving the Euler–Jacobi series is discussed in Paris and Kaminski (2001, §8.1.7).

When $1 < p < 2$, $E_p(x) \equiv 0$ and the expansion of $I_p(x)$ is algebraic with

$$I_p(x) \sim H_p(x) \qquad (1 < p < 2).$$

When $p = 2$ ($N = 0$), we find from (1.4.20) the values $\kappa = \frac{1}{2}$, $\vartheta = 0$, $X = x^2/4$ and the coefficients $A_k = 0$ for $k \geq 1$; see (1.4.27). From (1.4.30), the exponential expansion of $I_2(x)$ then consists of a single term and we find the readily verifiable result

$$I_2(x) \sim \tfrac{1}{2}\sqrt{\pi}e^{-x^2/4} + i\sum_{k=0}^{\infty} \frac{\Gamma(2k+1)}{k!\,x^{2p+1}}.$$

For $2 < p < 6$ ($N = 1$), there is just one contributory saddle t_0 and the expansion of $I_p(x)$ for this range of p is therefore given by (1.4.29), where $E_p(x) = e_0(e; p)$; that is,

$$E_p(x) = \frac{X_0^\vartheta}{p}e^{X_0}\sum_{k=1}^{\infty} A_{2k}X_0^{-k} \qquad (2 < p < 6).$$

For example, when $p = 3$, we have $\kappa = \frac{2}{3}$, $\vartheta = -\frac{1}{6}$; then, we find

$$H_3(x) = \sum_{k=0}^{\infty} \frac{i^{k+1}\Gamma(3k+1)}{k!\,x^{3k+1}}$$

and, from (1.4.25),

$$E_3(x) = \tfrac{1}{3}X^{-1/6}e^{Xe^{3\pi i/4}}\sum_{k=0}^{\infty} A_{2k}X^{-k}e^{-(6k+1)\pi i/8},$$

where $X = 2(x/3)^{3/2}$.

When $p = 6$, a second saddle t_1 crosses the positive imaginary t-axis with the consequence that, when $6 < p < 10$ ($N = 2$), there are now two exponentially small contributions which result from these saddles. In this case, we have $E_p(x) = e_0(x; p) + e_1(x; p)$, so that

$$E_p(x) = \frac{X_0^\vartheta}{p}e^{X_0}\sum_{k=0}^{\infty} A_{2k}X_0^{-k} - \frac{X_1^\vartheta}{p}e^{X_1}\sum_{k=0}^{\infty} A_{2k}X_1^{-k} \qquad (6 < p < 10).$$

As with the saddle t_0 in the case $p = 2$, the saddle t_1 makes only half the contribution to $I_p(x)$ when $p = 6$. Thus, when $p = 6$ (so that $\kappa = \frac{5}{6}$ and $\vartheta = -\frac{1}{3}$), we find

$$E_6(x) = \tfrac{1}{6} X^{-1/3} e^{X e^{3\pi i/5}} \sum_{k=0}^{\infty} A_{2k} X^{-k} e^{-(3k+1)\pi i/5}$$

$$+ \tfrac{1}{12} X^{-1/3} e^{-X} \sum_{k=0}^{\infty} (-i)^k A_k X^{-k/2},$$

where $X = 5(x/6)^{6/5}$. We remark that the second exponential e^{-X} is subdominant with respect to the real part of the first exponential $\exp\{X \cos \frac{3}{5}\pi\}$. The pattern that emerges in the exponentially small contribution $E_p(x)$ continues to hold for higher values of p. Thus, each time p increases by 4, a new saddle in (1.4.22) crosses over into the first quadrant of the t-plane and results in an additional exponentially small contribution to the asymptotics of $I_p(x)$.

The numerical interpretation of composite expansions such as (1.4.29) involving algebraic and exponentially small terms is discussed in §1.8. We point out that the integral $I_p(x)$ has been considered for complex values of x and integer values of p by Bakhoom (1933), who gave only the dominant exponential expansion. This expansion for complex x can also be obtained from the observation that $I_p(x)$ in (1.4.19) can be expressed as the series

$$I_p(x) = \frac{1}{p} \sum_{k=0}^{\infty} \frac{(ix)^k}{k!} \Gamma\left(\frac{k+1}{p}\right) = \frac{1}{p} U_{p,1}(ihx),$$

where the function $U_{p,1}(z)$ and its asymptotics for large complex z are considered in detail in Paris and Wood (1986, §3.2 and Eq. (3.10.12)). Related investigations can be found in Brillouin (1916), Burwell (1924) and Faxén (1921), who considered a transformation of the integral

$$\int_0^{\infty} t^{\alpha-1} e^{-\lambda(t^\mu - ct^m)} dt \qquad (\mu > m > 0, \ \mathrm{Re}\,(\alpha) > 0)$$

for large complex λ and finite complex parameter c. An extension to the asymptotics for large $|\lambda|$ of the n-dimensional Faxén integral given by

$$\int_0^{\infty} \cdots \int_0^{\infty} t_1^{\alpha_1-1} \ldots t_n^{\alpha_n-1} e^{-\lambda \psi(t_1,\ldots,t_n)} \, dt_1 \ldots dt_n,$$

where

$$\psi(t_1,\ldots,t_n) = \sum_{j=1}^{n} t_j^{\mu_j-1} - c\, t_1^{m_1} \ldots t_n^{m_n}$$

with $\mu_j > m_j > 0$, $\mathrm{Re}\,(\alpha_j) > 0$ $(1 \le j \le n)$, is described in Paris and Liakhovetski (2000); see also Paris (2010).

Finally, we remark that the integral

$$J(x) = \int_{-\infty}^{\infty} e^{-t^{2m}+ixt}\,dt \tag{1.4.32}$$

for positive integer m can be dealt with in a similar manner as $x \to +\infty$. In this case the path of integration can be made to pass over all the saddles (which are distributed symmetrically about the positive imaginary axis) in the upper half-plane, since the bottoms of the valleys V_0 and V_m coincide with the positive and negative real axes, respectively; see Paris and Kaminski (2001, p. 360). Since such paths avoid the neighbourhood of the origin, the expansion of $J(x)$ will consist entirely of exponentially small contributions given by

$$J(x) \sim \sum_{r=0}^{m-1} e_r(x; 2m) \qquad (x \to +\infty),$$

which, because of the symmetrical distribution of the saddles, is real. The integral (1.4.32) has also been considered by Senouf (1996) who obtained the expansion for $|x| \to \infty$ in the sector $|\arg x| < \frac{1}{2}\pi(1-(2n)^{-1})$ using a similar approach. A notable result of his analysis is the closed-form representation of the coefficients in the exponential expansions as sums involving combinatorial coefficients. The main aim of his paper was the derivation of asymptotic approximations for the (real) zeros of $J(x)$.

1.5 Uniform expansions

In integrals of the type

$$\int_C e^{\lambda\psi(t)} f(t)\,dt$$

it often happens that the phase function $\psi(t)$ or the amplitude function $f(t)$ depends on a parameter α, say. In certain situations the resulting asymptotic expansion for large λ can become invalid at a critical value $\alpha = \alpha_c$ of this parameter because the order estimate in (1.2.25) becomes invalid. In such cases, the expansion is said to be *non-uniform* in α. Common causes of non-uniformity in an asymptotic expansion are produced by a variety of coalescence phenomena, such as a saddle point coalescing either with an endpoint of the integration path C or with one or more other saddle points.

An example of the last-mentioned form of non-uniformity has already been seen in §1.4 when dealing with the expansion of the modified Bessel function $K_{i\nu}(x)$ for large x and ν. When $x < \nu$ or $x > \nu$, the active saddles are separated and the resulting expansions involve inverse powers of $\nu^{1/2}$; see (1.4.5) and (1.4.8). When $x = \nu$, the active saddles coalesce to form a double saddle and the resulting expansion involves inverse powers of $\nu^{1/3}$; see (1.4.6). Other causes of non-uniformity are produced by the presence of a pole of $f(t)$ approaching a saddle point and the Stokes phenomenon, which will be discussed separately in §1.7. Physical applications of such non-uniform behaviour arise in many short-wavelength phenomena, such as wave propagation and optical diffraction. A well-known example is the Pearcey integral that contains two real parameters, which when varied can result in both coalescence to form a caustic and the Stokes phenomenon; see §4.3.

It is clearly desirable to obtain a uniform expansion of the above integral that remains valid as α varies in a domain containing the critical value α_c. The price to be paid for such uniformity, however, is the increased complexity in the asymptotic sequence employed (which must encapsulate the change in form of the expansion as α passes through the value α_c) and in the associated coefficients. In this section, we wish to illustrate the different types of uniform expansion that result from the above-mentioned causes of non-uniformity. It is not our aim here to supply proofs of the validity of these expansions, for which we refer the reader to the literature. In Chapter 3, we shall consider these coalescence problems from the point of view of the Hadamard expansion procedure.

1.5.1 Saddle point near a pole

We consider the integral

$$I(\lambda; \alpha) = \int_C e^{-\lambda \psi(t)} f(t) dt, \qquad f(t) = \frac{F(t)}{t - \alpha}$$

for $\lambda \to +\infty$, where $\psi(t)$ has a simple saddle point at $t = t_s$ and $F(t)$ is regular at the simple pole $t = \alpha$. It is supposed that the integration path C may be deformed to coincide with the steepest descent path through the saddle point. If, during this deformation, the path passes over the pole then the contribution arising from the pole in the form of a residue must, of course, be added to $I(\lambda; \alpha)$. Application of the saddle-point method yields the approximation containing the factor $f(t_s)$ (see (1.2.30), with ψ replaced by $-\psi$)

$$I(\lambda; \alpha) \sim e^{-\lambda \psi(t_s)} f(t_s) \left(\frac{2\pi}{\lambda \psi''(t_s)} \right)^{1/2} \qquad (\lambda \to +\infty) \qquad (1.5.1)$$

which will provide an adequate description of the behaviour of $I(\lambda; \alpha)$ provided $|t_s - \alpha|$ is bounded away from zero. When $t_s \to \alpha$, however, the factor $f(t_s)$ becomes infinite and accordingly (1.5.1) cannot represent $I(\lambda; \alpha)$ uniformly for $|t_s - \alpha| \geq 0$. The procedure we describe here to obtain a uniformly valid expansion of $I(\lambda; \alpha)$ as t_s passes through $t = \alpha$ is based on that given in Jones (1972; 1997, p. 89); see also Wong (1989, p. 356).

If we set

$$\psi(t) - \psi(t_s) = u^2 \qquad (1.5.2)$$

then the integral becomes [19]

$$I(\lambda; \alpha) = e^{-\lambda\psi(t_s)} \int_{-\infty}^{\infty} e^{-\lambda u^2} \frac{F(t)}{t - \alpha} \frac{dt}{du} \, du.$$

The ambiguity in (1.5.2) can be removed by choosing

$$u = (t - t_s) \left\{ \frac{\psi(t) - \psi(t_s)}{(t - t_s)^2} \right\}^{1/2},$$

where the square root is positive for real positive argument on the steepest descent path. The pole is now situated at the point $u = u_\alpha$, given by

$$u_\alpha^2 = \psi(\alpha) - \psi(t_s). \tag{1.5.3}$$

From the above choice of u, $\mathrm{Im}(u_\alpha)$ is positive or negative according as the pole lies above or below the u-axis.

In the neighbourhood of the saddle point

$$\frac{F(t)}{t - \alpha} \frac{dt}{du} = \frac{c_{-1}}{u - u_\alpha} + G(u), \qquad G(u) = \sum_{k=0}^{\infty} c_k u^k \quad (|u| < u_0), \tag{1.5.4}$$

where c_{-1} denotes the residue. The radius u_0 of the circle of convergence of the expansion of $G(u)$ about the saddle $u = 0$ will be determined either by a distant saddle of $\psi(t)$ or additional singularities present in $F(t)$. Near t_s we have from (1.5.2)

$$u = \kappa(t - t_s) \left\{ 1 + \frac{\psi'''(t_s)}{6\psi''(t_s)}(t - t_s) + \cdots \right\}, \qquad \kappa := \{\tfrac{1}{2}\psi''(t_s)\}^{1/2}, \tag{1.5.5}$$

so that $(dt/du)_{t_s} = \kappa^{-1}$, and we obtain from (1.5.4)

$$c_{-1} = \lim_{\substack{t \to \alpha \\ u \to u_\alpha}} \frac{u - u_\alpha}{t - \alpha} F(\alpha) \left(\frac{dt}{du}\right)_\alpha = F(\alpha),$$

$$c_0 = \frac{F(\alpha)}{u_\alpha} + \frac{\kappa^{-1} F(t_s)}{t_s - \alpha}.$$

The coefficient c_0 has a removable singularity when $t_s = \alpha$; expansion of $F(\alpha)$ about t_s and use of (1.5.5) shows that

$$c_0 = \kappa^{-1} F(t_s) \left\{ \frac{F'(t_s)}{F(t_s)} - \frac{\psi'''(t_s)}{6\psi''(t_s)} \right\} + O(t_s - \alpha), \tag{1.5.6}$$

with the first term on the right-hand side of (1.5.6) giving the value of c_0 at coalescence.

Substitution of the expansion (1.5.4) into the above integral for $I(\lambda; \alpha)$ then yields

$$I(\lambda; \alpha) = e^{-\lambda\psi(t_s)} \int_{-\infty}^{\infty} \left\{ \frac{c_{-1}}{u - u_\alpha} + G(u) \right\} e^{-\lambda u^2} \, du.$$

---

[19] We assume here that the image of C can be deformed into the entire real axis in the u-plane.

We now make use of the standard integral given in Abramowitz and Stegun (1965, Eq. (7.1.4))

$$J_{\pm}(x, a) = \int_{-\infty}^{\infty} e^{-x(t^2 - a^2)} \frac{dt}{t - a} = \pm \pi i \, \mathrm{erfc} \, (\mp i a x^{\frac{1}{2}}) \qquad (1.5.7)$$

when $x > 0$, where the upper or lower signs are taken according as $\mathrm{Im}(a) > 0$ or $\mathrm{Im}(a) < 0$, respectively, and erfc denotes the complementary error function.[20] We then finally obtain the desired uniform expansion in the form

$$I(\lambda; \alpha) \sim e^{-\lambda \psi(\alpha)} F(\alpha) J_{\pm}(\lambda, u_\alpha) + e^{-\lambda \psi(t_s)} \sum_{k=0}^{\infty} c_{2k} \frac{\Gamma(k + \frac{1}{2})}{\lambda^{k + \frac{1}{2}}} \qquad (1.5.8)$$

as $\lambda \to +\infty$, where we have used (1.5.3) to write $\psi(t_s) + u_\alpha^2 = \psi(\alpha)$. The upper or lower signs are chosen according as $\mathrm{Im}(u_\alpha) > 0$ or $\mathrm{Im}(u_\alpha) < 0$, respectively. We repeat that, if the pole is crossed while displacing the integration path, the residue contribution must also be added to $I(\lambda; \alpha)$.

The expansion (1.5.8) describes the asymptotic behaviour of $I(\lambda; \alpha)$ uniformly as the pole at $t = \alpha$ changes its position relative to the saddle point. As $t_s \to \alpha$, the first term remains finite, even at coalescence. When the pole is sufficiently distant from the saddle point (so that u_α is bounded away from zero) we can employ the leading term of the asymptotic expansion of the complementary error function given by (Abramowitz and Stegun, 1965, p. 298)

$$e^{z^2} \mathrm{erfc} \, z \sim \frac{1}{\pi} \sum_{k=0}^{\infty} \frac{(-)^k \Gamma(k + \frac{1}{2})}{z^{2k+1}} \qquad (|z| \to \infty \text{ in } |\arg z| < \tfrac{3}{4}\pi) \qquad (1.5.9)$$

to find

$$J_{\pm}(x, a) \sim -\frac{1}{a} \sqrt{\frac{\pi}{x}} \, e^{xa^2} \qquad (\mathrm{Im}(a) \gtrless 0)$$

as $ax^{1/2} \to \infty$. Since the argument of erfc in $J_{\pm}(\lambda, u_\alpha)$ in (1.5.8) satisfies $|\arg(\mp i u_\alpha \lambda^{1/2})| < \tfrac{1}{2}\pi$ when $\mathrm{Im}(u_\alpha) > 0$ or $\mathrm{Im}(u_\alpha) < 0$, respectively, it follows that for $\lambda \to +\infty$ the first two terms of (1.5.8) yield

$$e^{-\lambda \psi(\alpha)} F(\alpha) J_{\pm}(\lambda, u_\alpha) + c_0 e^{-\lambda \psi(t_s)} \sqrt{\frac{\pi}{\lambda}} \sim e^{-\lambda \psi(t_s)} \sqrt{\frac{\pi}{\lambda}} \left\{ -\frac{F(\alpha)}{u_\alpha} + c_0 \right\}$$

$$= e^{-\lambda \psi(t_s)} \frac{F(t_s)}{t_s - \alpha} \left(\frac{2\pi}{\lambda \psi''(t_s)} \right)^{1/2}$$

in agreement with (1.5.1).

---

[20] We remark that $J_+(x, a) - J_-(x, a) = 2\pi i$, since the complementary error function satisfies the relation $\mathrm{erfc}(-z) = 2 - \mathrm{erfc}(z)$. This difference results from the contribution from the residue of the integrand at $t = a$ and shows that $J_{\pm}(x, a)$ are *not* the analytic continuation of each other.

An alternative discussion of the expansion of a Laplace-type integral with a nearly coincident singularity is given in Van der Waerden (1951); see also Lauwerier (1966, pp. 55–61).

Example 1.11 A particularly simple example of an integral with a coalescing saddle point and pole is given by

$$I = \int_{-\infty}^{\infty} e^{-\lambda(t^2 - 2it)} \frac{dt}{t^2 + a^2},$$

where we take $a > 0$ and $\lambda \to +\infty$. The exponential factor has a saddle point at $t_s = i$ and the amplitude function has simple poles at $t = \pm ia$. The steepest descent path through t_s is the horizontal line $t = i + u$, $u \in (-\infty, \infty)$. We are interested here in the case when $a \simeq 1$.

First suppose that $a < 1$. Displacement of the integration path over the pole at $t = ia$ to coincide with the steepest descent path through t_s yields

$$I = \frac{\pi}{a} e^{-a(2-a)\lambda} + e^{-\lambda} \int_{-\infty}^{\infty} e^{-\lambda u^2} \frac{du}{(u+i)^2 + a^2},$$

where the first term on the right-hand side is the residue at the pole. A partial-fraction decomposition of the amplitude function then shows that

$$I = \frac{\pi}{a} e^{-a(2-a)\lambda} + \frac{e^{-\lambda}}{2ia} \int_{-\infty}^{\infty} e^{-\lambda u^2} \left\{ \frac{1}{u + i\alpha_-} - \frac{1}{u + i\alpha_+} \right\} du, \qquad (1.5.10)$$

where $\alpha_{\pm} := 1 \pm a$ and each integral can be evaluated according to (1.5.7) with the lower signs. If, for convenience in presentation, we define $E_{\pm} := \exp(\lambda \alpha_{\pm}^2) \operatorname{erfc}(\alpha_{\pm} \lambda^{1/2})$ and use the reflection formula $\operatorname{erfc}(-z) = 2 - \operatorname{erfc}(z)$, then we find

$$I = \frac{\pi}{a} e^{-a(2-a)\lambda} + \frac{\pi}{2a} e^{-\lambda}(E_+ - E_-)$$

$$= \frac{\pi}{2a} e^{-a(2-a)\lambda} \operatorname{erfc}[(a-1)\lambda^{\frac{1}{2}}] + \frac{\pi}{2a} e^{-\lambda} E_+. \qquad (1.5.11)$$

For $a \lesssim 1$ and $\lambda \to +\infty$, we can expand the function E_+ by (1.5.9) to find the uniform expansion

$$I \sim \frac{\pi}{a} e^{-a(2-a)\lambda} \operatorname{erfc}[(a-1)\lambda^{\frac{1}{2}}] + \frac{e^{-\lambda}}{2a(1+a)} \sum_{k=0}^{\infty} \frac{(-)^k \Gamma(k+\frac{1}{2})}{(1+a)^{2k} \lambda^{k+\frac{1}{2}}}. \qquad (1.5.12)$$

When $a > 1$, the displacement of the integration path does not cross the pole at $t = ia$ and so there is no residue contribution. In the first integral in (1.5.10) this pole now lies above the path in the u-plane so that, from (1.5.9) with the upper signs, we obtain

$$I = \frac{\pi}{2a} e^{-\lambda} \left\{ e^{\lambda(1-a)^2} \operatorname{erfc}[(a-1)\lambda^{\frac{1}{2}}] + E_+ \right\}.$$

Upon expansion of E_+ this last result yields the same as (1.5.12).

In the special case $a = 1$, when the saddle and a pole coincide, the integration path has an indentation about $t = i$. We then find as $\lambda \to +\infty$

$$I \sim e^{-\lambda} \left(\frac{1}{2}\pi + \frac{1}{4} \sum_{k=0}^{\infty} \frac{(-)^k \Gamma(k + \frac{1}{2})}{2^{2k} \lambda^{k+\frac{1}{2}}} \right) \qquad (a = 1).$$

The first two terms of this expansion have been given by Lauwerier (1966, p. 61).

A similar treatment can be applied to the more general integral

$$I = \int_{-\infty}^{\infty} e^{-\lambda(t^2 - 2i\theta t)} \frac{dt}{\cosh t - \cos\omega} = e^{-\lambda\theta^2} \int_{-\infty}^{\infty} e^{-\lambda(t - i\theta)^2} \frac{dt}{\cosh t - \cos\omega}$$

as $\lambda \to +\infty$, where $0 < \theta < \pi$ and $0 < \omega < \pi$. We have $\psi(t) = (t - i\theta)^2$, which has a saddle point at $t_s = i\theta$, and the amplitude function has simple poles at $t = \pm i\omega \pm 2\pi ik$ $(k = 0, 1, 2, \ldots)$. The steepest descent path through t_s is again the horizontal line $t = i\theta + u$, $u \in (-\infty, \infty)$.

When $\omega \lesssim \theta$, we displace the integration path over the pole at $t = i\omega$ to coincide with the steepest descent path through t_s to find, upon evaluation of the residue,

$$e^{\lambda\theta^2} I = \frac{2\pi}{\sin\omega} e^{\lambda(\theta - \omega)^2} + \int_{-\infty}^{\infty} e^{-\lambda u^2} \frac{F(t)}{t - i\omega} du,$$

where

$$F(t) = \frac{t - i\omega}{\cosh t - \cos\omega}.$$

Straightforward calculations show that

$$F(t) = -i\operatorname{cosec}\omega + \frac{\cot\omega}{2\sin\omega}(t - i\omega) + O(t - i\omega)^2,$$

so that $F(i\omega) = -i\operatorname{cosec}\omega$ and $F'(t_s) = \cot\omega/(2\sin\omega) + O(\theta - \omega)$. We then obtain from (1.5.6)

$$c_{-1} = -i\operatorname{cosec}\omega, \qquad c_0 = \frac{\cot\omega}{2\sin\omega} + O(\theta - \omega).$$

When $\omega \lesssim \theta$, the first two terms in the uniform expansion (1.5.8) therefore yield

$$e^{\lambda\theta^2} I \sim \frac{2\pi}{\sin\omega} e^{\lambda(\theta - \omega)^2} - \pi i e^{\lambda(\theta - \omega)^2} F(i\omega) \operatorname{erfc}[(\theta - \omega)\lambda^{\frac{1}{2}}] + c_0 \sqrt{\frac{\pi}{\lambda}}$$

$$= \frac{\pi}{\sin\omega} e^{\lambda(\theta - \omega)^2} \operatorname{erfc}[(\omega - \theta)\lambda^{\frac{1}{2}}] + c_0 \sqrt{\frac{\pi}{\lambda}} \tag{1.5.13}$$

as $\lambda \to +\infty$. When $\omega \gtrsim \theta$, the path displacement does not cross the pole and we take the upper sign in (1.5.8). This results in the same expression as (1.5.13). When $\omega = \theta$ we have

$$e^{\lambda\theta^2} I \sim \frac{\pi}{\sin\theta} + c_0 \sqrt{\frac{\pi}{\lambda}} \qquad (\lambda \to +\infty).$$

1.5.2 Saddle point near an endpoint

In §1.2.3 we considered the expansion of Laplace-type integrals along steepest descent paths when the saddle point was either an interior point, an endpoint or outside the range of integration. These different cases are given by (1.2.19), (1.2.18) and (1.2.21), respectively. It is readily seen that each expansion is associated with a different asymptotic scale in the large parameter λ. Here we consider a uniform expansion that will describe the behaviour of the integral as the saddle point varies between the above three cases. The analysis presented is based on that given in Jones (1972; 1997, p. 89); see also Wong (1989, p. 360).

For simplicity, we consider the integral on the real axis

$$I(\lambda) = \int_0^\infty t^\mu f(t) e^{-\lambda \psi(t)} dt \qquad (\mu > -1) \tag{1.5.14}$$

where all quantities are real, $\lambda \to +\infty$ and $f(t)$ is regular at the origin. The integrand is supposed to have a single saddle point at $t = t_s$ situated on the real t-axis, which depends on an unspecified parameter contained in $\psi(t)$, and $\psi''(t_s) > 0$. In addition, it is assumed that $\psi(t)$ is such that $\psi(t) \to +\infty$ as $t \to +\infty$ and that the scaled amplitude function $f(t)$ is differentiable as many times as required.

We introduce the new variable u by

$$\psi(t) - \psi(t_s) = \tfrac{1}{2}(u - a)^2, \tag{1.5.15}$$

where the point $u = a$ corresponds to the saddle $t = t_s$. The points $u = 0$ and $t = 0$ are made to correspond by the choice

$$a = \pm\{2[\psi(0) - \psi(t_s)]\}^{1/2} \qquad (t_s \gtrless 0), \tag{1.5.16}$$

so that a has the same sign as t_s. Substitution of the new variable u into $I(\lambda)$ then produces

$$I(\lambda) = e^{-\lambda \psi(t_s)} \int_0^\infty u^\mu g_0(u) e^{-\frac{1}{2}\lambda(u-a)^2} du, \qquad g_0(u) = f(t)(t/u)^\mu \frac{dt}{du}.$$

We now write, following Bleistein (1966),

$$g_0(u) = A_0 + B_0(u - a) + u(u - a)G_0(u),$$

where the constants A_0 and B_0 are defined by [21]

$$A_0 = g_0(a) = f(t_s)(t_s/a)^\mu \{\psi''(t_s)\}^{-1/2},$$
$$B_0 = \frac{1}{a}\{g_0(a) - g_0(0)\} = \frac{1}{a}\{g_0(a) - f(0)(-a/\psi'(0))^{\mu+1}\}. \tag{1.5.17}$$

---

[21] In the neighbourhood of $t = t_s$ and $t = 0$ we find, by appropriate expansion of (1.5.15), that $\{\psi''(t_s)\}^{1/2}(t - t_s) \simeq u - a$ and $\psi'(0)t \simeq -ua$, respectively, which enables the corresponding values of dt/du to be determined.

The resulting integrals are expressed in terms of

$$\int_0^\infty u^\mu e^{-\frac{1}{2}\lambda(u-a)^2}\,du = \lambda^{-(\mu+1)/2}\int_0^\infty w^\mu e^{-\frac{1}{2}(w-a\sqrt{\lambda})^2}\,dw$$

$$= \lambda^{-(\mu+1)/2}W_\mu(a\lambda^{\frac{1}{2}})$$

and its derivative, where

$$W_\mu(x) = \int_0^\infty w^\mu e^{-\frac{1}{2}(w-x)^2}\,dw = \Gamma(\mu+1)e^{-x^2/4}U(\mu+\tfrac{1}{2},-x) \qquad (1.5.18)$$

is seen to be related to the parabolic cylinder function $U(a,z)$ defined in Abramowitz and Stegun (1965, p. 687).

Then we obtain

$$I(\lambda) = e^{-\lambda\psi(t_s)}\left\{A_0\frac{W_\mu(a\lambda^{\frac{1}{2}})}{\lambda^{\frac{1}{2}\mu+\frac{1}{2}}} + B_0\frac{W_\mu'(a\lambda^{\frac{1}{2}})}{\lambda^{\frac{1}{2}\mu+1}} + J_1(\lambda)\right\},$$

where

$$J_1(\lambda) = \int_0^\infty u^{\mu+1}(u-a)G_0(u)e^{-\frac{1}{2}\lambda(u-a)^2}\,du.$$

A straightforward integration by parts yields

$$J_1(\lambda) = \frac{1}{\lambda}\int_0^\infty u^\mu g_1(u)e^{-\frac{1}{2}\lambda(u-a)^2}\,du,$$

with $g_1(u)$ defined by

$$g_1(u) = u^{-\mu}\{u^{\mu+1}G_0(u)\}' = (\mu+1)G_0(u) + uG_0'(u).$$

The function $g_1(u)$ is now expanded as

$$g_1(u) = A_1 + B_1(u-a) + u(u-a)G_1(u)$$

and the process can be repeated to generate an expansion in the form

$$I(\lambda) = e^{-\lambda\psi(t_s)}\left\{\frac{W_\mu(a\lambda^{\frac{1}{2}})}{\lambda^{\frac{1}{2}\mu+\frac{1}{2}}}\sum_{k=0}^{n-1}\frac{A_k}{\lambda^k} + \frac{W_\mu'(a\lambda^{\frac{1}{2}})}{\lambda^{\frac{1}{2}\mu+1}}\sum_{k=0}^{n-1}\frac{B_k}{\lambda^k} + J_n(\lambda)\right\} \qquad (1.5.19)$$

for positive integer n. For $0 \le k \le n-1$, the coefficients A_k and B_k are determined from

$$g_k(u) = A_k + B_k(u-a) + u(u-a)G_k(u), \qquad g_{k+1}(u) = u^{-\mu}\{u^{\mu+1}G_k(u)\}',$$

$$A_k = g_k(a), \qquad B_k = \frac{1}{a}\{g_k(a) - g_k(0)\}$$

and the remainder

$$J_n(\lambda) = \frac{1}{\lambda^n}\int_0^\infty u^\mu g_n(u)e^{-\frac{1}{2}\lambda(u-a)^2}\,du.$$

The proof that (1.5.19) is a uniform asymptotic expansion is given in Wong (1989, p. 364).

It now remains to examine how the leading terms of (1.5.19), namely

$$I(\lambda) \sim e^{-\lambda\psi(t_s)} \left\{ A_0 \frac{W_\mu(a\lambda^{\frac{1}{2}})}{\lambda^{\frac{1}{2}\mu+\frac{1}{2}}} + B_0 \frac{W_\mu'(a\lambda^{\frac{1}{2}})}{\lambda^{\frac{1}{2}\mu+1}} \right\}, \tag{1.5.20}$$

change as the saddle point t_s moves through the endpoint at $t = 0$. For this purpose we shall require the following values of the parabolic cylinder function $W_\mu(x)$ (Abramowitz and Stegun, 1965, p. 687)

$$W_\mu(0) = \Gamma(\tfrac{1}{2}\mu + \tfrac{1}{2})2^{\frac{1}{2}\mu-\frac{1}{2}}, \qquad W_\mu'(0) = -\Gamma(\tfrac{1}{2}\mu + 1)2^{\frac{1}{2}\mu}, \tag{1.5.21}$$

together with the asymptotic behaviour[22]

$$W_\mu(x) \sim \begin{cases} (2\pi)^{\frac{1}{2}} x^\mu & (x \to +\infty) \\ \Gamma(\mu + 1)e^{-x^2/2}(-x)^{-\mu-1} & (x \to -\infty). \end{cases} \tag{1.5.22}$$

We now examine in turn the cases (i) $t_s > 0$, (ii) $t_s = 0$ and (iii) $t_s < 0$.

(i) $t_s > 0$ (and $a\lambda^{1/2} \gg 1$): Provided t_s is sufficiently distant from the endpoint $t = 0$, so that $a\lambda^{1/2} \gg 1$, the leading form of the right-hand side of (1.5.20) combined with the first result in (1.5.22) produces

$$e^{-\lambda\psi(t_s)} \left\{ A_0 a^\mu \sqrt{2\pi/\lambda} + O(\lambda^{-\frac{3}{2}}) \right\},$$

whence

$$I(\lambda) \sim e^{-\lambda\psi(t_s)} f(t_s) t_s^\mu \left(\frac{2\pi}{\lambda\psi''(t_s)} \right)^{1/2}.$$

This form agrees with that in (1.5.1), when $f(t)$ is replaced by $t^\mu f(t)$, and that obtained from (1.2.19).

(ii) $t_s = 0$ ($a = 0$): In this case the dominant term in (1.5.20) is again the first term and we find

$$I(\lambda) \sim e^{-\lambda\psi(0)} \frac{A_0\Gamma(\tfrac{1}{2}\mu + \tfrac{1}{2})}{2(\lambda/2)^{(\mu+1)/2}}$$

$$= \tfrac{1}{2} e^{-\lambda\psi(0)} f(0)\Gamma(\tfrac{1}{2}\mu + \tfrac{1}{2}) \left(\frac{2}{\lambda\psi''(0)} \right)^{(\mu+1)/2},$$

where we have employed the limiting value of $t_s/a = \{\psi''(0)\}^{-1/2}$. This can be seen to agree with the first term of (1.2.18) corresponding to the situation when the saddle point coincides with an endpoint.

---

[22] This follows from (1.5.18) and the relations between the parabolic cylinder functions $U(a, x)$ and $U(a, -x) = \pi V(a, x)/\Gamma(a + \tfrac{1}{2}) - \sin\pi a\, U(a, x)$, combined with the asymptotic behaviour of $U(a, x)$ and $V(a, x)$ for $x \to +\infty$ given in Abramowitz and Stegun (1965, p. 689).

(iii) $t_s < 0$ (and $|a|\lambda^{1/2} \gg 1$): When the saddle point is outside the interval of integration both t_s and a are negative. We shall suppose that t_s is sufficiently distant from $t = 0$, so that $|a|\lambda^{1/2} \gg 1$. In this case both terms in (1.5.20) contribute to the leading form. From the second result in (1.5.22) and (1.5.17), we obtain

$$I(\lambda) \sim \frac{e^{-\lambda\psi(t_s)-\frac{1}{2}a^2\lambda}}{\lambda^{(\mu+1)/2}}\Gamma(\mu+1)\{A_0(-a\lambda^{\frac{1}{2}})^{-\mu-1} + B_0\lambda^{-\frac{1}{2}}(-a\lambda^{\frac{1}{2}})^{-\mu}\}$$

$$= \frac{e^{-\lambda\psi(0)}}{(-a\lambda)^{\mu+1}}\Gamma(\mu+1)(A_0 - aB_0)$$

$$= e^{-\lambda\psi(0)}\frac{f(0)\Gamma(\mu+1)}{(\lambda\psi'(0))^{\mu+1}},$$

which agrees with the leading term in (1.2.21) corresponding to a linear endpoint.

We thus find that as the saddle point moves from well inside the interval of integration to well outside, there is a smooth transition in the leading form (1.5.20) between values that have been obtained in special cases in §1.2.3.

Example 1.12 The complementary incomplete gamma function $\Gamma(1 + x, y)$ is defined for $x > 0$, $y > 0$ by

$$\Gamma(1 + x, y) = \int_y^\infty e^{-t}t^x dt = e^{-y}\int_0^\infty e^{-t}(y + t)^x dt$$

$$= e^{-y}y^{1+x}\int_0^\infty e^{-x\{\xi t - \log(1+t)\}}dt, \qquad \xi = y/x.$$

We consider the behaviour of the integral

$$I = \int_0^\infty e^{-x\{\xi t - \log(1+t)\}}dt$$

for large positive and nearly equal values of x and y. Comparison with (1.5.14) shows that here we have $\mu = 0$, $f(t) = 1$ and $\psi(t) = \xi t - \log(1 + t)$. The function $\psi(t)$ has a saddle point at $t_s = \xi^{-1} - 1$, which lies inside the interval of integration when $\xi < 1$ and outside the interval when $\xi > 1$. As $\xi \to 1$, the saddle point coalesces with the endpoint $t = 0$.

From (1.5.16), it follows that

$$a = \pm\{2(\xi - 1 - \log \xi)\}^{1/2},$$

where the upper or lower sign is chosen according as $t_s > 0$ ($\xi < 1$) or $t_s < 0$ ($\xi > 1$), respectively. When $t_s = 0$ ($\xi = 1$) the quantity $a = 0$. From (1.5.17), the constants A_0 and B_0 are

$$A_0 = \frac{1}{\xi}, \qquad B_0 = \frac{1}{a\xi} + \frac{1}{\xi - 1} = \frac{2}{3} + O(\xi - 1).$$

Then, from (1.5.20), the first two terms of the uniform expansion are given by

$$I \sim e^{-x\psi(t_s)} \left\{ A_0 \frac{W_0(ax^{\frac{1}{2}})}{x^{\frac{1}{2}}} + B_0 \frac{W_0'(ax^{\frac{1}{2}})}{x} \right\},$$

where $\psi(t_s) = 1 - \xi + \log \xi$ and

$$W_0(ax^{\frac{1}{2}}) = e^{-a^2 x/4} U(\tfrac{1}{2}, -ax^{\frac{1}{2}}) = e^{x\psi(t_s)/2} U(\tfrac{1}{2}, -ax^{\frac{1}{2}}).$$

The leading term of this approximation therefore yields

$$\Gamma(1 + x, y) \sim e^{-y} y^{1+x} \frac{e^{-x\psi(t_s)/2}}{\xi \, x^{\frac{1}{2}}} U(\tfrac{1}{2}, -ax^{\frac{1}{2}})$$

$$= x^{\frac{1}{2}} (xy)^{x/2} e^{-(x+y)/2} U(\tfrac{1}{2}, -ax^{\frac{1}{2}})$$

for $x \to +\infty$, $y \to +\infty$ and finite y/x.

1.5.3 Two coalescing saddle points

We consider the contour integral of the form

$$I(\lambda; \sigma) = \frac{1}{2\pi i} \int_C e^{\lambda \psi(t;\sigma)} f(t) \, dt, \tag{1.5.23}$$

where λ is large and positive, $f(t)$ is analytic in t and the phase function[23] $\psi(t) \equiv \psi(t; \sigma)$ is an analytic function of both t and a parameter σ. We shall suppose that $\psi(t)$ has two distinct simple saddle points t_{s1} and t_{s2} that move in the complex t-plane as σ varies. At a critical value of this parameter, $\sigma = \sigma_c$, the two saddles coalesce to form a double saddle at t_c. When $\sigma \neq \sigma_c$, it is assumed that the path C can be deformed to pass over either one (say t_{s1}) or both saddles, so that from (1.2.19) we have the leading asymptotic behaviour of $I(\lambda; \sigma)$ as $\lambda \to +\infty$

$$I(\lambda; \sigma) \sim \sum_{j=1}^{2} \frac{\gamma_j}{2\pi i} e^{\lambda \psi(t_{sj})} f(t_{sj}) \left(\frac{2\pi}{-\lambda \psi''(t_{sj})} \right)^{1/2}, \tag{1.5.24}$$

where $\gamma_1 = \pm 1$ and $\gamma_2 = 0$ or ± 1 depending on C. On the other hand, when $\sigma = \sigma_c$, we have from (1.2.16)

$$I(\lambda; \sigma) \sim \frac{\delta}{2\pi i} e^{\lambda \psi(t_c)} f(t_c) \Gamma(\tfrac{4}{3}) \left(\frac{3!}{-\lambda \psi'''(t_c)} \right)^{1/3}, \tag{1.5.25}$$

where the value of the factor δ depends on which of the three steepest descent paths emanating from the double saddle t_c is selected. It is seen that there is a discontinuity in the asymptotic scale ($\lambda^{-1/2}$ to $\lambda^{-1/3}$) associated with these approximations and the former becomes singular as $\sigma \to \sigma_c$ (since $\psi''(t_{sj}) \to 0$). A uniform asymptotic

---

[23] To ease the presentation, the functional dependence of ψ on σ is not indicated.

expansion that interpolates between these two disparate forms as the parameter σ varies through σ_c is therefore desirable. However, this uniformity comes at the cost of additional complexity in the form of the approximating functions and the associated coefficients.

The simplest phase function that exhibits two coalescing saddle points is the cubic polynomial introduced by Chester, Friedman and Ursell (1957) in the form

$$\psi(t) = \tfrac{1}{3}u^3 - \alpha^2 u + \beta, \tag{1.5.26}$$

where $\alpha \equiv \alpha(\sigma)$ and $\beta \equiv \beta(\sigma)$ are to be determined by the requirement that the derivative

$$\frac{dt}{du} = \frac{u^2 - \alpha^2}{\psi'(t)} \tag{1.5.27}$$

must not vanish or become infinite in the relevant regions. This is achieved by making t_{s1} correspond to α and t_{s2} correspond to $-\alpha$, so that

$$\alpha^3 = \tfrac{3}{4}\{\psi(t_{s2}) - \psi(t_{s1})\}, \qquad \beta = \tfrac{1}{2}\{\psi(t_{s1}) + \psi(t_{s2})\}. \tag{1.5.28}$$

It was shown by Chester, Friedman and Ursell (1957) that one branch of the cubic transformation $t \mapsto u$ in (1.5.26) is one-to-one and analytic for all σ in a neighbourhood of σ_c. This last result, however, is only local; but it often transpires to be one-to-one and analytic in a domain containing the path of integration. Assuming this to be the case, we can substitute (1.5.26) into (1.5.23) to obtain

$$I(\lambda; \sigma) = \frac{e^{\lambda\beta}}{2\pi i} \int_{C'} e^{\lambda(\frac{1}{3}u^3 - \alpha^2 u)} f(t) \frac{dt}{du}\, du,$$

where C' denotes the image of C in the u-plane.

Following the procedure outlined in the last section, we now put

$$g_0(u) := f(t)\frac{dt}{du} = A_0 + B_0 u - (u^2 - \alpha^2)G_0(u),$$

where the constants $A_0 \equiv A_0(\sigma)$ and $B_0 \equiv B_0(\sigma)$ are given by

$$A_0 = \frac{1}{2}\{g_0(\alpha) + g_0(-\alpha)\}, \qquad B_0 = \frac{1}{2\alpha}\{g_0(\alpha) - g_0(-\alpha)\}. \tag{1.5.29}$$

Substitution of the above form for $g_0(u)$ into $I(\lambda; \sigma)$ then yields

$$I(\lambda; \sigma) = e^{\lambda\beta}\left\{\frac{A_0}{\lambda^{\frac{1}{3}}}U(\alpha^2\lambda^{\frac{2}{3}}) - \frac{B_0}{\lambda^{\frac{2}{3}}}U'(\alpha^2\lambda^{\frac{2}{3}}) + J_1(\lambda; \sigma)\right\},$$

where

$$U(x) = \frac{1}{2\pi i} \int_{C'} e^{\frac{1}{3}\tau^3 - x\tau}\, d\tau. \tag{1.5.30}$$

The remainder term is defined by

$$J_1(\lambda; \sigma) = \frac{1}{2\pi i} \int_{C'} e^{\lambda(\frac{1}{3}u^3 - \alpha^2 u)} (\alpha^2 - u^2) G_0(u)\, du$$

$$= \frac{\lambda^{-1}}{2\pi i} \int_{C'} e^{\lambda(\frac{1}{3}u^3 - \alpha^2 u)} g_1(u)\, du, \qquad g_1(u) := G_0'(u)$$

upon integration by parts. The function $g_1(u)$ is now expanded in the same manner as $g_0(u)$ and the process can be repeated. This generates an expansion in the form

$$I(\lambda; \sigma) = e^{\lambda \beta} \left\{ U(\alpha^2 \lambda^{\frac{2}{3}}) \sum_{k=0}^{n-1} \frac{A_k}{\lambda^{k+\frac{1}{3}}} - U'(\alpha^2 \lambda^{\frac{2}{3}}) \sum_{k=0}^{n-1} \frac{B_k}{\lambda^{k+\frac{2}{3}}} + J_n(\lambda; \sigma) \right\},$$

$$(1.5.31)$$

where, for $0 \le k \le n - 1$,

$$g_k(u) = A_k + B_k u - (u^2 - \alpha^2) G_k(u), \qquad g_{k+1}(u) = G_k'(u),$$

$$A_k = \frac{1}{2}\{g_k(\alpha) + g_k(-\alpha)\}, \qquad B_k = \frac{1}{2\alpha}\{g_k(\alpha) - g_k(-\alpha)\}$$

and

$$J_n(\lambda; \sigma) = \frac{\lambda^{-n}}{2\pi i} \int_{C'} e^{\lambda(\frac{1}{3}u^3 - \alpha^2 u)} g_n(u)\, du.$$

Note that the quantity β appearing in the exponential factor $e^{\lambda \beta}$ in (1.5.31) can, from (1.5.28), be interpreted as the *average* saddle height.

The leading coefficients A_0 and B_0 (when $\alpha \neq 0$) are defined in (1.5.29) in terms of $g_0(\pm\alpha) = f(t_{sj})t'_{u=\pm\alpha}$, where the upper or lower signs correspond to the choice $j = 1$ or $j = 2$, respectively. By differentiation of (1.5.27) in the form

$$\psi'(t)t'(u) = u^2 - \alpha^2, \tag{1.5.32}$$

where the prime denotes differentiation with respect to the argument concerned, we obtain

$$\psi''(t)(t'(u))^2 + \psi'(t)t''(u) = 2u,$$

$$\psi'''(t)(t'(u))^3 + 3\psi''(t)t'(u)t''(u) + \psi'(t)t'''(u) = 2,$$

so that

$$(t'(\pm\alpha))^2 = \pm 2\alpha/\psi''(t_{sj}) \qquad (j = 1, 2). \tag{1.5.33}$$

When $\alpha = 0$, we have

$$A_0 = g_0(0) = f(t_c)t'(0), \quad B_0 = g_0'(0) = f'(t_c)(t'(0))^2 + f(t_c)t''(0), \tag{1.5.34}$$

where[24]

$$(t'(0))^3 = 2/\psi'''(t_c), \qquad t''(0) = -\psi^{iv}(t_c)(t'(0))^2/(6\psi'''(t_c)).$$

The proof that (1.5.31) is a uniform asymptotic expansion for large λ is given in Chester, Friedman and Ursell (1957); see also Bleistein and Handelsman (1975, pp. 374–376) and Wong (1989, pp. 371–372). The integral $U(x)$ in (1.5.30) can be expressed in terms of an Airy function. However, the precise choice of Airy function depends on the integration path C', which in turn depends on the original path C in the t-plane and, when $t_{s1} \neq t_{s2}$, on the choice of the particular branch chosen for α in (1.5.28); see Bleistein and Handelsman (1975, p. 372) and Jones (1997, pp. 93–94) for a discussion of this point. There are three possible choices for C' as illustrated in Fig. 1.16: these are indicated by the three contours Γ_s ($s = 0, \pm 1$) which lie in sectors of angular width $2\pi/3$. If C' is taken to be the contour Γ_s then

$$U(x) = \omega^s \mathrm{Ai}\,(\omega^s x), \qquad \omega = e^{2\pi i/3} \;\; (s = 0, \pm 1). \tag{1.5.35}$$

We now examine how the leading terms of the expansion (1.5.31) change as the saddles move from being well separated to coalescence. For definiteness, we shall suppose that the integration path C and the branch of α are such that C' can be reconciled with the path Γ_0, so that

$$I(\lambda; \sigma) \sim e^{\lambda\beta} \left\{ \frac{A_0}{\lambda^{\frac{1}{3}}} \mathrm{Ai}(\alpha^2 \lambda^{\frac{2}{3}}) - \frac{B_0}{\lambda^{\frac{2}{3}}} \mathrm{Ai}'(\alpha^2 \lambda^{\frac{2}{3}}) \right\}. \tag{1.5.36}$$

For this we shall need the asymptotic behaviour (Abramowitz and Stegun, 1965, p. 448)

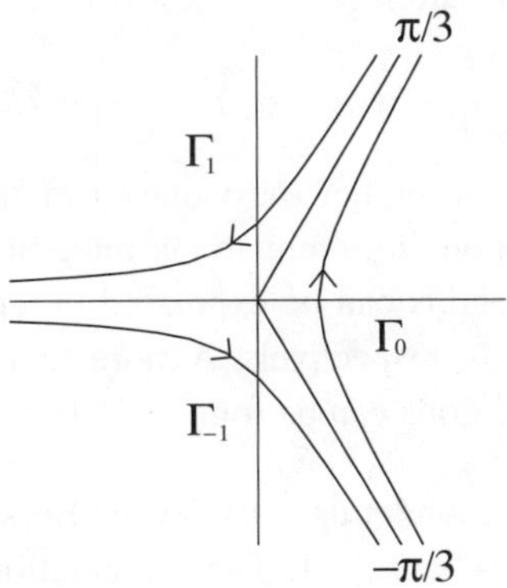

Figure 1.16 The integration paths Γ_s ($s = 0, \pm 1$) for $U(x)$.

---

[24] The value of $t''(0)$ follows from a third differentiation of (1.5.32) combined with the fact that $\psi'(t_c) = \psi''(t_c) = 0$.

$$\text{Ai}\,(z) \sim \frac{z^{-\frac{1}{4}}}{2\sqrt{\pi}}\exp(-\tfrac{2}{3}z^{3/2}), \qquad \text{Ai}'\,(z) \sim -\frac{z^{\frac{1}{4}}}{2\sqrt{\pi}}\exp(-\tfrac{2}{3}z^{3/2})$$

as $|z| \to \infty$ in $|\arg z| \le \pi - \delta$, $\delta > 0$, together with the values

$$\text{Ai}(0) = \Gamma(\tfrac{1}{3})/(2 \cdot 3^{1/6}\pi), \qquad \text{Ai}'(0) = -3^{1/6}\Gamma(\tfrac{2}{3})/(2\pi).$$

Substitution of these forms into the leading expression for $I(\lambda;\sigma)$ in (1.5.36) when the saddles are well separated then leads to

$$\begin{aligned}
I(\lambda;\sigma) &\sim \frac{e^{\lambda(\beta-\frac{2}{3}\alpha^3)}}{2\pi}\left(\frac{\pi}{\lambda\alpha}\right)^{1/2}(A_0 + B_0\alpha) \\
&= \frac{e^{\lambda\psi(t_{s1})}}{2\pi}f(t_{s1})\left(\frac{2\pi}{\lambda\psi''(t_{s1})}\right)^{1/2}
\end{aligned}$$

as $\lambda \to +\infty$ since, from (1.5.28) and (1.5.29),

$$\beta - \tfrac{2}{3}\alpha^3 = \psi(t_{s1}), \qquad A_0 + B_0\alpha = g_0(\alpha).$$

When $\sigma = \sigma_c$, we have $t_{s1} = t_{s2} = t_c$, $\beta = \psi(t_c)$ and $A_0 = g_0(0)$, so that the expression for $I(\lambda;\sigma)$ in (1.5.36) as $\lambda \to +\infty$ yields

$$\begin{aligned}
I(\lambda;\sigma) &= e^{\lambda\psi(t_c)}\left(\frac{A_0\text{Ai}(0)}{\lambda^{\frac{1}{3}}} + O(\lambda^{-\frac{2}{3}})\right) \\
&\sim e^{\lambda\psi(t_c)}f(t_c)\left(\frac{3!}{\lambda\psi'''(t_c)}\right)^{1/3}\frac{3^{\frac{1}{2}}\Gamma(\tfrac{4}{3})}{2\pi}.
\end{aligned}$$

These limiting forms are seen to agree with those in (1.5.24) and (1.5.25).

In situations where the phase function $\psi(t)$ in (1.5.23) has $m \ge 3$ saddle points that can coalesce (and, in general, with $m-1$ coalescence parameters $\vec{\sigma} = \{\sigma_1, \sigma_2, \ldots, \sigma_{m-1}\}$) the comparison integrals $W(x)$ are no longer Airy functions. The analogue of the transformation (1.5.26) now becomes

$$\psi(t;\vec{\sigma}) = \frac{u^{m+1}}{m+1} + \alpha_1 u + \alpha_2 u^2 + \cdots + \alpha_{m-1}u^{m-1} + \beta.$$

It is hardly surprising that the higher the value of m the more recondite become the properties of the corresponding comparison integrals. In the cases $m = 3$ and $m = 4$, the comparison integrals can be expressed in terms of the Pearcey and the so-called swallowtail integrals, respectively. A more detailed discussion of these and higher-order comparison integrals can be found in Olver *et al.* (2010, Ch. 36).

Example 1.13 As an illustration let us consider the Bessel function $J_\nu(\nu x)$ of large order and argument, where $0 < x \le 1$. The presentation is based on that given in Chester, Friedman and Ursell (1957) and Copson (1965, Chapter 10). From Watson (1952, §6.2), we have

$$J_\nu(\nu x) = \frac{1}{2\pi i}\int_{\infty-\pi i}^{\infty+\pi i} e^{\nu(\text{sech}\,\sigma\,\sinh t - t)}dt,$$

where $x = \operatorname{sech}\sigma$. Here $f(t) = 1$ and $\psi(t) \equiv \psi(t;\sigma) = \operatorname{sech}\sigma \sinh t - t$, so that the saddle points are situated at $t = \pm\sigma \pm 2\pi i k$ $(k = 0, 1, 2, \ldots)$. The relevant saddle points are $t_{s1} = \sigma$ and $t_{s2} = -\sigma$, which coalesce at $t = 0$ when $\sigma = 0$. The required cubic transformation is

$$\operatorname{sech}\sigma \sinh t - t = \tfrac{1}{3}u^3 - \alpha^2 u + \beta \tag{1.5.37}$$

and for the points $t = \pm\sigma$ to correspond to $u = \pm\alpha$, respectively, we require from (1.5.28) with $\sigma \geq 0$

$$\tfrac{2}{3}\alpha^3 = \sigma - \tanh\sigma \geq 0, \qquad \beta = 0.$$

For small σ it then follows that

$$\alpha = 2^{-1/3}\sigma\left(1 - \tfrac{2}{15}\sigma^2 + \tfrac{19}{525}\sigma^4 + \cdots\right).$$

The required root of (1.5.37) is given by the ordinary theory of cubic equations in the form [25]

$$u = 2\alpha \sin\tfrac{1}{3}\Phi, \qquad \tfrac{2}{3}\alpha^3 \sin\Phi = t - \operatorname{sech}\sigma \sinh t,$$

which is easily shown to satisfy $u = \pm\alpha$ when $t = \pm\sigma$. From this it follows that

$$u = \frac{(1 - \operatorname{sech}\sigma)}{\alpha^2}t + O(t^3) \qquad (t \to 0),$$

so that as $t \to 0$, $\sigma \to 0$ we have $dt/du \sim 2^{1/3}$. From (1.5.33) and the fact that $\psi''(\pm\sigma) = \pm\tanh\sigma$, we find $(dt(\pm\alpha)/du)^2 = 2\alpha/\tanh\sigma$ and this quantity approaches the value $2^{2/3}$ as $\sigma \to 0$. Hence, we require the positive root in (1.5.33) and so

$$t'(\pm\alpha) = +\left(\frac{2\alpha}{\tanh\sigma}\right)^{1/2}.$$

From (1.5.29) we then obtain $A_0 = (2\alpha/\tanh\sigma)^{1/2}$, $B_0 = 0$.

It remains to show that the steepest descent path C in the t-plane maps to a path in the u-plane of the type Γ_0 in Fig. 1.16. To see this we consider the intermediate transformations

$$z = t - \operatorname{sech}\sigma \sinh t, \qquad \tfrac{2}{3}\alpha^3 \sin\Phi = z, \qquad u = 2\alpha \sin\tfrac{1}{3}\Phi$$

and show in Fig. 1.17 the successive transformations of the upper half of the steepest descent path through $t = \sigma$ in the t-plane (solid curve) together with the part of the imaginary axis $0 \leq \operatorname{Im}(t) \leq \pi$ and the horizontal line (shown dashed) $\operatorname{Im}(t) = \pi$, $0 \leq \operatorname{Re}(t) < \infty$; the lower half of the integration path follows by symmetry. The

---

[25] The other two roots $u_\pm = -\alpha \sin\tfrac{1}{3}\Phi \pm \alpha\sqrt{3}\cos\tfrac{1}{3}\Phi$ do not satisfy the requirement $u = \pm\alpha$ when $t = \pm\sigma$, since $u_+ = \alpha$, $u_- = -2\alpha$ when $t = \sigma$ and $u_+ = 2\alpha$, $u_- = -\alpha$ when $t = -\sigma$.

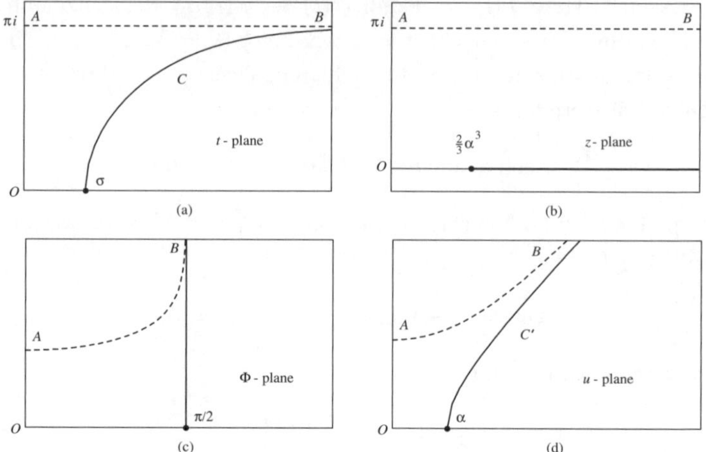

Figure 1.17 Successive transformations of the steepest descent path in (a) the t-plane, (b) the z-plane, (c) the Φ-plane and (d) the u-plane.

map of C in the u-plane is C', which is the steepest descent path through $u = \alpha$. The steepest paths through α are given by $\mathrm{Im}(\frac{1}{3}u^3 - \alpha^2 u) = 0$, or

$$y(y^2 - 3x^2 + 3\alpha^2) = 0, \qquad u = x + iy.$$

The steepest descent path is the branch of the hyperbola $y^2 - 3x^2 + 3\alpha^2 = 0$, which has endpoints at infinity on the rays $\arg u = \pm\pi/3$. Consequently, the path C' in the u-plane corresponds to the path Γ_0 and so, from (1.5.36), we find as $\nu \to +\infty$

$$J_\nu(\nu\operatorname{sech}\sigma) \sim \frac{A_0}{\nu^{\frac{1}{3}}} \operatorname{Ai}(\alpha^2\nu^{\frac{2}{3}}) = \left(\frac{2\alpha}{\tanh\sigma}\right)^{\frac{1}{2}} \frac{\operatorname{Ai}(\alpha^2\nu^{\frac{2}{3}})}{\nu^{\frac{1}{3}}}. \tag{1.5.38}$$

The validity of (1.5.38) has been established here only for σ sufficiently small. By means of a differential-equation approach, Olver (1954) obtained an expansion for all σ in domains extending to infinity and for complex values of x. In terms of $\alpha^2 = \zeta$, this more general expansion takes the form

$$J_\nu(\nu x) \sim \left(\frac{4\zeta}{1 - x^2}\right)^{\frac{1}{4}} \left\{ \frac{\operatorname{Ai}(\nu^{\frac{2}{3}}\zeta)}{\nu^{\frac{1}{3}}} \sum_{k=0}^{\infty} \frac{a_k(\zeta)}{\nu^{2k}} + \frac{\operatorname{Ai}'(\nu^{\frac{2}{3}}\zeta)}{\nu^{\frac{5}{3}}} \sum_{k=0}^{\infty} \frac{b_k(\zeta)}{\nu^{2k}} \right\};$$

see also Abramowitz and Stegun (1965, p. 368) and Olver *et al.* (2010, Eq. (10.20.4)). This expansion holds for $\nu \to +\infty$ uniformly with respect to x in the sector $|\arg x| \le \pi - \delta, \delta > 0$. The variable ζ is defined by

$$\tfrac{2}{3}\zeta^{\frac{3}{2}} = \log\left(\frac{1 + \sqrt{1 - x^2}}{x}\right) - \sqrt{1 - x^2} \qquad (0 < x \le 1),$$

$$\tfrac{2}{3}(-\zeta)^{\frac{3}{2}} = \sqrt{x^2 - 1} - \arccos(1/x) \qquad (1 \le x < \infty),$$

the branches being chosen so that ζ is real when $x > 0$. The coefficients $a_k(\zeta), b_k(\zeta)$ are complicated functions of ζ, with

$$a_0(\zeta) = 1, \qquad b_0(\zeta) = -\tfrac{5}{48}\zeta^{-2} + \zeta^{-\frac{1}{2}}\{\tfrac{5}{24}(1-x^2)^{-\frac{3}{2}} - \tfrac{1}{8}(1-x^2)^{-\frac{1}{2}}\}.$$

Although the coefficient functions $a_k(\zeta)$ $(k \geq 1)$ and $b_k(\zeta)$ $(k \geq 0)$ are analytic in the neighbourhood of the coalescence point $x = 1$ $(\zeta = 0)$, they are, in common with many uniform expansions, expressed in a form that possesses a removable singularity at this point. The leading term in this expansion is easily seen to correspond to the uniform approximation in (1.5.38).

1.6 Optimal truncation and superasymptotics

1.6.1 Optimal truncation

Let the function $f(z)$ possess the asymptotic expansion

$$f(z) \sim \sum_{k=0}^{\infty} c_k z^{-k} \tag{1.6.1}$$

as $z \to \infty$ in a sector $\mathbf{S}$ of the complex z-plane, where the c_k are coefficients independent of z. If we denote the error in approximating $f(z)$ by the sum of the first n terms by $R_n(z)$, then we have exactly

$$f(z) = \sum_{k=0}^{n-1} c_k z^{-k} + R_n(z) \qquad (z \in \mathbf{S}).$$

By Poincaré's definition of an asymptotic series (see Definition 1.1 in §1.2.2), the error (or remainder) term satisfies the bound $|R_n(z)| \leq A_n|z|^{-n}$, where A_n is an assignable constant, for z in the sector $\mathbf{S}$. The integer n is arbitrary, but fixed: truncation of the series (1.6.1) at the nth term consequently yields an error that decays algebraically as $z \to \infty$ in $\mathbf{S}$. When $|z|$ is large, the successive terms in a typical asymptotic expansion (1.6.1) initially start to decrease in absolute value, reach a minimum and thereafter increase without bound given the divergent character of the full expansion. If the series is truncated just before this minimum modulus term is reached, then this process is called *optimal truncation*, and the finite series that results is the *optimally truncated expansion*. As we shall see in a specific case below, the resulting rate of decay of $R_n(z)$ is then greatly enhanced.

As a specific example we consider the exponential integral

$$f(x) = xe^x E_1(x) = x \int_0^\infty \frac{e^{-xt}}{1+t}\, dt,$$

where $x > 0$. The reason for this choice is that this function provides one of the simplest examples of an asymptotic expansion complete with error bound. If we substitute the identity

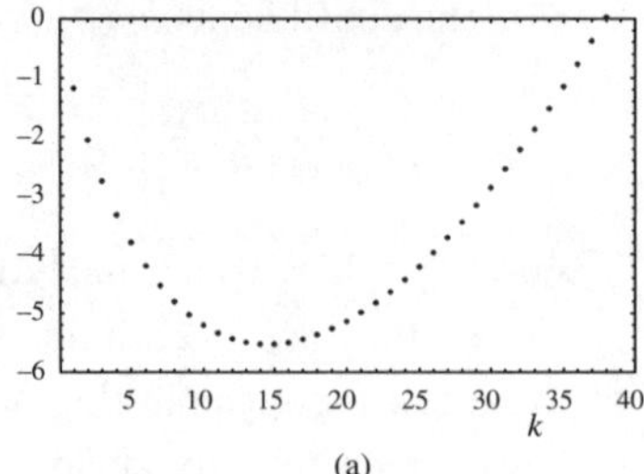 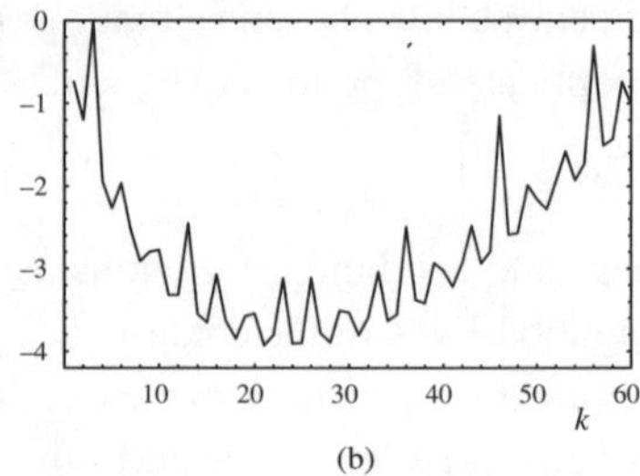

(a)   (b)

Figure 1.18 Magnitude of the terms (on a $\log_{10}$ scale) in the asymptotic series against ordinal number k for (a) the exponential integral in (1.6.2) when $x = 15$ and (b) the terms I_1 in (1.6.3) for $\mu = 3$, $m_1 = 1.5$, $m_2 = 1.047$ when $x = 3.6$. The points have been joined for clarity.

$$\frac{1}{1+t} = \sum_{k=0}^{n-1} (-)^k t^k + \frac{(-)^n t^n}{1+t} \qquad (t \neq -1)$$

followed by term-by-term integration, we obtain

$$f(x) = \sum_{k=0}^{n-1} (-)^k \frac{k!}{x^k} + R_n(x), \qquad R_n(x) = (-)^n \int_0^\infty \frac{t^n e^{-xt}}{1+t}\, dt. \qquad (1.6.2)$$

It is readily seen that

$$|R_n(x)| < \int_0^\infty t^n e^{-xt}\, dt = \frac{n!}{x^n},$$

so that this expansion enjoys the property that the remainder is bounded in magnitude by the first neglected term and has the same sign.[26] For a given value of x, the smallest term in absolute value of the series in (1.6.2) occurs when $k = \lfloor x \rfloor$ (except when x is an integer, in which case there are two equally small terms corresponding to $k = x - 1$ and $k = x$). Figure 1.18(a) illustrates the typical 'factorial divided by a power of x' behaviour of an asymptotic series.

If we denote the optimal truncation index by N, then we have

$$f(x) = \sum_{k=0}^{N-1} (-)^k \frac{k!}{x^k} + R_N(x).$$

It is clear that $R_N(x)$ is a discontinuous function of x, since N changes each time x passes through an integer value. Use of Stirling's formula to approximate $N!$ shows that

$$|R_N(x)| < \frac{N!}{x^N} \simeq (2\pi)^{\frac{1}{2}} \frac{e^{-N} N^{N+\frac{1}{2}}}{x^N} \simeq (2\pi x)^{\frac{1}{2}} e^{-x},$$

upon using the fact that $N \simeq x$. This reveals the fact that at optimal truncation the remainder term for $f(x)$ is of order $x^{1/2} e^{-x}$ as $x \to +\infty$ and consequently that

---

[26] It is sometimes (fallaciously) believed that all asymptotic expansions enjoy this property.

evaluation of $f(x)$ by this scheme will result in an error that is *exponentially small*. This level of asymptotic approximation has been given the neologism *superasymptotics* by Berry (1991a); alternatively, it is termed *asymptotics beyond all orders* by Kruskal and Segur (1991), since the small exponential that results is lost in the truncation errors $n!/x^n$ of ordinary Poincaré asymptotics.

It is worth pointing out that not all asymptotic series present the regular behaviour of the terms depicted in Fig. 1.18(a). For example, in certain expansions encountered in the asymptotic evaluation of two-dimensional Laplace-type integrals (Paris and Kaminski, 2001, §7.4), we find the compound expansion $I_1 + I_2$, where

$$I_r = \sum_{k=0}^{\infty} c_k^{(r)} z^{-(1+\mu k)/m_r} \qquad (r = 1, 2) \tag{1.6.3}$$

and, for positive parameters m_1, m_2 and μ satisfying $\Delta \equiv \mu - m_1 - m_2 > 0$,

$$c_k^{(1)} = \frac{1}{k!} \Gamma\left(\frac{m_2 + \mu k}{m_1}\right) \Gamma\left(\frac{m_1 - m_2}{m_1 \mu} - \frac{m_2 k}{m_1}\right)$$

with a similar expression for $c_k^{(2)}$ with m_1 and m_2 interchanged. If the parameters m_1, m_2 and μ are chosen such that the arguments of the second gamma functions in $c_k^{(1)}$ and $c_k^{(2)}$ are not close to a negative integer, then the variation of the absolute value of coefficients with ordinal number k will be similar to that shown in Fig. 1.18(a). If, however, the parameter values are chosen so that these arguments become close to a non-positive integer[27] for subsets of k values, then we find that the variation of the coefficients becomes irregular with a sequence of peaks of variable height superimposed on the basic structure of Fig. 1.18(a). Rearrangement of the coefficients $c_k^{(1)}$ yields the form

$$c_k^{(1)} = \frac{1}{k!} \frac{\Gamma\left(\dfrac{m_2 + \mu k}{m_1}\right)}{\Gamma\left(1 - \dfrac{m_1 - m_2}{m_1 \mu} + \dfrac{m_2 k}{m_1}\right)} \pi \operatorname{cosec} \pi\left(\frac{m_1 - m_2}{m_1 \mu} - \frac{m_2 k}{m_1}\right),$$

where, by application of Stirling's formula, the quotient of gamma functions can be seen to scale, for large k, like the single factorial

$$\Gamma\left(\frac{\Delta k}{m_1} + \frac{(m_1 - m_2)(1 - \mu)}{m_1 \mu}\right)$$

with ultimately positive argument since $\Delta > 0$; see Paris and Kaminski (2001, Lemma 2.2, p. 39). This shows that the terms $c_k^{(1)} z^{-(1+\mu k)/m_1}$ present the familiar 'factorial divided by a power' dependence as illustrated in Fig. 1.18(a), but decorated by the cosec term. An example of this behaviour is shown in Fig. 1.18(b). The

---

[27] If the parameter values are such that the argument of the second gamma function equals a non-positive integer for a subset of k values, it is found that this situation corresponds to the formation of double poles and logarithmic terms in the Mellin–Barnes integral description.

treatment of such series shows that if the terms of I_1 are truncated at a local peak (provided that the corresponding peak associated with the coefficients $c_k^{(2)}$ is included) increasingly accurate asymptotic approximations are obtained by steadily increasing the truncation indices in the series I_1 and I_2 until they correspond roughly to the global minimum of each curve.

1.6.2 Ursell's lemma

As a rule, it is found that similar exponential improvement at optimal truncation can be achieved in asymptotic expansions of other functions. In these cases the remainder can be written explicitly in the form of an integral and its value estimated to establish that it is exponentially small. There are few general results of this nature but the following result due to Ursell (1991) provides a significant advance in this direction.

Lemma 1.3 *Let $f(t)$ be analytic in the disc $|t| < R$ with the Maclaurin expansion*

$$f(t) = \sum_{n=0}^{\infty} c_n t^n \qquad (|t| \le R). \tag{1.6.4}$$

Let r be a fixed positive quantity such that $0 < r < R$ and suppose that $|f(t)| < K e^{\beta t}$ when $r \le t < \infty$, where K and β are positive constants. Then

$$I(x) = \int_0^{\infty} e^{-xt} f(t)\, dt = \sum_{n=0}^{n_*} \frac{c_n n!}{x^{n+1}} + R_{n_*}(x), \tag{1.6.5}$$

where $n_* = \lfloor rx \rfloor$ and $R_{n_*}(x) = O(e^{-rx})$ as $x \to +\infty$.

Before giving Ursell's proof of this result, we shall require two simple bounds for the incomplete gamma functions. These are

$$\left.\begin{array}{ll} \gamma(a+1, \chi) \le e^{-\chi} \chi^{a+1} & (a \ge \chi > 0) \\ \Gamma(a+1, \chi) \le 2e^{-\chi} \chi^{a+1} & (1 \le a \le \chi) \end{array}\right\} \tag{1.6.6}$$

where $a > 0$ and $\chi > 0$. The derivation of the first bound follows from

$$\gamma(a+1, \chi) = \int_0^{\chi} e^{-t} t^a dt = e^{-\chi} \chi^{a+1} \int_0^1 e^{u\chi} (1-u)^a du$$

$$\le e^{-\chi} \chi^{a+1} \qquad (a \ge \chi),$$

since $0 \le e^{u\chi}(1-u)^a \le 1$ on $[0, 1]$ when $a \ge \chi$. The second bound follows from

$$\Gamma(a+1, \chi) = \int_{\chi}^{\infty} e^{-t} t^a dt = e^{-\chi} \chi^{a+1} \int_0^{\infty} e^{-u\chi}(1+u)^a du$$

$$\le e^{-\chi} \chi^{a+1} \int_0^{\infty} \{e^{-u}(1+u)\}^a du$$

$$\le e^{-\chi} \chi^{a+1} \int_0^{\infty} e^{-u}(1+u)\, du = 2e^{-\chi} \chi^{a+1},$$

the first inequality holding when $a \leq \chi$ and the second inequality when $a \geq 1$, since the integrand $e^{-u}(1+u) \leq 1$ on $[0, \infty)$. The resulting bound therefore holds for $1 \leq a \leq \chi$.

Proof of Lemma 1.3 From (1.6.4), we have

$$I(x) = \int_0^\infty e^{-xt} f(t)\, dt = \sum_{n=0}^\infty c_n \int_0^r e^{-xt} t^n dt + I_1(x),$$

where $I_1(x) = \int_r^\infty e^{-xt} f(t)\, dt$ and the series converges since $\int_0^r e^{-xt} t^n dt < \int_0^r t^n dt = r^{n+1}/(n+1)$. Let us now split the series at $n = n_*$. Then

$$I(x) = \sum_{n \leq n_*} c_n \left\{ \int_0^\infty - \int_r^\infty \right\} e^{-xt} t^n dt + \sum_{n > n_*} c_n \int_0^r e^{-xt} t^n dt + I_1(x)$$

where, upon evaluation of the integral over $[0, \infty)$ as a factorial,

$$\sum_{n \leq n_*} c_n \int_0^\infty e^{-xt} t^n dt = \sum_{n \leq n_*} \frac{c_n n!}{x^{n+1}}.$$

The remainder $R_{n_*}(x) = I(x) - \sum_{n \leq n_*} c_n n!/x^{n+1}$ therefore satisfies the bound

$$|R_{n_*}(x)| \leq \sum_{n \leq n_*} |c_n| \int_r^\infty e^{-xt} t^n dt + \sum_{n > n_*} |c_n| \int_0^r e^{-xt} t^n dt + |I_1(x)|$$

$$= \sum_{n \leq n_*} \frac{|c_n|}{x^{n+1}} \Gamma(n+1, rx) + \sum_{n > n_*} \frac{|c_n|}{x^{n+1}} \gamma(n+1, rx) + |I_1(x)|.$$

Since $|f(t)| < K e^{\beta t}$ when $r \leq t < \infty$, then the third term on the right-hand side satisfies

$$|I_1(x)| = \int_r^\infty e^{-xt} |f(t)|\, dt < K \frac{e^{-r(x-\beta)}}{x-\beta} < K' \frac{e^{-rx}}{x},$$

where K' is a constant.

We now choose $n_* = \lfloor rx \rfloor$ and make use of the bounds in (1.6.6) to find

$$|R_{n_*}(x)| \leq e^{-rx} \left\{ \frac{|c_0| + K'}{x} + 2 \sum_{1 \leq n \leq rx} |c_n| r^{n+1} + \sum_{n > rx} |c_n| r^{n+1} \right\}$$

$$\leq e^{-rx} \left\{ \frac{|c_0| + K'}{x} + 2r \sum_{n=1}^\infty |c_n| r^n \right\}. \tag{1.6.7}$$

Since $\sum c_n R^n$ is convergent it follows that $\sum |c_n| r^n$ is convergent for $r < R$. Hence the quantity in braces in (1.6.7) is finite and the theorem follows. $\square$

We observe that the expansion

$$\sum_{n=0}^\infty \frac{c_n n!}{x^{n+1}}$$

is the standard asymptotic expansion of the integral $I(x)$ in (1.6.5) obtained by application of Watson's lemma; see (1.2.24). Ursell's lemma shows that if we take $1+\lfloor rx \rfloor$ terms in this expansion, then the resulting approximation to $I(x)$ is exponentially accurate, with the error being $O(e^{-rx})$ as $x \to +\infty$. As has been pointed out in Olver (1992), the above result can be extended straightforwardly to sectors of the complex x-plane, and also to the case where $f(t)$ possesses a branch point at $t = 0$ and the expansion (1.6.4) involves fractional powers. A heuristic treatment of this problem has also been considered by Jeffreys (1958).

1.7 The Stokes phenomenon

1.7.1 Description of the Stokes phenomenon

The Stokes phenomenon concerns the rapid change across certain rays in the complex plane exhibited by the coefficients multiplying exponentially subdominant terms in compound asymptotic expansions. The pervasive nature of this phenomenon indicates that its occurrence should have a fundamental origin. The root cause is a consequence of *asymptotically approximating a given function, which possesses a certain multi-valued structure, in terms of approximants of a different multi-valued structure.*

One of the simplest illustrations of this phenomenon for solutions of linear differential equations is Airy's equation $y'' = zy$, for which the solutions $y(z)$ are entire functions of z. For large $|z|$, the (formal) asymptotic solutions are given by[28]

$$u_{\pm}(z) = \frac{z^{-\frac{1}{4}}}{2\sqrt{\pi}} e^{\pm\zeta} \sum_{k=0}^{\infty} (\pm 1)^k \frac{c_k}{z^{3k/2}}, \qquad \zeta = \tfrac{2}{3} z^{\frac{3}{2}},$$

where

$$c_k = \frac{\Gamma(3k + \frac{1}{2})}{3^{2k} \pi^{\frac{1}{2}} (2k)!};$$

see Bender and Orszag (1978, pp. 100–101) for a derivation of this result. The functions $u_{\pm}(z)$ are multi-valued functions with a branch point at $z = 0$. Consequently, any linear combination $Au_+(z) + Bu_-(z)$, where A and B are constants, cannot represent asymptotically a solution of Airy's equation *uniformly* as one describes a complete circuit about $z = 0$, since $y(ze^{2\pi i}) = y(z)$ but $u_{\pm}(ze^{2\pi i}) \neq u_{\pm}(z)$. It follows that the constants A and B must change somewhere in a sector of angular width 2π. This is the Stokes phenomenon.

To be more specific, let us consider the solution denoted by the Airy function $\mathrm{Ai}(z)$. In the neighbourhood of $\arg z = 0$, this solution is associated with the

---

[28] The factor $1/(2\sqrt{\pi})$ has been added for convenience.

constants $A = 0$ and $B = 1$. Then, in the sector $|\arg z| \leq \pi - \delta$, we have the expansion

$$\mathrm{Ai}(z) \sim u_-(z) \qquad (|\arg z| \leq \pi - \delta), \tag{1.7.1}$$

where δ is an arbitrary small positive quantity; compare (1.3.21) *et seq.* Use of the circuit relation (Abramowitz and Stegun, 1965, p. 446)

$$\mathrm{Ai}(z) + \omega^{-1}\mathrm{Ai}(\omega^{-1}z) + \omega^{-2}\mathrm{Ai}(\omega^{-2}z) = 0, \qquad \omega = e^{2\pi i/3}$$

shows that in the adjacent sector $\frac{1}{3}\pi + \delta \leq \arg z \leq \frac{5}{3}\pi - \delta$ we obtain the compound asymptotic expansion

$$\mathrm{Ai}(z) \sim u_-(z) + i u_+(z) \qquad (\tfrac{1}{3}\pi + \delta \leq \arg z \leq \tfrac{5}{3}\pi - \delta). \tag{1.7.2}$$

A conjugate expansion holds in the sector $-\frac{5}{3}\pi + \delta \leq \arg z \leq -\frac{1}{3}\pi - \delta$. These expansions (in the Poincaré sense) are uniformly valid as $|z| \to \infty$ with respect to $\arg z$ in their respective sectors of validity. In their common sector of validity, $\frac{1}{3}\pi + \delta \leq \arg z \leq \pi - \delta$, these expansions differ by the inclusion of the series $u_+(z)$. The resulting contradiction, however, is only apparent, since $u_+(z)$ is uniformly exponentially small for large $|z|$ in this common sector and so is asymptotically smaller than any term in the dominant series $u_-(z)$.

Then, as $\arg z$ increases from 0 to π, the coefficient A multiplying the asymptotic series $u_+(z)$ changes from 0 to i; an analogous change occurs in the conjugate sector. Stokes (1857), who was the first to observe this phenomenon, argued that this change should take place across the rays $\arg z = \pm\frac{2}{3}\pi$, corresponding to where the difference between the two exponential factors $e^{\pm\zeta}$ is greatest and where the terms c_k in the dominant expansion $u_-(z)$ all have the same sign, so that the divergence of this series is most severe. The rays[29] $\arg z = \pm\frac{2}{3}\pi$ are called the *Stokes lines* for the function $\mathrm{Ai}(z)$, while the rays $\arg z = \pm\frac{1}{3}\pi$ and $\pm\pi$ are called *anti-Stokes lines* where the exponentials in $u_\pm(z)$ are oscillatory and of the same order; see Fig. 1.19(a).

Since the discovery of this phenomenon, the conventional view has been that of a discontinuous change in the constant (called a *Stokes multiplier*) associated with the subdominant asymptotic expansion. The fact that an analytic function could possess such a discontinuous representation in the neighbourhood of an irregular singular point, together with the inherent vagueness associated with the precise location in the complex plane of these jumps, has resulted in the Stokes phenomenon being considered somewhat mysterious. This viewpoint, however, was completely overturned when Berry (1989) argued that the coefficient of the subdominant expansion should be regarded not as a discontinuous constant but, for fixed $|z|$, as a *continuous* function of $\theta = \arg z$. He showed that, when viewed on an appropriately magnified scale, the change in the subdominant multiplier across a Stokes line is in fact continuous.

---

[29] In some references the names of the Stokes and anti-Stokes lines are interchanged.

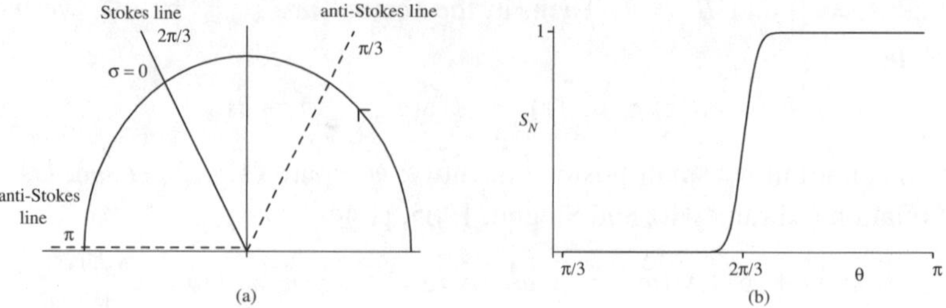

Figure 1.19 (a) The Stokes (solid) and anti-Stokes (dashed) lines in the upper half-plane for the Airy function Ai(z) and (b) the behaviour of S_N in (1.7.4) in the sector $\pi/3 < \arg z < \pi$ when $|z| = 20$.

For a wide class of functions, the functional form of this rapid but smooth transition possesses a universal structure approximated by an error function, whose argument is an appropriate variable describing transition across the Stokes line.

In order to detect the appearance of such exponentially small terms, it is necessary to compute the dominant series to at least a comparable accuracy. This is achieved by optimal truncation of the series $u_-(z)$ (that is, truncation just before the least term of the expansion; see §1.6.1) after N terms, where $N = \lfloor \frac{4}{3}|z|^{3/2} \rfloor$, and truncation of $u_+(z)$ at its first term. The Stokes multiplier S_N is then *defined* by

$$\mathrm{Ai}(z) = \frac{z^{-\frac{1}{4}}}{2\sqrt{\pi}} e^{-\zeta} \sum_{k=0}^{N-1} (-)^k \frac{c_k}{z^{3k/2}} + \frac{i z^{-\frac{1}{4}}}{2\sqrt{\pi}} e^{\zeta} S_N. \tag{1.7.3}$$

Clearly S_N is a piecewise continuous function of $|z|$ (as a result of the relation between N and $|z|$) and is a continuous function of arg z. Since the term containing S_N is proportional to the leading term of the expansion of $u_+(z)$, we see by comparison with (1.7.1) and (1.7.2) that as we traverse the Stokes line $\theta = \frac{2}{3}\pi$ (with fixed large $|z|$) the value of S_N changes from 0 to 1. By resumming the divergent tail of the series $u_-(z)$ by Borel summation, Berry (1989, 1991a) obtained the approximate functional form of S_N given by

$$S_N \simeq \tfrac{1}{2} + \tfrac{1}{2}\mathrm{erf}\,\sigma, \tag{1.7.4}$$

where erf denotes the error function. The variable σ is specified in terms of the so-called singulant $F(z) = -\frac{4}{3}z^{3/2}$ (which is the difference between the exponents of the two exponentials in $u_{\mp}(z)$) by

$$\sigma = \frac{\mathrm{Im}\,F}{(2\mathrm{Re}\,F)^{1/2}} = \sin \tfrac{3}{2}\phi \left(\frac{|F|}{2\cos \tfrac{3}{2}\phi} \right)^{1/2}, \tag{1.7.5}$$

where ϕ is the angular separation measured from the Stokes line under consideration, namely $\phi = \theta - \frac{2}{3}\pi$. Between the two anti-Stokes lines adjacent to the Stokes line $\theta = \frac{2}{3}\pi$, the argument of the error function increases from $\sigma = -\infty$ on $\theta = \frac{1}{3}\pi$ to $\sigma = +\infty$ on $\theta = \pi$, so that S_N increases from 0 to 1 in this sector. The angular width of the transition across the Stokes line decreases as $|z|$ increases and is easily seen to scale like $|z|^{-3/4}$. For example, when $|z| = 20$, the Stokes multiplier S_N varies from 0.005 (the transition is just commencing) to 0.995 (the transition is virtually completed) in an angular zone of about $9°$ either side of $\theta = \frac{2}{3}\pi$; see Fig. 1.19(b).

The details of Berry's arguments are summarised in the survey paper Paris and Wood (1995); see also Olver (1990) for a more detailed presentation. Alternative, descriptive derivations of the Stokes phenomenon have been given by Berry (1991c), Nishikov and Ritus (1992), Olde Daalhuis *et al.* (1995) and López (2007), and an interesting discussion from a physical point of view can be found in Berry and Howls (1993). Survey papers on divergent series and hyperasymptotic expansions are given in Boyd (1999) and Kowalenko (2001).

1.7.2 A rigorous approach

A rigorous approach to the Stokes phenomenon for the confluent hypergeometric function $U(a, b, z)$ (which includes Ai(z) as a particular case) has been been given in Olver (1990, 1991a) starting from a Laplace integral representation; see also Paris and Kaminski (2001, §6.2) for a different account in terms of Mellin–Barnes integrals. He established that an exponentially improved expansion for this function, valid for $|z| \to \infty$ in the sector $|\arg z| \leq \frac{5}{2}\pi - \delta$, is more accurately expressed in terms of the incomplete gamma function $\Gamma(1 - v, z)$ (the so-called *terminant function* introduced by Dingle (1973, Ch. 23)) given by [30]

$$T_v(z) = \frac{e^{\pi i v}\Gamma(v)}{2\pi i}\,\Gamma(1 - v, z), \qquad (1.7.6)$$

rather than the error function. The form of this expansion when $|z| \to \infty$ is given by

$$z^a U(a, a - b + 1, z) = \sum_{k=0}^{N-1}(-)^k\frac{(a)_k(b)_k}{k!\,z^k} + R_N(z), \qquad (1.7.7)$$

where

$$R_N(z) = \frac{2\pi i e^{-\pi i \vartheta}}{\Gamma(a)\Gamma(b)}z^\vartheta e^z \sum_{j=0}^{M-1}\frac{(1 - a)_j(1 - b)_j}{j!\,z^j}\,T_{N+\vartheta-j}(z) + R_{N,M}(z), \quad (1.7.8)$$

$\vartheta = a + b - 1$, $N = |z| - \vartheta + \alpha$, with $|\alpha|$ bounded, M is an abitrary positive integer and the remainder term $R_{N,M}(z)$ satisfies the order estimates

$$R_{N,M}(z) = \begin{cases} O(e^{-|z|}z^{\vartheta - M}) & |\arg z| \leq \pi \\ O(e^z z^{\vartheta - M}) & \pi \leq |\arg z| \leq \frac{5}{2}\pi - \delta. \end{cases}$$

When $v \sim |z|$, the Laplace integral representation for $T_v(z)$ is associated with a pole that coincides with a saddle point when $\arg z = \pm\pi$ (Olver, 1991b), with

---

[30] An equivalent representation involves the exponential integral $E_v(z)$, since $E_v(z) = z^{v-1}\Gamma(1 - v, z)$.

the behaviour of $T_\nu(z)$ therefore involving the error function; cf. (1.5.7). The precise form of this dependence is given by[31]

$$T_\nu(z) \sim \tfrac{1}{2} + \tfrac{1}{2}\mathrm{erf}\,\{c(\phi)(\tfrac{1}{2}|z|)^{\frac{1}{2}}\}$$

near $\arg z = \pi$, where the quantity $c(\phi)$ is defined by

$$\tfrac{1}{2}c^2(\phi) = 1 + i\phi - e^{i\phi}, \qquad \phi = \arg z - \pi$$

and corresponds to the branch that behaves like $c(\phi) \simeq \phi$ near $\arg z = \pi$. Consequently, the terminant function $T_\nu(z)$ implicitly contains the error function-behaviour required to describe the smooth transition across the Stokes lines $\arg z = \pm \pi$ in accord with Berry's formula (1.7.4), with an appropriately defined value of the transition variable σ.

Finally, we mention an interesting peculiarity of the gamma function $\Gamma(z)$. The Stokes lines for this function are $\arg z = \pm\tfrac{1}{2}\pi$ and it is found that not one but *infinitely many* subdominant exponential terms switch on, each associated with its own Stokes multiplier, as these rays are crossed; see Berry (1991b), Paris and Wood (1992) and Paris and Kaminski (2001, §6.4). The appearance of these exponentials is essential since, as the negative z-axis is approached, they combine to generate the poles of $\Gamma(z)$. A similar phenomenon has been shown to occur for the Hurwitz zeta function $\zeta(s, a)$ in the neighbourhood of the Stokes lines $\arg a = \pm\tfrac{1}{2}\pi$ (Paris, 2005).

1.7.3 The Stokes phenomenon and steepest descents

The Stokes phenomenon in the treatment of the Laplace-type integral

$$\int_C e^{\lambda\psi(t)} f(t)\,dt$$

by the method of steepest descents manifests itself in the following ways. As the phase of λ varies, a path of steepest descent from a saddle point of the phase function $\psi(t)$ can connect abruptly with a neighbouring subdominant saddle point at a critical value of $\arg \lambda$. The ray on which this occurs is a Stokes line in the complex λ-plane and the appearance of the subdominant contribution across this ray is the Stokes phenomenon. A different situation in which the appearance of a subdominant contribution can arise occurs with integrals where the path C has a finite endpoint. In such cases, there can be an abrupt change in the integration path leading from the finite endpoint to join the steepest descent path through a dominant saddle point as $\arg \lambda$ passes through a critical value. This is called an *endpoint Stokes phenomenon*.

We illustrate these two different situations with reference to the Airy function $\mathrm{Ai}(z)$ and an integral connected with the parabolic cylinder function. The Airy function $\mathrm{Ai}(z)$ is defined by

---

[31] In Olver (1991a, 1991b) and Olver *et al.* (2010, §2.11) the function $T_\nu(z)$ is called
$F_\nu(z) \equiv ie^{-\pi i\nu}T_\nu(z)$.

$$\mathrm{Ai}(z) = \frac{1}{2\pi i} \int_C e^{\frac{1}{3}t^3 - zt}\, dt, \qquad z = |z|e^{i\theta},$$

where C denotes any path that starts and ends at infinity in the sectors $-\frac{1}{2}\pi < \arg t < -\frac{1}{6}\pi$ and $\frac{1}{6}\pi < \arg t < \frac{1}{2}\pi$, respectively. The saddle points are situated at $t_s = \pm z^{1/2}$ and as θ varies at fixed $|z|$, they rotate in the complex plane.[32] The heights of the saddles are $\exp\{\mp\frac{2}{3}z^{3/2}\}$, so that the saddle at $z^{1/2}$ is subdominant in the sector $|\theta| < \frac{1}{3}\pi$ and dominant in the sectors $\frac{1}{3}\pi < |\theta| < \pi$. The paths of steepest descent and ascent through the saddles are shown in Fig. 1.20 for different θ in the range $0 \le \theta \le \pi$. When $\theta < \frac{2}{3}\pi$, it is seen that the integration path C can be made to coincide with the steepest descent path through the saddle at $z^{1/2}$; but that when $\theta = \frac{2}{3}\pi$ the steepest descent path through $z^{1/2}$ connects with that through the subdominant saddle at $-z^{1/2}$. This is an example of the Stokes phenomenon. The steepest descent paths for values of θ close to the Stokes line $\theta = \frac{2}{3}\pi$ are shown in Fig. 1.20(e,f).

It should be pointed out that the Stokes phenomenon in Laplace-type integrals is not confined to situations where the asymptotic variable is complex as in the above example. When the exponential factor in the integrand contains two or more real parameters, a similar connection of neighbouring saddle points can occur for critical values of these parameters. A classical example of this is the Pearcey integral which depends on two real variables; see Paris and Kaminski (2001, §8.3) and also §4.3.

Our second example is furnished by the integral

$$\int_0^\infty t^{\nu-1} e^{-\frac{1}{2}t^2 + zt}\, dt, \tag{1.7.9}$$

when $\mathrm{Re}(\nu) > 0$, which is equal to a multiple of the parabolic cylinder function $U(\nu - \frac{1}{2}, -z) = D_{-\nu}(-z)$ and provides a simple illustration of an endpoint Stokes phenomenon. There is a single saddle point at $t_s = z$ and the steepest descent path through t_s is a line parallel to the real t-axis. When $\theta = 0$, the integration path coincides with part of the steepest descent path. When $\theta \neq 0$, the integration path may be deformed to pass over the steepest descent path from the origin out to infinity and thence along the steepest path through the saddle point. Thus, as θ passes through the value $\theta = 0$, there is a change in the form of the subdominant (algebraic) contribution from the neighbourhood of the origin. This change is illustrated in Fig. 1.21.

1.8 Hyperasymptotics

1.8.1 Overview

The term *hyperasymptotics* was adopted in Berry (1991a) and Berry and Howls (1990, 1991) to refer to systematic improvements in the exponentially small

---

[32] A rescaling of the variable $t \to z^{1/2}\tau$ produces the exponential factor $\exp\{z^{3/2}(\tau^3/3 - \tau)\}$. In this case, the saddle points remain fixed at $\tau = \pm 1$ as θ varies.

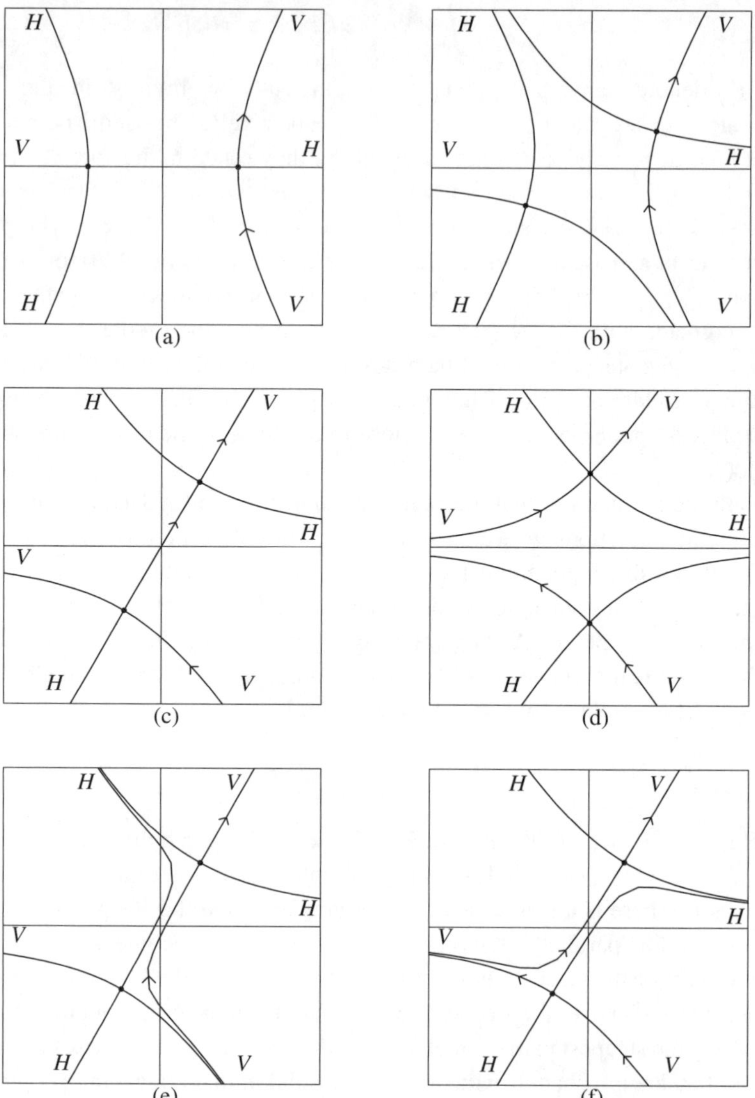

Figure 1.20 The paths of steepest descent and ascent for Ai(z) when (a) $\theta = 0$, (b) $\theta = \frac{1}{3}\pi$, (c) $\theta = \frac{2}{3}\pi$, (d) $\theta = \pi$, (e) $\theta = \frac{2}{3}\pi - \epsilon$ and (f) $\theta = \frac{2}{3}\pi + \epsilon$, where $\epsilon = 0.02\pi$. The valleys (V) and the hills (H) at infinity are indicated and the arrows indicate the integration path.

remainder in an optimally truncated asymptotic expansion. Thus, hyperasymptotics goes beyond the level of exponential improvement at optimal truncation. The earliest attempt at a systematic approach to this problem is usually attributed to Stieltjes (1886) in his doctoral dissertation. However, there is an example of such a procedure by Stokes that predates this work by about 30 years; see Paris (1996).

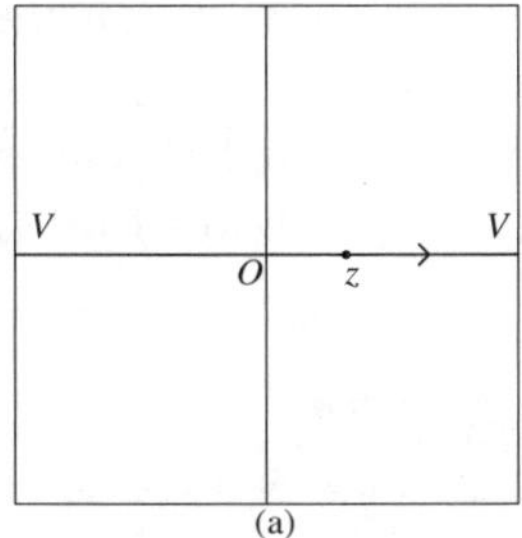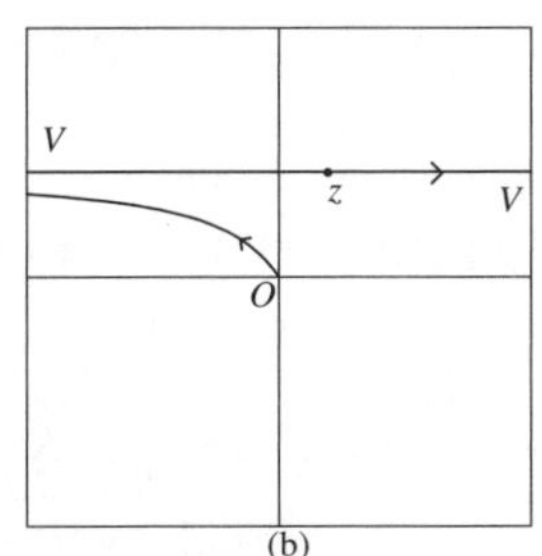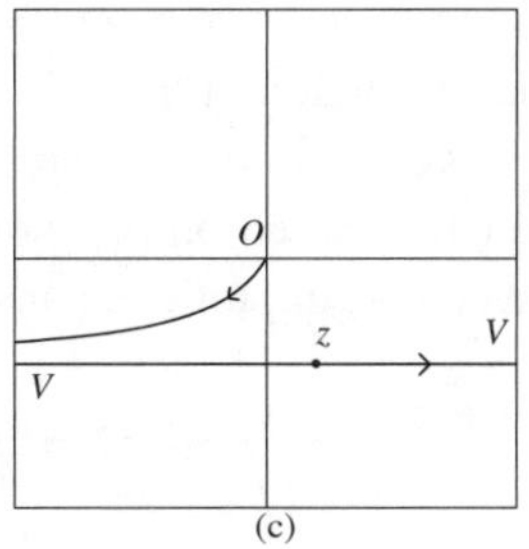

Figure 1.21 The integration path along steepest descent paths through the saddle point $t_s = z$ for the integral in (1.7.9) when (a) $\theta = 0$, (b) $\theta > 0$ and (c) $\theta < 0$. The valleys (V) at infinity are indicated.

When investigating the phenomenon that now bears his name (see §1.7), Stokes (1857) was led to verify his assertions with a numerical example applied to what we now call the Airy function Ai(z). He compared his asymptotic predictions for the change in the Stokes multiplier $30°$ either side of the so-called Stokes line $\arg z = \frac{2}{3}\pi$ with exact values of Ai(z) for $|z| = 4.16$ laboriously computed from convergent series expansions. To detect the change in such exponentially small terms requires the calculation of the dominant series to at least a comparable accuracy. Stokes did this by first employing optimal truncation of the dominant asymptotic expansion of Ai(z) (that is, just before the least term in modulus in the expansion) to produce an exponentially small remainder. And secondly, not content with this level of exponential improvement, he proceeded to resum the divergent tail of the asymptotic expansion by what is effectively an Euler transformation, thereby further increasing the accuracy of his calculations. It should be emphasised that all this was carried out well before Poincaré's definition of an asymptotic expansion (Poincaré, 1886) and Stieltjes' above-mentioned work on divergent series of the same year. Indeed, this calculation was later described by Lord Kelvin (1903) in his memorial appreciation of Stokes as 'mathematical supersubtlety'.

1.8.2 Airey's converging factors

Following on from the pioneering work of Stieltjes, Airey (1937) developed a formal theory of *converging factors*. The basic idea of this theory is to truncate an asymptotic expansion at, or near, its least term (optimal truncation) and to express the remainder as a multiple of the first term neglected of the series; thus

$$f(z) = \sum_{k=0}^{N-1} \frac{a_k}{z^k} + C_N(z)\frac{a_N}{z^N},$$

where $N \equiv N(|z|)$ is the optimal truncation index. The term $a_N z^{-N}$ will, in general, be exponentially small for large $|z|$ at optimal truncation, but if the converging factor $C_N(z)$ can be calculated this will lead to hyperasymptotic improvement in the determination of $f(z)$.

As an example, we present the converging factor theory for the incomplete gamma function based on the work of Rosser (1955). It is a well-known result that $\Gamma(a, z)$ has the asymptotic expansion

$$\Gamma(a, z) = \int_z^\infty t^{a-1} e^{-t} dt \sim z^{a-1} e^{-z} \sum_{k=0}^\infty (-)^k \frac{(1-a)_k}{z^k}$$

for fixed a as $|z| \to \infty$ in the sector $|\arg z| \leq \frac{3}{2}\pi - \delta, \delta > 0$; see Abramowitz and Stegun (1965, p. 263). We introduce the converging factor $C_n(z)$ by

$$\Gamma(a, z) = z^{a-1} e^{-z} \left\{ \sum_{k=0}^{n-1} (-)^k \frac{(1-a)_k}{z^k} + (-)^n C_n(z) \frac{(1-a)_n}{z^n} \right\}, \qquad (1.8.1)$$

where, for the moment, n is an unspecified positive integer. Then, with $m = n - a$ and $\beta = m/z$, we have formally

$$\begin{aligned} C_n(z) &= \frac{z^n}{(1-a)_n} \sum_{k=n}^\infty (-)^{k-n} \frac{(1-a)_k}{z^k} \\ &= 1 - \frac{n-a+1}{z} + \frac{(n-a+1)(n-a+2)}{z^2} - \cdots \\ &= 1 - \beta(1 + m^{-1}) + \beta^2(1 + m^{-1})(1 + 2m^{-1}) - \cdots . \end{aligned} \qquad (1.8.2)$$

If we denote the expression on the right-hand side of (1.8.2) by $F(\beta)$ and carry out formal operations with power series in β, we obtain

$$(1 + \beta)F(\beta) + \frac{\beta}{m} \frac{d}{d\beta} (\beta F(\beta)) = 1. \qquad (1.8.3)$$

We now suppose that n is chosen to be the optimal truncation index $N = \lfloor |z| + a \rfloor$ or, if $|z| + a$ is an integer, $N = |z| + a - 1$. Consequently $m = N - a \gg 1$ for large $|z|$ and $F(\beta)$ may be assumed to possess an expansion in inverse powers of m in the form

$$F(\beta) = F_0(\beta) + \frac{F_1(\beta)}{m} + \frac{F_2(\beta)}{m^2} + \cdots . \qquad (1.8.4)$$

Substitution of this expansion into (1.8.3) then yields

$$F_0(\beta) = \frac{1}{1+\beta}, \qquad F_k(\beta) = -\frac{\beta}{1+\beta} \frac{d}{d\beta} (\beta F_{k-1}(\beta)) \quad (k \geq 1).$$

The first few coefficients with $k \geq 1$ are therefore given by

$$F_1(\beta) = -\frac{\beta}{(1+\beta)^3}, \qquad F_2(\beta) = -\frac{\beta^2(\beta-2)}{(1+\beta)^5}, \qquad F_3(\beta) = -\frac{\beta^3(\beta^2 - 8\beta + 6)}{(1+\beta)^7},$$

$$F_4(\beta) = -\frac{\beta^4(\beta^3 - 22\beta^2 + 58\beta - 24)}{(1+\beta)^9},$$

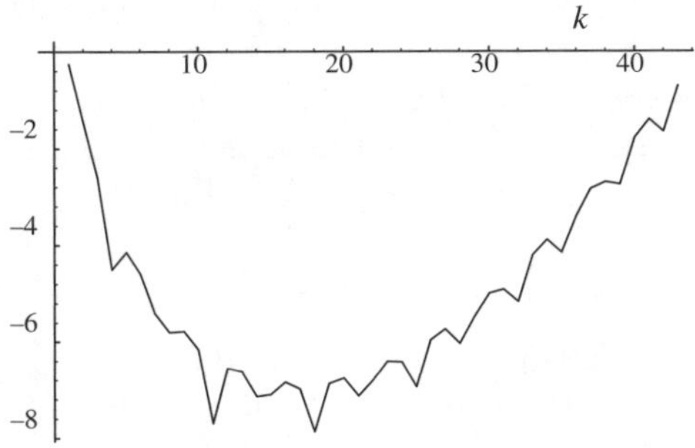

Figure 1.22 The behaviour of the terms (on a $\log_{10}$ scale) against ordinal number k for the converging factor $C_N(z)$ in (1.8.5) when $z = 5$. The points have been joined for clarity.

$$F_5(\beta) = -\frac{\beta^5(\beta^4 - 52\beta^3 + 328\beta^2 - 444\beta + 120)}{(1+\beta)^{11}}, \dots$$

The above procedure is, of course, formal since the series (1.8.2) is divergent. Nevertheless, the improvement in the accuracy in the computation of $\Gamma(a, z)$ can be striking. In Rosser (1951), it is established that, in certain cases, the Euler transformation yields Airey's converging factor and that, in those situations where the Euler transformation can be justified, this affords a justification of Airey's method.

To illustrate, we take the case with $a = 0$ and consider the function $ze^z\Gamma(0, z) = ze^z E_1(z)$, where $E_1(z)$ is the exponential integral. From (1.8.1) we have

$$\mathcal{E}(z) \equiv ze^z E_1(z) = \sum_{k=0}^{N-1} (-)^k \frac{k!}{z^k} + \frac{(-)^N N!}{z^N} C_N(z), \qquad (1.8.5)$$

where the optimal truncation index $N = \lfloor |z| \rfloor$ (or $N = |z| - 1$ when $|z|$ is an integer). We let $z = 5e^{i\theta}$, so that $N = m = 4$ and set $\beta = \frac{4}{5}e^{-i\theta}$. When $\theta = 0$, the optimally truncated series yields an absolute error of 2.011×10^{-2} for this value of z. The corresponding terms in (1.8.4) appear to possess an asymptotic structure as shown in Fig. 1.22, albeit of the irregular type illustrated in Fig. 1.18(b). When the series (1.8.4) is truncated near its global minimum at $k_0 = 20$, the resulting absolute error in $\mathcal{E}(z)$ is found to be 3.262×10^{-10}, an increase in 8 digits of precision. The results for different values of θ, together with those corresponding to $|z| = 8$, are presented in Table 1.3; the truncation index k_0 used in the computation of the converging factor $C_N(z)$ is also indicated. It is worth pointing out that the irregular behaviour of the terms in the converging factor present in Fig. 1.22 becomes smoothed out once $|\theta| \gtrsim \pi/20$.

1.8.3 Dingle's converging factors

An alternative theory of converging factors was given by Dingle (1973, Ch. 23) which followed on from the work of Airey. His expansions are expressed in terms of

Table 1.3 *The absolute errors in $ze^z E_1(z)$ at optimal truncation (second column) and with the converging factor (third column) truncated at $k = k_0$ when $z = |z|e^{i\theta}$*

θ	|Error|	|Error|	k_0
$|z| = 5$			
0	2.011×10^{-2}	3.262×10^{-10}	20
$\pi/8$	2.046×10^{-2}	2.762×10^{-8}	13
$\pi/4$	2.157×10^{-2}	5.385×10^{-7}	10
$|z| = 8$			
0	2.011×10^{-3}	1.470×10^{-15}	25
$\pi/8$	1.261×10^{-3}	2.181×10^{-13}	25
$\pi/4$	1.332×10^{-3}	3.378×10^{-11}	20

certain functions related to the incomplete gamma function (or exponential integral) which he called *basic terminants*. Foremost among these is the function defined by

$$\Lambda_\nu(z) = \frac{1}{\Gamma(\nu)} \int_0^\infty \frac{t^{\nu-1}e^{-t}}{1+t/z}dt \qquad (\mathrm{Re}(\nu) > 0, \quad |\arg z| < \pi),$$

which can be shown to equal (Olver, 1990)

$$\Lambda_\nu(z) = z^\nu e^z \Gamma(1 - \nu, z) = \frac{2\pi i e^{-\pi i \nu}}{\Gamma(\nu)} z^\nu e^z T_\nu(z), \tag{1.8.6}$$

where $T_\nu(z)$ is given in (1.7.6).

Dingle's procedure was to resum the divergent tail of an asymptotic expansion by means of Borel summation. In the case of some of the simpler transcendental functions, such as the error function, Fresnel integrals and the incomplete gamma function, Borel summation yielded an explicit representation for the remainder term in terms of $\Lambda_\nu(z)$ or one of his other basic terminants. For higher transcendental functions, such as the confluent hypergeometric functions, it is first necessary to employ a resurgence-type relation for the coefficients in the tail of the expansion before the Borel summation process to yield a series involving the $\Lambda_\nu(z)$ functions.

Taking as an example the confluent hypergeometric function $U(a, a - b + 1, z)$ mentioned in §1.7.2, we have from (1.7.7)

$$z^a U(a, a - b + 1, z) = \sum_{k=0}^{N_0-1} (-)^k \frac{(a)_k (b)_k}{k! z^k} + R_0(z; N_0), \tag{1.8.7}$$

where N_0 is, for the moment, an arbitrary truncation index and the remainder $R_0(z; N_0)$ is given (formally) by the divergent tail

$$R_0(z; N_0) = \sum_{k=N_0}^\infty (-)^k \frac{(a)_k (b)_k}{k! z^k}. \tag{1.8.8}$$

As $|z| \to \infty$, the index N_0 will be chosen to scale like $|z|$ and so can be taken to be a large positive integer. This means that in the sum (1.8.8) we can employ the expansion of the quotient of gamma functions $(a)_k(b)_k/k!$ given in Paris and Kaminski (2001, p. 53) and Olver (1995). If we define the coefficients

$$A_k = \frac{(a)_k(b)_k}{k!}, \qquad A'_k = \frac{(1-a)_k(1-b)_k}{k!},$$

where $(a)_k = \Gamma(a+k)/\Gamma(a)$, then we obtain the resurgence-like expansions for $k \to \infty$

$$\left.\begin{aligned}
A_k &\sim D_1 \sum_{j=0}^{\infty} (-)^j A'_j \Gamma(k + \vartheta - j) \\
A'_k &\sim D'_1 \sum_{j=0}^{\infty} (-)^j A_j \Gamma(k - \vartheta - j),
\end{aligned}\right\} \tag{1.8.9}$$

where $\vartheta = a+b-1$, $D_1 \equiv D_1(a,b) = \{\Gamma(a)\Gamma(b)\}^{-1}$ and $D'_1 = D_1(1-a, 1-b)$. It should be remarked that in these large-order expansions the coefficients A_k involve the complementary coefficients A'_k, and vice versa.

Formal substitution of the first result in (1.8.9) into (1.8.8), followed by reversal of the order of summation and replacement of the factorial $\Gamma(k + \vartheta - j)$ by its Euler integral representation (which constitutes the Borel summation process), yields

$$\begin{aligned}
R_0(z; N_0) &\sim D_1 \sum_{j=0}^{\infty} (-)^j A'_j \sum_{k=N_0}^{\infty} (-z)^{-k} \Gamma(k + \vartheta - j) \\
&= D_1 \sum_{j=0}^{\infty} (-)^j A'_j \int_0^{\infty} t^{\vartheta - j - 1} e^{-t} \sum_{k=N_0}^{\infty} (-t/z)^k dt \\
&= \frac{D_1}{(-z)^{N_0}} \sum_{j=0}^{\infty} (-)^j A'_j \int_0^{\infty} t^{N_0 + \vartheta - j - 1} \frac{e^{-t}}{1 + t/z} dt \\
&= \frac{D_1}{(-z)^{N_0}} \sum_{j=0}^{\infty} (-)^j A'_j \Gamma(N_0 + \vartheta - j) \Lambda_{N_0 + \vartheta - j}(z). \tag{1.8.10}
\end{aligned}$$

When N_0 is set equal to the optimal truncation index $N_0 = |z| + O(1)$ in (1.8.10), we obtain one of Dingle's expansions for the remainder. If the terminant function $\Lambda_\nu(z)$ is expressed in terms of $T_\nu(z)$ by (1.8.6), then the above resummation of the remainder $R_0(z; N_0)$ agrees with that given in (1.7.8) when the upper limit in the sum is replaced by ∞.

1.8.4 A formal discussion of hyperasymptotics

The re-expansion in (1.8.10) represents the first stage of the hyperasymptotic process. The terms in $R_0(z; N_0)$ are found similarly to decrease in magnitude before finally diverging in asymptotic fashion. If this series is now in its turn also optimally truncated, the new remainder resulting from this re-expansion process is found to be exponentially small compared to $R_0(z; N_0)$ as $|z| \to \infty$ in a certain sector. This process can be repeated to produce a sequence of re-expanded remainder terms, each of which is exponentially smaller than its predecessor. This process of successive exponential improvement has been termed *hyperasymptotics* by Berry and Howls (1990) and Berry (1991a). This improvement in accuracy comes at a cost however: the terms in the re-expanded remainders involve progressively more complicated terminant functions, known as *hyperterminants*, which have to be computed to the required precision.

To illustrate the hyperasymptotic procedure, we continue with our paradigmatic illustration of the confluent hypergeometric function $U(a, a - b + 1, z)$ and give only an outline description based on that presented in Berry and Howls (1990). A more detailed discussion of this function using Mellin–Barnes integrals is given in Paris and Kaminski (2001, §6.3). Let us introduce the constants $D_j \equiv D_j(a, b)$ for $j = 1, 2, \ldots$ by

$$D_{2j} = \pi^{-2j}(\sin \pi a \, \sin \pi b)^j, \qquad D_{2j+1} = \frac{D_{2j}}{\Gamma(a)\Gamma(b)},$$

where $D_0 = 1$ and $D_1 = \{\Gamma(a)\Gamma(b)\}^{-1}$. We now truncate the sum for the remainder term $R_0(z; N_0)$ in (1.8.10) after N_1 terms and write

$$R_0(z; N_0) = D_1 \sum_{k=0}^{N_1-1} (-)^k A_k' \mathbf{T}_{k1} + R_1(z; N_0, N_1),$$

where (formally)

$$R_1(z; N_0, N_1) = D_1 \sum_{k=N_1}^{\infty} (-)^k A_k' \mathbf{T}_{k1}$$

with

$$\mathbf{T}_{k1} = \frac{(-)^{N_0}}{z^{N_0}} \int_0^\infty t^{N_0+\vartheta-k-1} \frac{e^{-t}}{1+t/z} dt. \tag{1.8.11}$$

Substitution of the second result in (1.8.9), followed by reversal of the order of summation and Borel summation as in the first stage, then yields

$$R_1(z; N_0, N_1) \sim D_2 \sum_{j=0}^{\infty} (-)^j A_j \sum_{k=N_1}^{\infty} (-)^k \Gamma(k - \vartheta - j) \mathbf{T}_{k1}$$

$$= D_2 \sum_{j=0}^{\infty} (-)^j A_j \int_0^\infty t^{-\vartheta-j-1} e^{-t} \sum_{k=N_1}^{\infty} (-t)^k \mathbf{T}_{k1} \, dt$$

$$= D_2 \sum_{k=0}^{\infty} (-)^k A_k \mathbf{T}_{k2}, \tag{1.8.12}$$

where

$$\mathbf{T}_{k2} = \frac{(-)^{N_0+N_1}}{z^{N_0}} \int_0^\infty t_1^{N_0-N_1+\vartheta-1} \frac{e^{-t_1}}{1+t_1/z} dt_1 \int_0^\infty t_2^{N_1-\vartheta-k-1} \frac{e^{-t_2}}{1+t_2/t_1} dt_2$$

and we have made use of the fact that $D_1 D_1' = D_2$.

The result in (1.8.12) represents the second stage in the hyperasymptotic expansion process. This procedure of truncation, substitution of the resurgence-like expansions in (1.8.9) and Borel summation can be repeated to produce the hyperasymptotic sequence

$$z^a U(a, a-b+1, z) = \sum_{k=0}^{N_0-1} (-)^k A_k z^{-k} + S_1(z) + S_2(z) + \cdots, \qquad (1.8.13)$$

where the hyperseries $S_j(z)$ are defined by

$$S_j(z) = D_j \sum_{k=0}^{N_j-1} (-)^k A_k^{(j)} \mathbf{T}_{kj} \qquad (j = 1, 2, \ldots)$$

with $A_k^{(j)} = A_k$ (j even), A_k' (j odd). The sequence of hyperterminants $\mathbf{T}_{kj}$ is given by the n-fold multiple integrals

$$\mathbf{T}_{kj} = \mathbf{T}_{kj}(z; N_0, N_1, \ldots, N_{j-1})$$

$$= z^{-N_0} \prod_{r=1}^{j} (-)^{N_r-1} \int_0^\infty \frac{t_r^{\alpha_r-1} e^{-t_r}}{1+t_r/t_{r-1}} dt_r \qquad (t_0 = z, N_j = k), \quad (1.8.14)$$

where, for convenience, we have set $\alpha_r = N_{r-1} - N_r + (-)^{r-1}\vartheta$. The convergence of these integrals requires $N_r < N_{r-1}$, so that successive hyperseries contain fewer terms. The evaluation of the hyperterminants has been discussed by Olde Daalhuis (1996, 1997).

1.8.5 Truncation schemes

The principle adopted by Berry and Howls (1990) in choosing the truncation indices N_j was that the successive hyperseries $S_j(z)$, including the zeroth level sum, are truncated near their least terms. For the zeroth series in (1.8.13) we readily find $N_0 = |z| + O(1)$. To estimate N_1, we find from (1.8.9) that $A_k' \sim \Gamma(k - \vartheta)$ for large k, while the terminant $\mathbf{T}_{k1}$ can be estimated by replacing in (1.8.11) the slowly varying denominator by its value at the maximum of the rest of the integrand, namely at $t_1^* = N_0 + \vartheta - k - 1$, to obtain

$$\mathbf{T}_{k1} \simeq (-z)^{-N_0} \frac{\Gamma(N_0 - k + \vartheta)}{1 + t_1^*/z}.$$

Then the least value of the kth term $A_k' \mathbf{T}_{k1}$ is determined by the minimum of the product of factorials $\Gamma(k - \vartheta)\Gamma(N_0 - k + \vartheta)$, which is easily seen to correspond to

the choice $k = N_1 = \frac{1}{2}|z| + O(1)$. An estimate for the remainder when the expansion
(1.8.13) is truncated at the optimally truncated first level is given by the magnitude
of the least term in $S_1(z)$, which is controlled by

$$|z|^{-N_0}\Gamma(N_1 - \vartheta - 1)\Gamma(N_0 - N_1 + \vartheta - 1) = O(e^{-|z|}2^{-|z|})$$

upon use of Stirling's formula and the above values of N_0, N_1. The successive
truncation indices can be determined in a similar manner through use of the estimate

$$\mathbf{T}_{kj} \simeq z^{-N_0} \prod_{r=1}^{j}(-)^{N_{r-1}} \frac{\Gamma(\alpha_r)}{1 + \alpha_r/\alpha_{r-1}},$$

to find the truncation scheme given by

$$N_j = 2^{-j}|z| + O(1). \tag{1.8.15}$$

The last term of the jth hyperseries is associated with the estimate

$$O(e^{-|z|}2^{-\lambda_j|z|}), \tag{1.8.16}$$

where

$$\lambda_j = 1 + \tfrac{1}{2} + \tfrac{1}{4} + \cdots + 2^{1-j} = 2 - 2^{1-j} \qquad (j \geq 1).$$

The truncation index at level j is seen to be approximately half that of the preceding
level, so that this scheme comes to a halt when the last hyperseries contains a single
term. Since $\lambda_j < 2$, the exponential improvement is bounded by

$$e^{-|z|}2^{-2|z|} = e^{-|z|(1+2\log 2)};$$

the maximal exponential improvement with the Berry–Howls scheme is therefore
limited to a factor $e^{-\gamma|z|}$, where $\gamma = 1 + 2\log 2 = 2.386\ldots$.

   A rigorous treatment using Stieltjes integrals of the hyperasymptotic expansion of
solutions of second-order linear differential equations with a singularity at infinity of
rank 1 (of which the confluent hypergeometric function is an example) was given by
Olde Daalhuis and Olver (1995); see also Boyd (1990) for the case of the modified
Bessel function. By estimating the remainder in the expansion at level $m \geq 1$ as a
function of the truncation indices $N_0, N_1, \ldots, N_m$, they showed that greater accuracy
could be attained with the truncation scheme

$$N_j = (m + 1 - j)|z| + O(1) \qquad (0 \leq j \leq m), \tag{1.8.17}$$

where the remainder at the jth level is controlled by $\exp\{-(j + 1)|z|\}$. These esti-
mates are also derived using a Mellin–Barnes integral approach in the context of the
confluent hypergeometric function in Paris and Kaminski (2001, §6.3). The treatment of the solutions of second-oder differential equations of arbitrary integer rank
has been considered by Murphy and Wood (1997).

   This result of Olde Daalhuis and Olver is of interest for two reasons. First, with
the optimal truncation scheme in (1.8.17) the remainder at each level is exponentially

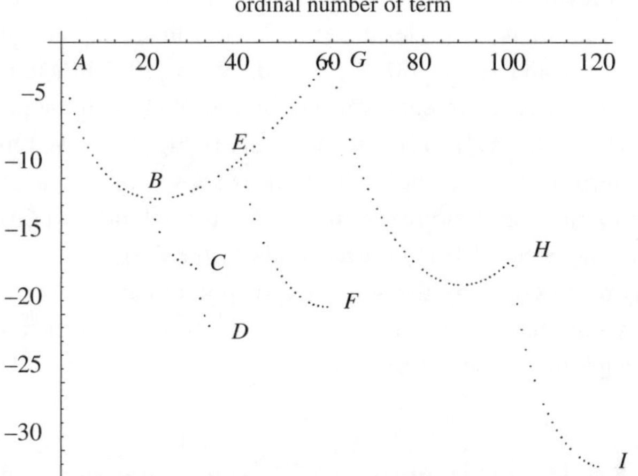

Figure 1.23 The magnitude of the terms (on a $\log_{10}$ scale) against ordinal number k for different hyperasymptotic truncation schemes.

smaller than its predecessor by the factor $e^{-|z|}$ *uniformly* throughout the sector $|\arg z| \le \pi$ and the remainder at the mth level is controlled by $e^{-(m+1)|z|}$. And second, this level of accuracy is achieved by employing *post-optimal* truncation of the hyperseries in all but the mth level. Thus, if we take $m = 2$ and consider the expansion (1.8.13) truncated after $S_2(z)$, then $N_0 = 3|z| + O(1)$, $N_1 = 2|z| + O(1)$ and $N_2 = |z| + O(1)$. This results in the zeroth series (the basic Poincaré asymptotic series) being summed to about $[3|z|]$ terms, which is three times its optimal truncation value.[33] Similarly, the first hyperseries in (1.8.13) is summed to about $[2|z|]$ terms with the final hyperseries then being truncated after about $[|z|]$ terms. This truncation scheme shows that by extracting information contained in the divergent tails of the lower-order hyperseries it is possible to obtain even greater accuracy than that using the scheme in (1.8.15). A similar observation was made by Byatt-Smith (1998) in the context of the hyperasymptotic expansion of the exponential integral.

To illustrate the two different truncation schemes in (1.8.15) and (1.8.17) we present the results of numerical computations for the case $a = -\frac{1}{2}v$, $b = \frac{1}{2} - \frac{1}{2}v$ so that

$$U(-\tfrac{1}{2}v, \tfrac{1}{2}, z) = 2^{-\frac{1}{2}v} e^{\frac{1}{2}z} D_v(\sqrt{2z}),$$

where $D_v(x)$ denotes the Weber parabolic cylinder function; see Abramowitz and Stegun (1965, p. 510). The absolute value of each term in the series in (1.8.13) up to the level $j = 2$ is shown in Fig. 1.23 as a function of ordinal number for the particular values $v = \frac{1}{2}$ and $z = 20$.

---

[33] A problem can arise if the terms in the zeroth series have an absolute value much greater than unity, since this can lead to a loss of accuracy when working with fixed-decimal arithmetic; see Olde-Daalhuis and Olver (1995, §8) for a discussion of this point.

For the zeroth level we set $N_0 = 20$, which terminates the Poincaré asymptotic series in (1.8.13) at, or near, its least term. For the first and second levels we set $N_0 = 40$, $N_1 = 20$ and $N_0 = 60$, $N_1 = 40$, $N_2 = 20$ following the truncation scheme in (1.8.17). These different levels are represented by the sequences of points labelled AB, $ABEF$ and $ABEGHI$, respectively in Fig. 1.23. We remark that at the second level no term in the sequence $ABEG$ rises above unity in magnitude, thereby avoiding the numerical problem mentioned in Olde Daalhuis and Olver (1995, §8). The truncation scheme in (1.8.15) corresponds to the choice $N_0 = 20$, $N_1 = 10$, $N_2 = 5$ and is represented by the sequence of points labelled $ABCD$. From this example, it is clear that greater accuracy results from the truncation scheme in (1.8.17) but at a greater computational cost.

1.8.6 Hyperasymptotics of Laplace-type integrals

The evaluation of the hyperasymptotic expansion of Laplace-type integrals possessing several saddle points was pioneered by Berry and Howls (1991) and we conclude this chapter with an outline of this elaborate theory. The treatment differs from that in the previous section as it employs the iteration of an exact resurgence formula[34] for the remainder of the truncated Poincaré asymptotic series rather than Borel summation.

Let λ be a large complex parameter with $\theta = \arg \lambda$. We consider integrals of the form

$$\int_{C_n} e^{-\lambda \psi(t)} f(t)\, dt,$$

where the functions $\psi(t)$ and $f(t)$ are analytic in a region Δ_n to be specified below. It is assumed that the phase function $\psi(t)$ possesses several *simple* saddle points which we denote by t_{sr} $(r = 1, 2, \ldots)$. The integration path $C_n \equiv C_n(\theta)$ is chosen to be a path of steepest descent through the saddle labelled t_{sn} and will be assumed to pass to infinity down two valleys of the real part of the exponential factor which issue from t_{sn}. It is clear that, as θ varies, the path C_n will also vary.

Reference to (1.2.19), with $\psi(t)$ replaced by $-\psi(t)$, shows that the leading behaviour of the above integral is controlled by $\exp\{-\lambda\psi(t_{sn})\}/\lambda^{1/2}$ as $|\lambda| \to \infty$ in a certain sector $\mathbf{S}$. Accordingly, we remove this rapidly varying factor and consider the slowly varying part $I^{(n)}(\lambda)$ defined by

$$I^{(n)}(\lambda) = \lambda^{\frac{1}{2}} \int_{C_n} e^{-\lambda\{\psi(t) - \psi(t_{sn})\}} f(t)\, dt. \tag{1.8.18}$$

With the change of variable $u = \lambda\{\psi(t) - \psi(t_{sn})\}$ it follows that u is real on the path C_n and increases monotonically from 0 on both halves $C_n^{\pm}$ as one moves away from

---

[34] The *principle of resurgence* was initiated in the works of Dingle (1973) and Écalle (1981).

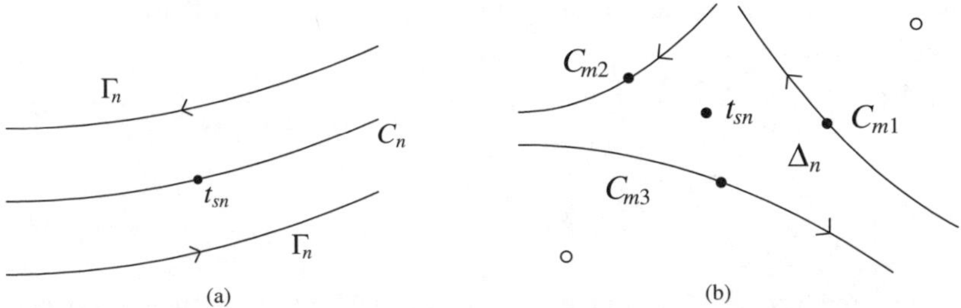

Figure 1.24 (a) The contour Γ_n that surrounds the steepest descent path C_n in the positive sense. (b) Three adjacent contours C_m relative to the saddle t_{sn} and the region Δ_n. The open dots denote more remote, non-adjacent saddles.

the saddle t_{sn}; see §1.2.2. Assuming that $|\psi(t)| \to \infty$ as $|t| \to \infty$ on C_n, we have, upon noting that $dt/du = \{\lambda\psi'(t)\}^{-1}$,

$$I^{(n)}(\lambda) = \lambda^{-\frac{1}{2}} \int_0^\infty e^{-u} \left\{ \frac{f(t^+)}{\psi'(t^+)} - \frac{f(t^-)}{\psi'(t^-)} \right\} du, \qquad t^\pm \equiv t^\pm(u/\lambda),$$

where $t^\pm$ refers to the values of t on the two halves $C_n^\pm$, respectively; compare (1.2.19).

The quantity in curly brackets can be expressed as an integral by an extension of Cauchy's integral to find (Berry and Howls, 1991)

$$\lambda^{-\frac{1}{2}} \left\{ \frac{f(t^+)}{\psi'(t^+)} - \frac{f(t^-)}{\psi'(t^-)} \right\} = \frac{u^{-\frac{1}{2}}}{2\pi i} \int_{\Gamma_n} \frac{f(t)\{h_n(t)\}^{\frac{1}{2}}}{h_n(t) - u/\lambda} dt,$$

where [35]

$$h_n(t) = \psi(t) - \psi(t_{sn}), \tag{1.8.19}$$

and Γ_n consists of two infinite contours that enclose C_n in an anti-clockwise loop within Δ_n; see Fig. 1.24(a). This leads to the *exact* representation for $I^{(n)}(\lambda)$ in the form

$$I^{(n)}(\lambda) = \frac{1}{2\pi i} \int_0^\infty e^{-u} u^{-\frac{1}{2}} \int_{\Gamma_n} \frac{f(t)\{h_n(t)\}^{\frac{1}{2}}}{h_n(t) - u/\lambda} dt\, du. \tag{1.8.20}$$

We now employ the result for non-negative integer N

$$\frac{1}{1-x} = \sum_{k=0}^{N-1} x^k + \frac{x^N}{1-x} \qquad (x \neq 1)$$

[35] The square root factor $\{h_n(t)\}^{1/2}$ is specified by having zero phase on the path C_n^+ when $\theta = 0$.

to expand the denominator appearing in the integral over t in (1.8.20). We find

$$I^{(n)}(\lambda) = \frac{1}{2\pi i} \sum_{k=0}^{N-1} \frac{1}{\lambda^k} \int_0^\infty e^{-u} u^{k-\frac{1}{2}} du \int_{\Gamma_n} \frac{f(t)}{\{h_n(t)\}^{k+\frac{1}{2}}} dt + R(\lambda; N)$$

$$= \sum_{k=0}^{N-1} \frac{a_k^{(n)}}{\lambda^k} + R(\lambda; N), \tag{1.8.21}$$

where the integral over u is evaluated as a factorial and, in the integral over t, the contour Γ_n has been shrunk to a closed contour encircling the saddle point t_{sn} to yield the coefficients

$$a_k^{(n)} = \frac{\Gamma(k + \frac{1}{2})}{2\pi i} \int^{(t_{sn}+)} \frac{f(t)}{\{\psi(t) - \psi(t_{sn})\}^{k+\frac{1}{2}}} dt;$$

compare (1.2.17) when the order of the saddle $m = 2$. The first few values of these coefficients can be obtained from (1.2.15). The remainder $R(\lambda; N)$ in the Poincaré asymptotic series for $I^{(n)}(\lambda)$ truncated after N terms is then given exactly by

$$R(\lambda; N) = \frac{\lambda^{-N}}{2\pi i} \int_0^\infty e^{-u} u^{N-\frac{1}{2}} \int_{\Gamma_n} \frac{f(t)}{h_n(t) - u/\lambda} \{h_n(t)\}^{-N+\frac{1}{2}} dt\, du. \tag{1.8.22}$$

The contour Γ_n is now deformed in a special manner which requires the domain Δ_n to be specified as follows. As θ varies, the steepest descent path C_n through the saddle t_{sn} will vary in a continuous manner, with its endpoints still passing to infinity, until it connects with other neighbouring saddles at t_{sm} ($m \neq n$) to produce a Stokes phenomenon. For these special values of θ, the steepest descent path turns sharply through a right angle at t_{sm} to continue its descent into a valley at infinity. In general, this will only arise for certain neighbouring saddles that Berry and Howls call the *adjacent saddles*; more distant saddles will not connect with t_{sn} in this manner. If we introduce the so-called singulants F_{nm} (Dingle, 1973, p. 147) by

$$\psi(t_{sm}) - \psi(t_{sn}) = F_{nm}, \quad \arg F_{nm} = \phi_{nm}, \tag{1.8.23}$$

the connections with the adjacent saddles occur when λF_{nm} is positive real, so that the phase θ takes the values $\theta = -\phi_{nm}$, for each m. The steepest descent paths C_m through the saddles t_{sm}, associated with these values of θ, then satisfy $\lambda\{\psi(t) - \psi(t_{sm})\} \geq 0$, so that on each path C_m we have

$$\arg\{\psi(t) - \psi(t_{sm})\} = \phi_{nm}. \tag{1.8.24}$$

The paths C_m are called the *adjacent contours*;[36] the domain Δ_n is defined by that region bounded by these paths as illustrated schematically in Fig. 1.24(b), where it is understood that the C_m are to be traversed in an *anti-clockwise sense* with respect to Δ_n. We shall now suppose that the sector $\mathbf{S}$ in the λ-plane is chosen such that

---

[36] It is assumed that each adjacent contour C_m contains only a single saddle point t_{sm}.

the steepest descent path C_n does not pass through a saddle other than t_{sn}, which requires θ to be restricted to an interval

$$-\phi_{nm_1} < \theta < -\phi_{nm_2},$$

where t_{sm_1} and t_{sm_2} are saddles adjacent to t_{sn}.

The contour Γ_n is now deformed by expanding it onto the union of the arcs at infinity and the contours C_m through the adjacent saddles t_{sm}. We repeat that these latter paths are steepest descent paths associated with the special values of $\theta = \arg \lambda = -\phi_{nm}$. This expansion process is justified provided that (i) the quantity $|f(t)|/|\psi(t)|^{N+1/2}$ decays at infinity faster than $|z|^{-1}$, in order to produce a zero contribution from the infinite arcs, (ii) there is no other zero of the denominator of (1.8.22), and (iii) that $\psi(t)$ and $f(t)$ are analytic in the domain Δ_n. These points are discussed in Berry and Howls (1991) and Boyd (1993). With these provisos, we have

$$R(\lambda; N) = \frac{\lambda^{-N}}{2\pi i} \int_0^\infty e^{-u} u^{N-\frac{1}{2}} \sum_m \int_{C_m} \frac{f(t)}{h_n(t) - u/\lambda} \{h_n(t)\}^{-N+\frac{1}{2}} dt\, du,$$

$$(1.8.25)$$

where the sum over m denotes the contribution from the adjacent contours.

Berry and Howls now introduced the change of variable

$$u = \{\psi(t) - \psi(t_{sn})\} \frac{v}{F_{nm}} = v + \{\psi(t) - \psi(t_{sm})\} \frac{v}{F_{nm}},$$

where, from (1.8.23) and (1.8.24), the new variable $v \geq 0$ on the adjacent paths C_m. Substitution of this transformation into (1.8.25) then yields

$$R(\lambda; N) = \frac{\lambda^{-N}}{2\pi i} \sum_m \frac{1}{F_{nm}^{N+\frac{1}{2}}} \int_0^\infty \frac{e^{-v} v^{N-\frac{1}{2}}}{1 - v/(\lambda F_{nm})} \int_{C_m} e^{-v h_m(t)/F_{nm}} f(t)\, dt\, dv$$

$$= \frac{1}{2\pi i} \sum_m \frac{1}{(\lambda F_{nm})^N} \int_0^\infty \frac{e^{-v} v^{N-1}}{1 - v/(\lambda F_{nm})} I^{(m)}\left(\frac{v}{F_{nm}}\right) dv, \qquad (1.8.26)$$

where $h_m(t)$ is defined as in (1.8.19), with n replaced by m, and we have identified the integral over t in terms of $I^{(m)}(v/F_{nm})$ by (1.8.18). The result in (1.8.26) provides an exact form for the remainder in the method of steepest descents expressed as a sum involving integrals through the adjacent saddles. Equations (1.8.21) and (1.8.26) then yield an exact resurgence formula for $I^{(n)}(\lambda)$.

In the special case $N = 0$, we obtain the result

$$I^{(n)}(\lambda) = \frac{1}{2\pi i} \sum_m \int_0^\infty \frac{e^{-v} v^{-1}}{1 - v/(\lambda F_{nm})} I^{(m)}\left(\frac{v}{F_{nm}}\right) dv, \qquad (1.8.27)$$

provided $f(t)/\{\psi(t)\}^{1/2}$ decays sufficiently rapidly at infinity. The form of the remainder in (1.8.26), together with the result (1.8.27), have been employed by Boyd (1993) to obtain new error bounds for asymptotic expansions derived using the method of steepest descents. These bounds are simpler than those in the literature (Wong, 1980; Olver, 1997, p. 135) and a significant difference is that *global* properties of the integrand – not just those on the steepest descent path – play an essential role.

The resurgence relation (1.8.26) can now be iterated to produce the hyperasymptotic expansion of $I^{(n)}(\lambda)$. The first stage in this process takes the form

$$I^{(n)}(\lambda) = \sum_{k=0}^{N_0-1} \frac{a_k^{(n)}}{\lambda^k} + \frac{1}{2\pi i} \sum_m \frac{1}{(\lambda F_{nm})^{N_0}} \int_0^\infty \frac{e^{-v_0} v_0^{N_0-1}}{1 - v_0/(\lambda F_{nm})} I^{(m)}\left(\frac{v_0}{F_{nm}}\right) dv_0,$$

where each term $I^{(m)}$ in the integrand can be expanded as an asymptotic expansion of the form (1.8.21), namely

$$I^{(m)}\left(\frac{v_0}{F_{nm}}\right) = \sum_{k=0}^{N_1-1} \frac{a_k^{(m)}}{(v_0/F_{nm})^k} + \frac{1}{2\pi i} \sum_{m'} \left(\frac{F_{nm}}{v_0 F_{mm'}}\right)^{N_1}$$
$$\times \int_0^\infty \frac{e^{-v_1} v_1^{N_1-1}}{1 - v_1 F_{nm}/(v_0 F_{mm'})} I^{(m')}\left(\frac{v_1}{F_{mm'}}\right) dv_1$$

and we have allowed different truncation indices N_0 and N_1. The sum over m' refers to those saddles that are adjacent to the saddles t_{sm} from the first stage: this will include the original saddle t_{sn} together with (possibly) more remote saddles that are not adjacent to t_{sn}. The quantities $F_{mm'}$ are the singulants between the saddles t_{sm} and $t_{sm'}$ and are defined in an analogous manner to (1.8.23). This then yields

$$I^{(n)}(\lambda) = \sum_{k=0}^{N_0-1} \frac{a_k^{(n)}}{\lambda^k} + \frac{\lambda^{-N_0}}{2\pi i} \sum_m \sum_{k=0}^{N_1-1} \frac{a_k^{(m)}}{F_{nm}^{N_1-k}} \int_0^\infty \frac{e^{-v_0} v_0^{N_0-k-1}}{1 - v_0/(\lambda F_{nm})} dv_0$$

$$+ R_1(\lambda; N_0, N_1), \qquad\qquad (1.8.28)$$

where

$$R_1(\lambda; N_0, N_1) = \left(\frac{1}{2\pi i}\right)^2 \sum_m \sum_{m'} \frac{1}{(\lambda F_{nm})^{N_0}} \left(\frac{F_{nm}}{F_{mm'}}\right)^{N_1}$$
$$\times \int_0^\infty \frac{e^{-v_0} v_0^{N_0-N_1-1}}{1 - v_0/(\lambda F_{nm})} \int_0^\infty \frac{e^{-v_1} v_1^{N_1-1}}{1 - v_1 F_{nm}/(v_0 F_{mm'})} I^{(m')}\left(\frac{v_1}{F_{mm'}}\right) dv_1 dv_0.$$

The integral appearing in (1.8.28) is the first-level hyperterminant and is analogous to that in (1.8.10). Continuation of this iteration process yields the hyperasymptotic expansion of $I^{(n)}(\lambda)$ given in Berry and Howls (1991), with the hyperterminant at the rth stage involving an r-fold integral and associated with the truncation indices $N_0, N_1, \ldots, N_r$. The convergence of the hyperterminants requires $N_r < N_{r-1}$, so that the number of terms in successive hyperseries decreases and the process comes to a halt: the indices are chosen to correspond to optimal truncation at each stage. The repeated sampling of the different saddles at successive levels of the hyperasymptotic expansion has been termed 'multiple scattering' by Berry and Howls. In general, this results in an extremely intricate sequence of contributions to the expansion, although in the case of the familiar Airy function, which is associated with only two saddle points, this consists of simply bouncing back-and-forth between the two saddles.

2

Hadamard expansion of Laplace integrals

2.1 Introduction

In a short paper in the early years of the last century, Hadamard (1908) gave a novel, absolutely convergent expansion for the zeroth-order modified Bessel function $I_0(x)$ when $x > 0$. The terms in this expansion (which we present for general order ν) involve the well-known large-variable asymptotic coefficients $a_k(\nu)(2x)^{-k}$ for the Bessel function smoothed by the normalised incomplete gamma function $P(a, x) := \gamma(a, x)/\Gamma(a)$ in the form

$$I_\nu(x) = \frac{e^x}{\sqrt{2\pi x}} \sum_{k=0}^{\infty} \frac{a_k(\nu)}{(2x)^k} \, P(k + \nu + \tfrac{1}{2}, 2x) \qquad (2.1.1)$$

when $\mathrm{Re}(\nu) > -\tfrac{1}{2}$, where the coefficients $a_k(\nu)$ are given by

$$a_k(\nu) = \frac{(\tfrac{1}{2} + \nu)_k(\tfrac{1}{2} - \nu)_k}{k!}, \qquad (2.1.2)$$

with $(a)_k = \Gamma(k + a)/\Gamma(a)$ being the Pochhammer symbol. The Hadamard series (2.1.1) was described by Watson (1952, p. 204) to be of 'considerable theoretical importance'. Despite this theoretical interest, however, little use seems to have been made of (2.1.1), presumably on account of its slow rate of convergence of algebraic order (see below) and the fact that the necessary computational power and interest in hyperasymptotic evaluation are relatively recent developments.

More generally, we define a Hadamard expansion to be a compound series expansion of the form

$$\sum_{n=0}^{\infty} e^{-\Omega_n x} S_n(x), \qquad (2.1.3)$$

where each term $S_n(x)$ is a series given by

$$S_n(x) = \sum_{k=0}^{\infty} \frac{c_{k,n}}{x^{k+\mu_n}} \, P(k + \mu_n, \omega_n x).$$

100

The $c_{k,n}$ are coefficients, μ_n are parameters and the normalised incomplete gamma function $P(a, x)$ is defined by

$$P(a, x) = \frac{1}{\Gamma(a)} \int_0^1 e^{-t} t^{a-1} dt = \frac{x^a e^{-x}}{\Gamma(1+a)} \, {}_1F_1(1; 1+a; x) \qquad (2.1.4)$$

where the integral holds for $\mathrm{Re}(a) > 0$ and ${}_1F_1$ denotes the confluent hypergeometric function; further properties of $P(a, x)$ are given in Appendix A. Expansions of the type (2.1.3) may have as few as a single series $S_0(x)$ (as in the Bessel function above) but frequently have infinitely many series. The index n is called the level of the expansion, and typically the factors Ω_n (termed the expansion points) and the associated quantities ω_n are related by $\Omega_n = \sum_{r=0}^{n-1} \omega_r$, or an analogous expression, for $n \geq 1$ with Ω_0 often set equal to zero or some other suitable initial value. The quantities ω_n are positive so that the expansion (2.1.3) displays a growing exponential separation between successive levels.

In addition, the series $S_n(x)$ associated with each level in the expansion (2.1.3) are usually absolutely convergent subject possibly to some restriction on the parameters; see Appendix B for a discussion of the convergence of a Hadamard series. The expansion (2.1.3) is then a convergent series expansion which for large positive x exhibits progressively exponentially smaller terms as the number of levels increases. The presence of the incomplete gamma functions in the series $S_n(x)$ does not result in high computational cost, since they can easily be generated by recursion. More interestingly, in view of the asymptotic behaviour $P(a, x) \sim 1$ as $x \to +\infty$ with a fixed, we see that the early terms in the series $S_n(x)$ reduce to a more commonplace Poincaré asymptotic series. As a consequence, the zeroth constituent series $S_0(x)$ in a Hadamard expansion effectively *contains within it the type of asymptotic power series produced by many asymptotic methods.*

In this chapter, we shall show how the Hadamard expansion of functions defined by Laplace integrals can be employed in high-precision evaluation, either by a simple rearrangement of the terms in the tail of the series to produce a rapidly convergent series or by choosing appropriate integration intervals. We shall use Bessel functions as convenient illustrations of the numerical accuracy achievable but our theory will be presented more generally for the confluent hypergeometric functions and so include many of the commonly used functions of mathematical physics.

2.2 The Hadamard series for $I_\nu(x)$

2.2.1 Derivation of the single-stage expansion

The derivation of the Hadamard series (2.1.1) starts with the integral representation (Abramowitz and Stegun, 1965, p. 376; Temme, 1996, p. 237), valid when $\mathrm{Re}(\nu) > -\frac{1}{2}$,

$$I_\nu(x) = \frac{(\frac{1}{2}x)^\nu}{\sqrt{\pi}\,\Gamma(\nu + \frac{1}{2})} \int_{-1}^1 (1 - t^2)^{\nu - \frac{1}{2}} e^{-xt}\,dt = \frac{(2x)^\nu e^x}{\sqrt{\pi}\,\Gamma(\nu + \frac{1}{2})} \int_0^1 e^{-2xt} f(t)\,dt,$$

$$(2.2.1)$$

where we have made the change of variable $t \to 2t - 1$ in the first integral and set $f(t) = \{t(1 - t)\}^{\nu - 1/2}$. The amplitude function $f(t)$ has (provided $\nu \neq \frac{1}{2}, \frac{3}{2}, \ldots$) branch-point singularities at $t = 0$ and $t = 1$. If we introduce the series expansion of $f(t)$ about $t = 0$ valid in $|t| < 1$, we find upon inversion of the order of summation and integration

$$I_\nu(x) = \frac{(2x)^\nu e^x}{\sqrt{\pi}\,\Gamma(\nu + \frac{1}{2})} \sum_{k=0}^\infty (-)^k \binom{\nu - \frac{1}{2}}{k} \int_0^1 e^{-2xt} t^{k+\nu - \frac{1}{2}}\,dt. \qquad (2.2.2)$$

The usual asymptotic procedure for estimating the behaviour of the above integral as $x \to +\infty$ is to replace the integration path $[0, 1]$ by $[0, \infty)$, on the grounds that the relative error introduced is exponentially small of order $O(x^{k+\nu - 1/2} e^{-2x})$ for each integer value of k. The resulting integral can then be evaluated as a factorial to yield the well-known asymptotic expansion

$$I_\nu(x) \sim \frac{e^x}{\sqrt{2\pi x}} \frac{1}{\Gamma(\nu + \frac{1}{2})} \sum_{k=0}^\infty \frac{(-)^k}{(2x)^k} \binom{\nu - \frac{1}{2}}{k} \int_0^\infty e^{-u} u^{k+\nu - \frac{1}{2}}\,du$$

$$= \frac{e^x}{\sqrt{2\pi x}} \sum_{k=0}^\infty \frac{a_k(\nu)}{(2x)^k}, \qquad (2.2.3)$$

where the coefficients $a_k(\nu)$ are defined in (2.1.2). The coefficients in this expansion exhibit the familiar 'factorial divided by a power' dependence of an asymptotic series (see §1.6.1), since application of Stirling's formula readily shows that

$$a_k(\nu) \sim \Gamma(k) \qquad (k \to \infty). \qquad (2.2.4)$$

This divergence is a direct consequence of having extended the interval of integration beyond the domain of validity of the Maclaurin series for $f(t)$.

If, on the other hand, we evaluate the integral in (2.2.2) in terms of the normalised incomplete gamma function defined in (2.1.4) we obtain the convergent expansion [1] (when $\mathrm{Re}(\nu) > -\frac{1}{2}$)

$$I_\nu(x) = \frac{e^x}{\sqrt{2\pi x}} \sum_{k=0}^\infty \frac{a_k(\nu)}{(2x)^k} P(k + \nu + \tfrac{1}{2}, 2x). \qquad (2.2.5)$$

For positive variables, the function $P(a, x)$ exhibits a 'cut-off' structure illustrated in Fig. 2.1(a), since we have the leading behaviour

---

[1] When ν takes on half-integer values the series (2.2.5) reduces to a finite sum, since $a_k(\pm m \pm \frac{1}{2}) = (-m)_k (m + 1)_k / k! = (-)^k \Gamma(1 + m + k)/(k!\,\Gamma(1 + m - k))$ $(m = 0, 1, 2, \ldots)$ which vanishes for $k \geq m + 1$.

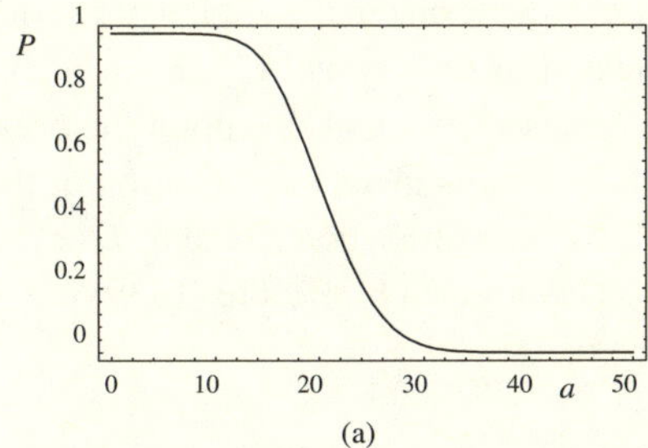 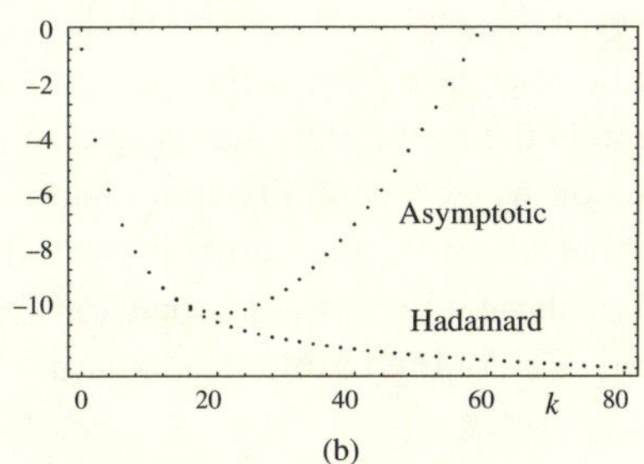

Figure 2.1 (a) The graph of $P(a, x)$ as a function of a when $x = 20$. (b) The behaviour of the terms (on a $\log_{10}$ scale) in the expansion (2.2.5) for $(2\pi x)^{1/2} e^{-x} I_\nu(x)$ against ordinal number k when $x = 10$, $\nu = \frac{1}{4}$ compared with the asymptotic series (2.2.3).

$$P(a, x) \sim \begin{cases} 1 & (x \to \infty) \\[2mm] \dfrac{x^a e^{-x}}{\Gamma(1+a)} & (a \to \infty); \end{cases} \tag{2.2.6}$$

see Appendix A for details. The presence of the incomplete gamma function in (2.2.5) acts as a 'smoothing' factor on the coefficients $a_k(\nu)$, since the behaviour of $P(a, x)$ changes from approximately unity when $a \lesssim x$ to a rapid decay to zero when $a \gtrsim x$. Consequently the early terms ($k < 2x$) in (2.2.5) behave like those of the Poincaré asymptotic series in (2.2.3) for large positive x. Once $k \simeq 2x$, which corresponds roughly to the transition point $a = x$ of the incomplete gamma function $P(a, x)$, the terms in (2.2.5) continue to decrease, but at a much slower rate. This is in contrast to the terms of the associated asymptotic series which start to diverge.

This slow decay follows from the large-a behaviour of $P(a, x)$ in (2.2.6) to produce the behaviour of the kth term in (2.2.5) given by

$$\frac{a_k(\nu)}{(2x)^k} P(k + \nu + \tfrac{1}{2}, 2x) \sim (2x)^{\nu + \frac{1}{2}} e^{-2x} \frac{a_k(\nu)}{\Gamma(k + \nu + \frac{3}{2})} \tag{2.2.7}$$

as $k \to \infty$. Application of the standard result $\Gamma(k+a)/\Gamma(k+b) \sim k^{a-b}$ as $k \to \infty$ and (2.2.4) then yields the order estimate

$$\frac{a_k(\nu)}{\Gamma(k + \nu + \frac{3}{2})} = O(k^{-\nu - 3/2}) \qquad (k \to \infty). \tag{2.2.8}$$

The Hadamard series (2.1.1) therefore has late terms ($k \gg 2x$) that decay like $k^{-\nu - 3/2}$. Although this algebraic decay is sufficient to secure the absolute convergence of the series when $\mathrm{Re}(\nu) > -\frac{1}{2}$, it is clearly too slow in this form for computational purposes.

In Fig. 2.1(b) we contrast the behaviour of the terms in the Poincaré series (2.2.3) with those in the Hadamard series (2.2.5). It is seen that the terms of the latter series decay in typical asymptotic fashion down to the ordinal index $k \simeq [2x]$ (where square brackets denote the integer part), which corresponds to the optimal truncation point of the asymptotic series (2.2.3). For $k > [2x]$, the terms in the Hadamard series continue to decrease at the algebraic rate indicated in (2.2.7) and (2.2.8), whereas those in the asymptotic series start to diverge. We also remark at this point the presence of the exponentially small factor e^{-2x} in (2.2.7): this gives an indication of the level at which the late terms in the tail of the Hadamard series contribute to $I_\nu(x)$.

An alternative form of expansion for $I_\nu(x)$ has been given by Kibble (1939) in the form of a doubly infinite Hadamard series

$$I_\nu(x) = \frac{e^x}{\sqrt{2\pi x}} \sum_{k=-\infty}^{\infty} \frac{(a+\nu)_k(a-\nu)_k}{\Gamma(k+a+\frac{1}{2})(2x)^{k+a-\frac{1}{2}}} P(k+\nu+a, 2x), \qquad (2.2.9)$$

for $\mathrm{Re}(\nu) > -\frac{1}{2}$, where a is an arbitrary free parameter. It is readily shown that when $\frac{1}{2} - a$ is any integer, this result reduces to that given in (2.2.5). Some generalisations of the Hadamard expansion for the modified Bessel function have also been given in Yang and Srivastava (2004).

2.2.2 The modified Hadamard series

One way to overcome the slow convergence of (2.2.5) is by a simple rearrangement of the tail of the series which will then enable high-precision evaluation of $I_\nu(x)$; another approach is discussed in §2.2.4. Let M denote a positive integer to be suitably chosen and write (2.2.5) as

$$I_\nu(x) = \frac{e^x}{\sqrt{2\pi x}} \left\{ \sum_{k=0}^{M-1} \frac{a_k(\nu)}{(2x)^k} P(k+\nu+\tfrac{1}{2}, 2x) + T_M(x) \right\}, \qquad (2.2.10)$$

where the tail $T_M(x)$ is given by

$$T_M(x) = \sum_{k=M}^{\infty} \frac{a_k(\nu)}{(2x)^k} P(k+\nu+\tfrac{1}{2}, 2x) \qquad (\mathrm{Re}(\nu) > -\tfrac{1}{2}). \qquad (2.2.11)$$

If we replace the normalised incomplete gamma function P by its representation in terms of the confluent hypergeometric function in (2.1.4), we find

$$\begin{aligned}
T_M(x) &= (2x)^{\nu+\frac{1}{2}} e^{-2x} \sum_{k=M}^{\infty} \frac{a_k(\nu)}{\Gamma(k+\nu+\frac{3}{2})} {}_1F_1(1; k+\nu+\tfrac{3}{2}; 2x) \\
&= (2x)^{\nu+\frac{1}{2}} e^{-2x} \sum_{k=M}^{\infty} a_k(\nu) \sum_{r=0}^{\infty} \frac{(2x)^r}{\Gamma(k+r+\nu+\frac{3}{2})} \\
&= e^{-2x} \sum_{r=0}^{\infty} \sigma_r \xi^{r+\nu+\frac{1}{2}}, \qquad \xi = \frac{2x}{M}, \qquad (2.2.12)
\end{aligned}$$

where the interchange in the order of summation is justified by absolute convergence when $\text{Re}(v) > -\frac{1}{2}$. The coefficients $\sigma_r \equiv \sigma_r(M)$ are given by

$$\sigma_r M^{-r-v-\frac{1}{2}} = \sum_{k=M}^{\infty} \frac{a_k(v)}{\Gamma(k+r+v+\frac{3}{2})} = \left\{ \sum_{k=0}^{\infty} - \sum_{k=0}^{M-1} \right\} \frac{a_k(v)}{\Gamma(k+r+v+\frac{3}{2})}$$

$$= \frac{\Gamma(r+v+\frac{1}{2})}{r!\,\Gamma(r+2v+1)} - \sum_{k=0}^{M-1} \frac{a_k(v)}{\Gamma(k+r+v+\frac{3}{2})} \qquad (2.2.13)$$

when $\text{Re}(v) > -\frac{1}{2}$, where the infinite sum has been evaluated in terms of a Gauss hypergeometric function of unit argument as

$$\frac{1}{\Gamma(r+v+\frac{3}{2})} \, {}_2F_1(\tfrac{1}{2}+v, \tfrac{1}{2}-v; r+v+\tfrac{3}{2}; 1) = \frac{\Gamma(r+v+\frac{1}{2})}{r!\,\Gamma(r+2v+1)}$$

upon application of the well-known summation formula (Abramowitz and Stegun, 1965, p. 556)

$$ {}_2F_1(a, b; c; 1) = \frac{\Gamma(c)\Gamma(c-a-b)}{\Gamma(c-a)\Gamma(c-b)} \qquad (\text{Re}(c-a-b) > 0), \qquad (2.2.14)$$

provided $c \neq 0, -1, -2, \dots$. This result enables σ_r to be expressed in terms of the difference between a quotient of gamma functions and the sum of the first M terms of the series, thereby removing the slow convergence present in the first sum in (2.2.13).

The *modified* Hadamard series for the Bessel function $I_v(x)$ is then, from (2.2.10) and (2.2.12), given by

$$I_v(x) = \frac{e^x}{\sqrt{2\pi x}} \left\{ \sum_{k=0}^{M-1} \frac{a_k(v)}{(2x)^k} P(k+v+\tfrac{1}{2}, 2x) + e^{-2x} \sum_{r=0}^{\infty} \sigma_r \xi^{r+v+\frac{1}{2}} \right\}, \qquad (2.2.15)$$

where ξ is defined in (2.2.12). It is important to stress that no approximation has been introduced in this representation and that the expansion is still absolutely convergent provided $\text{Re}(v) > -\frac{1}{2}$. The form of the tail in (2.2.12) has made apparent the subdominant exponential factor e^{-2x}, indicating the level of precision at which the tail contributes to the expansion. We also remark that the variable in the rearranged tail is $\xi = 2x/M$, which will be less than unity when the truncation index M is chosen to be greater than the optimal truncation index $[2x]$ of the associated Poincaré series (2.2.3); compare §1.8.5.

We illustrate the behaviour of the terms in the modified expansion (2.2.15) for $(2\pi x)^{1/2} e^{-x} I_v(x)$ in Fig. 2.2, where the magnitude of the terms (on a $\log_{10}$ scale) in the finite main sum and the tail is shown as a function of the overall ordinal number[2] k for two different values of the truncation index M. A remarkable feature is that, for

---

[2] The terms corresponding to $r \geq 0$ in the tail $T_M(x)$ are labelled $k \geq M$.

Table 2.1 *Values of the coefficients σ_r for different r and M when $v = \frac{1}{4}$*

r	$M = 20$	$M = 40$	$M = 60$
0	3.00280×10^{-1}	3.00199×10^{-1}	3.00169×10^{-1}
10	2.16436×10^{-3}	6.21257×10^{-3}	9.12606×10^{-3}
20	4.00661×10^{-6}	1.29046×10^{-4}	4.94580×10^{-4}
40	2.31919×10^{-14}	9.02199×10^{-10}	7.46402×10^{-8}
60	2.03704×10^{-25}	5.39409×10^{-17}	2.67243×10^{-13}

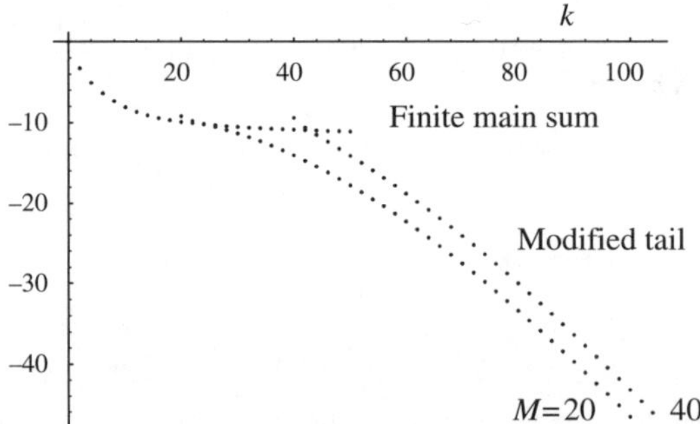

Figure 2.2 The magnitude of the terms (on a $\log_{10}$ scale) in the modified expansion (2.2.10) for $(2\pi x)^{1/2} e^{-x} I_v(x)$ against ordinal number k for different truncation index M when $x = 10$, $v = \frac{1}{4}$.

$M \gtrsim [2x]$, the terms in the tail are found to decay monotonically at a rate *comparable with that of the optimally truncated asymptotic series* in (2.2.3). In addition, it may be remarked that the leading term of $T_M(x)$ is somewhat larger than the Mth term of the finite main sum. Thus the modified Hadamard series for $I_v(x)$ provides a description that not only possesses terms in the finite main sum that behave like the asymptotic series (2.2.3) down to its optimal truncation point, but also has a tail that continues to decay at a rapid rate (when $M \gtrsim [2x]$). We thus have a means of hyperasymptotic evaluation of $I_v(x)$ when $x > 0$. The role of the terminant function in the usual hyperasymptotic theory (see §1.8) is here played by the coefficients σ_r, which are defined as a deleted Gauss hypergeometric function in (2.2.13) or, more generally, in terms of a one-dimensional integral (see below). We mention that a different procedure, using Mellin–Barnes regularisation, of high-precision calculation of the Bessel functions of large argument has recently been given in Kowalenko (2002).

The results of calculations using the modified expansion (2.2.15) are presented in Tables 2.1 and 2.2. We show the values of the coefficients σ_r and the absolute error in the computation of $(2\pi x)^{1/2} e^{-x} I_v(x)$ for different truncation index M, with the tail in (2.2.12) truncated after N terms. It is seen that, for fixed N, the achievable

Table 2.2 *The absolute error in the computation of* $(2\pi x)^{1/2}e^{-x}I_v(x)$ *for different truncation indices M and N when $x = 10$, $v = \frac{1}{4}$*

N	$M = 20$	$M = 30$	$M = 40$	$M = 50$
0	6.83×10^{-10}	2.21×10^{-10}	1.14×10^{-10}	7.05×10^{-11}
10	6.08×10^{-12}	9.03×10^{-14}	4.15×10^{-15}	3.64×10^{-16}
20	7.03×10^{-15}	1.01×10^{-17}	6.84×10^{-20}	1.19×10^{-21}
30	9.81×10^{-19}	2.09×10^{-22}	2.82×10^{-25}	1.22×10^{-27}
40	2.24×10^{-23}	9.52×10^{-28}	3.19×10^{-31}	4.07×10^{-34}
50	1.08×10^{-28}	1.14×10^{-33}	1.13×10^{-37}	4.82×10^{-41}

accuracy increases as M increases. Note also that even when $N = 0$ the expansion still receives an exponentially subdominant contribution from the tail corresponding to the leading term in $T_M(x)$. On the other hand, if we adopt the strategy of fixing the overall number $M + N$ of terms used in the expansion, then the greatest accuracy occurs when $M \simeq [2x]$, corresponding to the optimal truncation index of the Poincaré series (2.2.3).

In Paris (2001a), it was shown that the coefficients σ_r appearing in (2.2.12) can be expressed alternatively as a rapidly convergent $_3F_2$ hypergeometric function of unit argument in the form

$$\sigma_r = C_r \, _3F_2 \left(\begin{array}{c} \frac{1}{2}+v, \frac{1}{2}-v, 1; \\ M+1, r+v+\frac{3}{2}; \end{array} 1 \right), \qquad (2.2.16)$$

where

$$C_r = \frac{M^{r+v+\frac{1}{2}}}{\Gamma(r+v+\frac{3}{2})} \frac{a_M(v)}{(r+v+\frac{1}{2})_M}.$$

The hypergeometric function in (2.2.16) has the desirable property that its evaluation becomes more efficient as M and r increase, with $_3F_2(1) \to 1$ as either (or both) $M \to \infty$ and $r \to \infty$. The coefficients $C_r = O(1)$ as $M \to \infty$ with r fixed and decay factorially as $r \to \infty$ with M fixed; the coefficients σ_r therefore behave in a similar manner in these two limits.

We note that the coefficients σ_r can also be expressed in terms of an integral by

$$\sigma_r = M^{r+v+\frac{1}{2}} \left\{ \frac{1}{r!} \int_0^1 (1-t)^r \hat{f}(t)\, dt - \sum_{k=0}^{M-1} \frac{a_k(v)}{\Gamma(k+r+v+\frac{3}{2})} \right\}, \qquad (2.2.17)$$

where

$$\hat{f}(t) = \frac{f(t)}{\Gamma(\frac{1}{2}+v)}.$$

This follows from the Maclaurin expansion for $\hat{f}(t)$ given by

$$\hat{f}(t) = \frac{\{t(1-t)\}^{\nu-\frac{1}{2}}}{\Gamma(\nu+\frac{1}{2})} = \frac{1}{\Gamma(\nu+\frac{1}{2})} \sum_{k=0}^{\infty} (-)^k \binom{\nu-\frac{1}{2}}{k} t^{k+\nu-\frac{1}{2}}$$

$$= \frac{\cos\pi\nu}{\pi} \sum_{k=0}^{\infty} \frac{\Gamma(k+\nu-\frac{1}{2})}{k!} t^{k+\nu-\frac{1}{2}}$$

valid in $|t| < 1$, so that

$$\frac{1}{r!} \int_0^1 (1-t)^r \hat{f}(t)\, dt = \frac{\cos\pi\nu}{\pi} \sum_{k=0}^{\infty} \frac{\Gamma(k+\nu-\frac{1}{2})}{r!\,k!} \int_0^1 (1-t)^r t^{k+\nu-\frac{1}{2}}\, dt$$

$$= \frac{\cos\pi\nu}{\pi} \sum_{k=0}^{\infty} \frac{\Gamma(k+\nu-\frac{1}{2})}{k!} \frac{\Gamma(k+\nu+\frac{1}{2})}{\Gamma(k+r+\nu+\frac{3}{2})}$$

$$= \sum_{k=0}^{\infty} \frac{a_k(\nu)}{\Gamma(k+r+\nu+\frac{3}{2})},$$

where we have evaluated the integral in terms of the beta function. Although this form is not really required in this example, it will be found indispensable when dealing with Laplace integrals where the analogue of the infinite sum in (2.2.13) cannot be evaluated in closed form.

2.2.3 The Hadamard series for complex z

In this section we let $z = xe^{i\theta}$ denote a complex variable, where $x > 0$ and $\theta = \arg z$. The expansions in (2.2.5) and (2.2.15) are in fact valid for complex values[3] of the variable z since the integral representation employed in its derivation is valid without restriction on $\arg z$. However, if we attempt to use (2.2.10) for complex z it is found that the incomplete gamma function $P(a, z)$ (with $a > 0$) no longer exhibits the simple cut-off property, but develops an oscillatory structure with an increasingly pronounced maximum in the neighbourhood of the point $a = |z|$ as $|\arg z|$ increases. This loss of the simple cut-off property when z is complex results in a deterioration of the achievable accuracy when computing $I_\nu(z)$ for a given truncation index M. This becomes particularly severe, for example, when attempting to compute the Bessel function of the first kind $J_\nu(x) = e^{-\frac{1}{2}\pi i\nu} I_\nu(ix)$ from (2.2.10) by putting $z = ix$ $(x > 0)$.

To avoid the presence of terms involving the incomplete gamma function of complex argument in (2.2.5) when x is replaced by z, we employ a device (a path rotation argument) used by Hadamard (1908) for the modified Bessel function $I_0(z)$ in the

---

[3] In practice, however, it is sufficient to consider only the sector $|\arg z| \le \frac{1}{2}\pi$, since we have the continuation formula $I_\nu(z) = e^{\pm\pi i\nu} I_\nu(ze^{\mp\pi i})$.

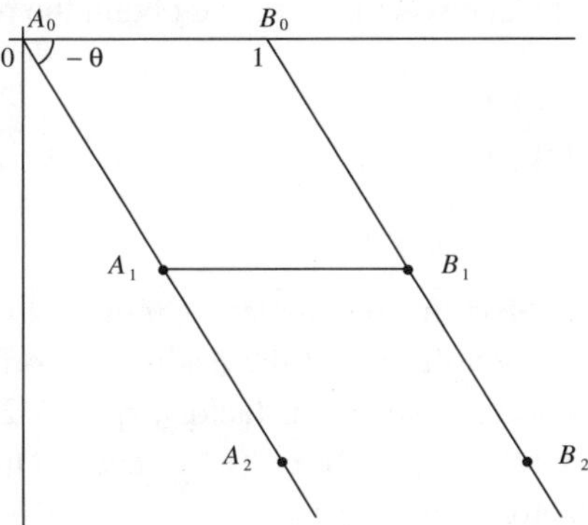

Figure 2.3 The integration path $A_0 A_1 B_1 B_0$.

case $\theta = \frac{1}{2}\pi$. The integration path $[0, 1]$ in (2.2.1) is taken along the sides of the rhombus $A_0 A_1 B_1 B_0$, with the sides $A_0 A_1$ and $B_0 B_1$ inclined at an angle $-\theta$ to the positive real axis and with vertices at the points 0, $e^{-i\theta}$, $1 + e^{-i\theta}$ and 1; see Fig. 2.3. On the sides $A_0 A_1$ and $B_0 B_1$ we put $t = e^{-i\theta} u$ and $t = 1 + e^{-i\theta} u$ $(0 \leq u \leq 1)$, respectively. The integral along $A_0 A_1$ then involves the *real* exponential factor $\exp(-xu)$ and so can be evaluated as described in §2.2.1 to find

$$\int_0^{e^{-i\theta}} e^{-2zt} f(t)\, dt = \sum_{k=0}^{\infty} (-)^k \binom{\nu - \frac{1}{2}}{k} e^{-(k+\nu+\frac{1}{2})i\theta} \int_0^1 e^{-2xu} u^{k+\nu-\frac{1}{2}}\, du,$$

so that

$$\frac{(2z)^\nu e^z}{\sqrt{\pi}\,\Gamma(\nu + \frac{1}{2})} \int_{A_0}^{A_1} e^{-2zt} f(t)\, dt = \frac{e^z}{\sqrt{2\pi z}} \sum_{k=0}^{\infty} \frac{a_k(\nu)}{(2z)^k} P(k + \nu + \tfrac{1}{2}, 2x). \qquad (2.2.18)$$

Proceeding in a similar manner for the integral[4] along $B_0 B_1$ we find

$$\frac{(2z)^\nu e^z}{\sqrt{\pi}\,\Gamma(\nu + \frac{1}{2})} \int_{B_1}^{B_0} e^{-2zt} f(t)\, dt = \frac{e^{z \pm \pi i(\nu + \frac{1}{2})}}{\sqrt{2\pi z}} \sum_{k=0}^{\infty} \frac{(-)^k a_k(\nu)}{(2z)^k} P(k + \nu + \tfrac{1}{2}, 2x).$$

The contribution from the path $A_1 B_1$ is obtained by putting $t = e^{-i\theta} + u$ $(0 \leq u \leq 1)$ to yield

$$R(z) = \frac{(2z)^\nu e^{z-2x}}{\sqrt{\pi}\,\Gamma(\nu + \frac{1}{2})} \int_0^1 e^{-2zu} f(u + e^{-i\theta})\, du. \qquad (2.2.19)$$

---

[4] Along the path $B_0 B_1$ we have $1 - t = e^{(\pm \pi - \theta)i} u$, where the upper or lower sign is chosen according as $\theta > 0$ or $\theta < 0$, respectively.

Then, by application of Cauchy's theorem, we obtain the result

$$
I_\nu(z) = \frac{e^z}{\sqrt{2\pi z}} \sum_{k=0}^{\infty} \frac{a_k(\nu)}{(2z)^k} \left\{ 1 + (-)^k e^{-2z \pm \pi i (\nu + \frac{1}{2})} \right\} P(k + \nu + \tfrac{1}{2}, 2x) + R(z).
$$

$$(2.2.20)$$

The Hadamard series involve the incomplete gamma functions of real variable $x = |z|$, which consequently will exhibit the desired cut-off property. The remainder integral $R(z)$ contains the exponential factor $\exp(z - 2x)$, which represents a more subdominant contribution as $|z| \to \infty$ (when $\theta \neq 0$) than the second series in (2.2.20) containing the factor e^{-2z}.

The integral for $R(z)$ in (2.2.19) is similar to that in (2.2.1), so that the above expansion procedure can be repeated to produce further subdominant exponential contributions. This is achieved by using the lines $A_0 A_1$ and $B_0 B_1$ produced as integration paths. The amplitude function in (2.2.19) is then expanded about the points A_1 and B_1 and the path $A_1 B_1$ is deformed into the path[5] $A_1 A_2 B_2 B_1$. The domain of convergence about B_1 is controlled by the branch point at $t = 1$, whereas that about A_1 is controlled by the branch point at $t = 1$ when $|\theta| \leq \frac{1}{3}\pi$ and by that at $t = 0$ when $\frac{1}{3}\pi \leq |\theta| \leq \frac{1}{2}\pi$. This process of successive deformation can be viewed in the t-plane as descending an 'exponential ladder' inclined at an angle $-\theta$ to the positive real axis. The spacing between the 'rungs' of this ladder is determined by the radius of convergence of the Taylor expansion of $f(t)$ about the sequence of points $A_1, A_2, A_3, \ldots$. The remainder at the nth stage corresponds to the contribution from the nth rung $A_n B_n$, which is associated with the exponential level $\exp(z - 2\Omega_n x)$, where $\Omega_n = A_0 A_n$. This procedure is described more fully in Paris (2001b).

An inconvenience with such an approach is that the lengths between successive rungs of the ladder depend on θ: only when $\theta = \frac{1}{2}\pi$ are the rungs at their maximum separation. In addition, our strict adherence to producing Hadamard series with real argument has led to a representation for $I_\nu(z)$ in terms of an infinite number of Hadamard series, instead of the single series in (2.2.5). In the neighbourhood of $\theta = 0$ (which is a Stokes line) the separation between the different levels $A_n B_n$ ($n \geq 1$) shrinks to zero as $\theta \to 0$. This fact has the consequence that the different exponential levels in the expansion become progressively less separated and, from a practical point of view, become indistinguishable from one another in this limit. On the Stokes line $\theta = 0$ we are left with just the contribution from $A_0 B_0$ which yields (2.2.5).

An alternative expansion for $I_\nu(z)$ which overcomes this cancellation effect in the neighbourhood of the Stokes line, but which no longer involves all incomplete gamma functions of real argument, is to deform the integration path into the path

---

[5] For convenience, the paths $A_n B_n$ ($n \geq 1$) are taken parallel to the real axis.

$A_0 A_1 B_0$. The contribution from $A_0 A_1$ is as in (2.2.18) with the integral along the segment $A_1 B_0$ being

$$\int_{e^{-i\theta}}^{1} e^{-2zt} f(t)\, dt = -e^{-2z} \int_{0}^{e^{-i\theta}-1} e^{-2zu} f(1+u)\, du,$$

where we have put $t = 1 + u$. Since

$$f(1+u) = -\{u(1+u)\}^{\nu-\frac{1}{2}} e^{\pm \pi i (\nu + \frac{1}{2})},$$

where the upper or lower sign is taken according as $\theta > 0$ or $\theta < 0$ respectively, we find upon expansion of the factor $(1+u)^{\nu-1/2}$ that

$$\frac{(2z)^\nu e^z}{\sqrt{\pi}\,\Gamma(\nu+\frac{1}{2})} \int_{A_1}^{B_0} e^{-2zt} f(t)\, dt = \frac{e^{-z\pm\pi i(\nu+\frac{1}{2})}}{\sqrt{2\pi z}} \sum_{k=0}^{\infty} \frac{(-)^k a_k(\nu)}{(2z)^k} P(k+\nu+\tfrac{1}{2}, 2x-2z).$$

This last expansion is valid provided[6] $|\theta| \le \frac{1}{3}\pi$, since in this sector we have $|u| \le 1$ on the integration path.

Hence we obtain the expansion

$$I_\nu(z) = \frac{e^z}{\sqrt{2\pi z}} \sum_{k=0}^{\infty} \frac{a_k(\nu)}{(2z)^k} P(k+\nu+\tfrac{1}{2}, 2x)$$

$$+ \frac{e^{-z\pm\pi i(\nu+\frac{1}{2})}}{\sqrt{2\pi z}} \sum_{k=0}^{\infty} \frac{(-)^k a_k(\nu)}{(2z)^k} P(k+\nu+\tfrac{1}{2}, 2x-2z) \qquad (|\arg z| \le \tfrac{1}{3}\pi).$$

$$(2.2.21)$$

We remark that when $\theta = 0$ the second series in (2.2.21) vanishes, since $P(a, 0) = 0$, to yield the single Hadamard series in (2.2.5). Although the second series involves the incomplete gamma functions of the complex argument $2x - 2z$, it is found that the terms decay rapidly *without any dramatic reduction* when $|\theta|$ is small; see Paris (2001b, §4).

2.2.4 Multi-stage expansions for $I_\nu(z)$

The slow algebraic convergence rate of the Hadamard series (2.2.5) is a consequence of having integrated over the full interval of convergence of the series expansion of the amplitude function $f(t)$, which in the case of $I_\nu(z)$ is the interval $[0, 1]$. In this sense, the interval $[0, 1]$ can be said to be a *maximal integration interval* and the resulting series (2.2.5) is a single-stage expansion. When dealing with Laplace-type integrals with coalescing saddle points in §3.4.1, the ideas of using non-maximal intervals of integration combined with what we shall call *forward-reverse expansion* will be found to lead to more rapidly convergent Hadamard series. Forward

---

[6] Alternatively, the fact that $a_k(\nu) \sim \Gamma(k)$ for large k and use of (2.2.6) show that the expansion converges in $|e^{-i\theta} - 1| \le 1$; that is, in the sector $|\theta| \le \pi/3$.

(resp. reverse) expansion consists of expansion of $f(t)$ about a point $t = t_0$ and, in the case of real variables, integration along the t-axis to the right (resp. left) of t_0. It can be seen that in obtaining (2.2.5) we have used forward expansion about $t = 0$ using the maximal integration interval $[0, 1]$.

We now apply these ideas to $I_\nu(z)$ as follows, where z is a complex variable. We subdivide the integration path in (2.2.1) into the intervals $[0, \frac{1}{2}]$ and $[\frac{1}{2}, 1]$ to find

$$I_\nu(z) = \frac{(2z)^\nu e^z}{\sqrt{\pi}\,\Gamma(\nu + \frac{1}{2})} \left\{ \int_0^{\frac{1}{2}} e^{-2zt} f(t)\, dt + e^{-2z} \int_0^{\frac{1}{2}} e^{2zt} f(t)\, dt \right\}, \qquad (2.2.22)$$

where, in the second integral over $[\frac{1}{2}, 1]$, we have replaced t by $1 - t$ and used the fact that $f(1 - t) = f(t)$. Upon insertion of the series expansion of $f(t)$ about $t = 0$, the first and second integrals then correspond to forward-reverse expansion over the non-maximal interval of length $\frac{1}{2}$. The first integral in (2.2.22) is evaluated as in (2.2.5). The second can be evaluated with the help of the result

$$\frac{1}{\Gamma(a)} \int_0^1 t^{a-1} e^{zt}\, dt = \frac{e^{\pm\pi i a}}{z^a \Gamma(a)} \int_0^{-z} \tau^{a-1} e^{-\tau}\, d\tau = z^{-a} e^{\pm\pi i a}\, P(a, -z), \qquad (2.2.23)$$

where, since the τ-plane is cut along the negative real axis, the upper or lower signs are chosen according as $\arg z > 0$ or $\arg z < 0$, respectively. Then (2.2.22) immediately yields the 2-stage expansion

$$I_\nu(z) = \frac{e^z}{\sqrt{2\pi z}} \left\{ \sum_{k=0}^\infty \frac{a_k(\nu)}{(2z)^k}\, P(k + \nu + \tfrac{1}{2}, z) \right.$$

$$\left. + e^{-2z \pm \pi i(\nu+\frac{1}{2})} \sum_{k=0}^\infty \frac{(-)^k a_k(\nu)}{(2z)^k}\, P(k + \nu + \tfrac{1}{2}, -z) \right\}, \qquad (2.2.24)$$

where the signs [7] are chosen as mentioned above.

From (2.2.6), the behaviour of the late terms in both series in (2.2.24) (after removal of the factor $e^z/\sqrt{(2\pi z)}$ but inclusion of the factor e^{-2z} multiplying the second series) is seen to be

$$z^{\nu+\frac{1}{2}} e^{-z} O(2^{-k} k^{-\nu-\frac{3}{2}}) \qquad (k \to \infty). \qquad (2.2.25)$$

This shows that by using non-maximal intervals both series in (2.2.24) now converge at a much faster, geometric rate. In addition, both series in (2.2.24) converge for all finite values of ν, so that the condition $\mathrm{Re}(\nu) > -\frac{1}{2}$ imposed on (2.2.5) can be removed by appeal to analytic continuation and (2.2.24) holds without restriction on the index ν. An advantage with the I Bessel function is the symmetry of the amplitude function $f(t) = \{t(1 - t)\}^{\nu-1/2}$ about $t = 0$ and $t = 1$. This results in both the series in (2.2.24) having the same coefficients $a_k(\nu)$, a fact that will not

---

[7] In *Mathematica*, the value of $P(a, z)$ for $z < 0$ and non-integer a is the value on the upper side of the branch cut $[0, -\infty)$ on the z-axis.

be shared, in general, by the confluent hypergeometric function ${}_1F_1(a; b; z)$ which involves different coefficients about $t = 0$ and $t = 1$; see §2.2.6.

This process of subdivision of the integration path in (2.2.1) can be continued to produce Hadamard series possessing even faster rates of convergence. For example, we can divide the integration path into the intervals $[0, \frac{1}{4}]$, $[\frac{1}{4}, \frac{3}{4}]$ and $[\frac{3}{4}, 1]$, and expand $f(t)$ about the points $t = 0, \frac{1}{2}, 1$. For the first and last intervals, we find the same contribution as (2.2.24) but with the argument of the normalised incomplete gamma functions replaced by $\frac{1}{2}z$. The contribution from the central interval is

$$
\int_{\frac{1}{4}}^{\frac{3}{4}} e^{-2zt} f(t)\, dt = e^{-z} \int_{-\frac{1}{4}}^{\frac{1}{4}} e^{-2zu} (\tfrac{1}{4} - u^2)^{\nu - \frac{1}{2}}\, du
$$

$$
= 2^{1-2\nu} e^{-z} \sum_{k=0}^{\infty} (-)^k \binom{\nu - \frac{1}{2}}{k} \int_{-\frac{1}{4}}^{\frac{1}{4}} e^{-2zu} (2u)^{2k}\, du
$$

$$
= 2^{-2\nu} e^{-z} \sum_{k=0}^{\infty} \frac{b_k(\nu)}{z^{2k+1}} \Delta P(2k + 1, \tfrac{1}{2}z),
$$

where we have put $u = t - \frac{1}{2}$ and employed (2.2.23). The coefficients $b_k(\nu)$ are given by

$$
b_k(\nu) = \frac{(\frac{1}{2} - \nu)_k (2k)!}{k!}
$$

and we have defined the difference between $P(m + 1, \pm z)$ for non-negative integer m by

$$
\Delta P(m + 1, z) = P(m + 1, z) - P(m + 1, -z). \tag{2.2.26}
$$

The incomplete gamma functions $P(2k + 1, \pm\frac{1}{2}z)$ result from forward-reverse expansion about the point $t = \frac{1}{2}$, respectively. Hence, the 3-stage Hadamard expansion for $I_\nu(z)$ takes the form

$$
I_\nu(z) =
$$

$$
\frac{e^z}{\sqrt{2\pi z}} \left\{ \sum_{k=0}^{\infty} \frac{a_k(\nu)}{(2z)^k} P(k + \nu + \tfrac{1}{2}, \tfrac{1}{2}z) + \frac{2e^{-z}(\frac{1}{2}z)^{\nu + \frac{1}{2}}}{\Gamma(\nu + \frac{1}{2})} \sum_{k=0}^{\infty} \frac{b_k(\nu)}{z^{2k+1}} \Delta P(2k + 1, \tfrac{1}{2}z) \right.
$$

$$
\left. + e^{-2z \pm \pi i (\nu + \frac{1}{2})} \sum_{k=0}^{\infty} \frac{(-)^k a_k(\nu)}{(2z)^k} P(k + \nu + \tfrac{1}{2}, -\tfrac{1}{2}z) \right\}, \tag{2.2.27}
$$

where the choice of signs is as in (2.2.24).

Application of (2.2.6) shows that the first and third series in (2.2.27) now have late terms that decay like

$$
\binom{1}{e^{-2z}} \frac{a_k(\nu)}{(2z)^k} P(k + \nu + \tfrac{1}{2}, \pm \tfrac{1}{2}z) = z^{\nu + \frac{1}{2}} \binom{e^{-\frac{1}{2}z}}{e^{-\frac{3}{2}z}} O(4^{-k} k^{-\nu - \frac{3}{2}}), \tag{2.2.28}
$$

and the two constituent parts of the middle series decay like

$$e^{-z}(\tfrac{1}{2}z)^{\nu+\frac{1}{2}}\,\frac{b_k(\nu)}{z^{2k+1}}\,P(2k+1,\pm\tfrac{1}{2}z)=z^{\nu+\frac{1}{2}}\begin{pmatrix}e^{-\frac{3}{2}z}\\e^{-\frac{1}{2}z}\end{pmatrix}O(4^{-k}k^{-\nu-\frac{3}{2}})\qquad(2.2.29)$$

as $k\to\infty$, where the upper and lower signs and components of the column vectors are taken together. The rate of convergence of each series is now controlled by $4^{-k}k^{-\nu-3/2}$, which is twice the rate of that in the 2-stage expansion in (2.2.24). The expansion (2.2.27) can again be extended without restriction on ν by analytic continuation. It follows that this process of subdivision could be continued to produce even more rapidly convergent series, but this would come at the cost of increased computation of the associated incomplete gamma functions and the evaluation of the coefficients in the expansion of the amplitude function $f(t)$ about the internal points.

In Fig. 2.4 we illustrate the decay of the terms for $(2\pi z)^{1/2}e^{-z}I_\nu(z)$ when $z>0$ in the different series appearing in the 2-stage and 3-stage expansions in (2.2.24) and (2.2.27) – including the exponential factors e^{-z} and e^{-2z} – against ordinal number k. The terms in the asymptotic series in (2.2.3) are also shown for comparison. In the case of the 2-stage expansion, the terms in the first series initially closely follow those of the asymptotic series but start to deviate once $k\simeq 20$. The terms in the second series are initially smaller than those of the first series by roughly the factor e^{-z}, although both series ultimately possess the same dominant behaviour as indicated in (2.2.25). In the case of the 3-stage expansion, the first and second series behave in a similar fashion, whereas the behaviour of the third series differs roughly by the factor e^{-z}; see (2.2.28) and (2.2.29). It is clear from these results that modification of the tails of the Hadamard series, as was done for the single-stage expansion in (2.2.15), is not necessary and so considerably reduces the computational effort.

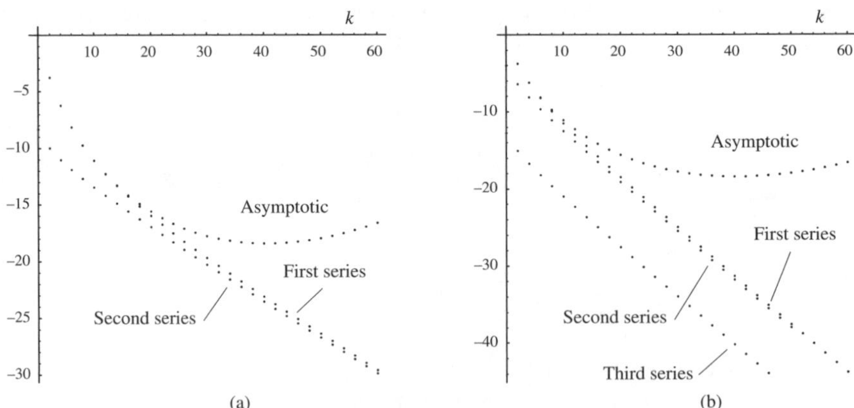

Figure 2.4 The behaviour of the terms (on a $\log_{10}$ scale) in the constituent series in (a) the 2-stage expansion (2.2.24) and (b) the 3-stage expansion (2.2.27) for the Bessel function $(2\pi z)^{1/2}e^{-z}I_\nu(z)$ when $z=20$, $\nu=\frac{3}{4}$. The terms in the asymptotic expansion (2.2.3) are also indicated.

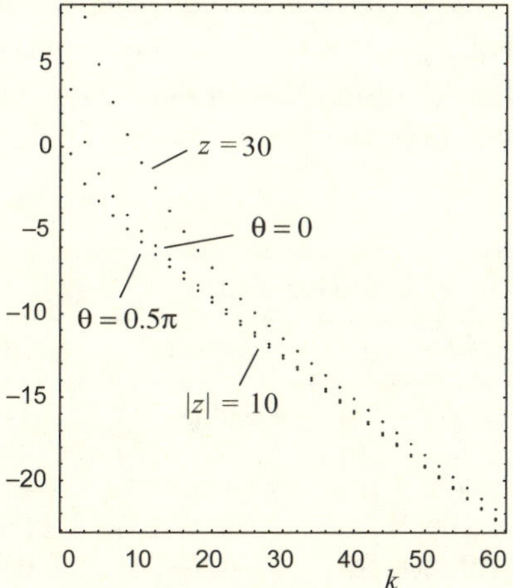

Figure 2.5 The behaviour of the terms (on a $\log_{10}$ scale) of the first series in (2.2.24) for the Bessel function $I_\nu(z)$ when $z = 10e^{i\theta}$ for $\theta = 0$, $\frac{1}{2}\pi$ and when $z = 30$, $\nu = \frac{3}{4}$.

The behaviour of the terms in the first series in (2.2.24) – including the prefactor $e^z/\sqrt{(2\pi z)}$ – when z is complex is illustrated in Fig. 2.5; the behaviour of the terms in the second series is similar. From (2.2.25), this series possesses terms with the large-k behaviour $z^\nu O(2^{-k}k^{-\nu-3/2})$, which depends only weakly on z. The figure demonstrates that the late terms are indeed relatively insensitive to both the phase and the absolute value of z. In Table 2.3 we show the absolute error in the computation of $I_\nu(z)$ for complex z as a function of $\theta = \arg z$. The series in (2.2.24) and (2.2.27) have been truncated after M terms, with the third series in (2.2.27) truncated at $k \leq M$ commensurate with the desired accuracy. The error associated with the optimally truncated asymptotic expansion in (2.2.3) is also indicated for comparison.[8] A significant feature of these results is that the absolute error *decreases* as $|\theta|$ increases towards $\frac{1}{2}\pi$. This is in contrast to the single-stage expansion (2.2.5), where it was found necessary to modify the integration path in order to preserve the accuracy as $|\theta| \to \frac{1}{2}\pi$.

The computation of $I_\nu(z)$ in the sectors $\frac{1}{2}\pi < |\arg z| \leq \pi$ can be accomplished by means of the continuation formula in Abramowitz and Stegun (1965, p. 376)

$$I_\nu(z) = e^{\pm\frac{1}{2}\pi i \nu} I_\nu(ze^{\mp\pi i}).$$

Thus, computation in these sectors can be reduced to the computation of $I_\nu(z)$ in the sector[9] $|\arg z| \leq \frac{1}{2}\pi$. Also, since

---

[8] We remark that the increase in accuracy of the asymptotic result (2.2.3) as θ increases is due to the Stokes phenomenon on $\arg z = 0$. A more precise treatment taking into account the smooth appearance of the subdominant second series in (2.2.32) across $\arg z = 0$ (see §1.7) would yield a more uniform accuracy.

[9] If ν is real this can be further reduced to the sector $0 \leq \arg z \leq \frac{1}{2}\pi$.

Table 2.3 *Values of the absolute error in the computation of $I_\nu(z)$ for different z and ν in the sector $0 \le \arg z \le \frac{1}{2}\pi$ using the 2-stage expansion (2.2.24) and the 3-stage expansion (2.2.27); the absolute error of the optimally truncated asymptotic expansion (2.2.3) is given for comparison*

$z = 10e^{i\theta}, \quad \nu = 3/4$

θ/π	Asymptotic[a] (2.2.3)	2-stage expansion		3-stage expansion	
		$M = 40$	$M = 60$	$M = 40$	$M = 60$
0	3.877(−6)	2.107(−16)	8.190(−23)	3.149(−27)	1.160(−39)
0.1	1.087(−6)	2.085(−16)	8.151(−23)	2.454(−27)	9.054(−40)
0.2	1.590(−7)	2.031(−16)	8.052(−23)	1.192(−27)	4.418(−40)
0.3	1.246(−8)	1.969(−16)	7.934(−23)	3.869(−28)	1.444(−40)
0.4	6.039(−10)	1.922(−16)	7.842(−23)	9.015(−29)	3.386(−41)
0.5	3.803(−11)	1.905(−16)	7.807(−23)	1.328(−29)	4.780(−42)

$z = 30e^{i\theta}, \quad \nu = 9/4$

θ/π	Asymptotic (2.2.3)	2-stage expansion		3-stage expansion	
		$M = 80$	$M = 100$	$M = 60$	$M = 80$
0	5.352(−15)	2.366(−29)	9.507(−36)	7.999(−36)	2.416(−48)
0.1	3.485(−16)	2.300(−29)	9.342(−36)	3.787(−36)	1.149(−48)
0.2	2.719(−18)	2.148(−29)	8.946(−36)	4.338(−37)	1.334(−49)
0.3	2.476(−21)	1.994(−29)	8.517(−36)	1.499(−38)	4.677(−51)
0.4	4.497(−25)	1.890(−29)	8.211(−36)	2.185(−40)	6.911(−53)
0.5	1.661(−29)	1.855(−29)	8.101(−36)	3.371(−42)	1.058(−54)

[a] We have adopted the convention of writing $x(-y)$ for $x \times 10^{-y}$.

$$J_\nu(z) = e^{\mp\frac{1}{2}\pi i\nu} I_\nu(\pm iz), \qquad (2.2.30)$$

it follows that computation of the Bessel function of the first kind $J_\nu(z)$ in $-\pi \le \arg z \le \pi$ can also be achieved by computation of $I_\nu(z)$ in the sector $|\arg z| \le \frac{1}{2}\pi$. Consequently, a separate treatment for $J_\nu(z)$ on the lines of the expansions in (2.2.24) and (2.2.27) is not necessary.

We make a remark about use of (2.2.24) and (2.2.27) for negative values of ν. From (2.2.25), it is clear that the different Hadamard series converge more rapidly when $\nu > 0$ and detailed calculations confirm that there is a progressive loss of accuracy at fixed truncation indices as ν decreases through negative values. In addition, the behaviour of the incomplete gamma functions $P(a, \pm x)$ presents a series of growing oscillations when a is large and negative (see Appendix A), which is detrimental to a high-precision computational scheme. Consequently, although valid for $\nu < 0$, the expansions (2.2.24) and (2.2.27) will lose accuracy for sufficiently large $-\nu$.

Finally, we mention the recent theory developed by Borwein, Borwein and Crandall (2008) and Borwein, Borwein and Chan (2008) for the evaluation of the Bessel functions employing what they call 'exp-arc integrals'. In Appendix C, it is shown in the particular case of the Bessel function $I_\nu(z)$ how their theory is related to the Hadamard expansion process.

2.2.5 The Stokes phenomenon

In this section we examine more closely the application of the 2-stage expansion for $I_\nu(z)$ in (2.2.24) in the neighbourhood of the Stokes line $\arg z = 0$. This expansion takes the form

$$
I_\nu(z) = \frac{e^z}{\sqrt{2\pi z}} \sum_{k=0}^{\infty} \frac{a_k(\nu)}{(2z)^k} P(k + \nu + \tfrac{1}{2}, z)
$$

$$
+ \frac{e^{-z \pm \pi i (\nu + \frac{1}{2})}}{\sqrt{2\pi z}} \sum_{k=0}^{\infty} \frac{(-)^k a_k(\nu)}{(2z)^k} P(k + \nu + \tfrac{1}{2}, -z), \qquad (2.2.31)
$$

where the coefficients $a_k(\nu)$ are defined in (2.1.2) and the upper or lower signs are chosen according as $\arg z > 0$ or $\arg z \leq 0$, respectively. We recall that the late terms in both series are associated with the geometric decay factor 2^{-k} and that the expansion is valid for all finite ν. It may be remarked that if the incomplete gamma functions are formally replaced by unity, then (2.2.31) reduces to the standard compound asymptotic expansion in the form (Watson, 1952, p. 203)

$$
I_\nu(z) \sim \frac{e^z}{\sqrt{2\pi z}} \sum_{k=0}^{\infty} \frac{a_k(\nu)}{(2z)^k} + \frac{e^{-z \pm \pi i (\nu + \frac{1}{2})}}{\sqrt{2\pi z}} \sum_{k=0}^{\infty} \frac{(-)^k a_k(\nu)}{(2z)^k} \qquad (2.2.32)
$$

for $|z| \to \infty$ in the sector $|\arg z| \leq \pi - \delta < \pi$.

We now concentrate on the detailed structure of the terms in (2.2.31) as the Stokes line $\theta = \arg z = 0$ is crossed at *fixed* large $|z|$. To do this, we separate out from the right-hand side of (2.2.31) the dominant Poincaré asymptotic series truncated after m terms, where m will be chosen to correspond to the optimal truncation index given by $m = [2|z|]$. The Stokes multiplier $F(\theta)$ is then defined by

$$
I_\nu(z) = \frac{e^z}{\sqrt{2\pi z}} \sum_{k=0}^{m-1} \frac{a_k(\nu)}{(2z)^k} + \frac{e^{-z}}{\sqrt{2\pi z}} F(\theta);
$$

compare (1.7.3). Comparison with (2.2.32) shows that $F(\theta) \simeq e^{\pi i (\nu + \frac{1}{2})}$ when $\theta > 0$ and $F(\theta) \simeq e^{-\pi i (\nu + \frac{1}{2})}$ when $\theta < 0$. We write $F(\theta)$ in the form

$$
F(\theta) = e^z \{ \Sigma_1 + \Sigma_2 + \Sigma_3 \}, \qquad (2.2.33)
$$

where, from (2.2.31),

$$\Sigma_1 = e^z \sum_{k=0}^{m-1} \frac{a_k(\nu)}{(2z)^k} \{P(k + \nu + \tfrac{1}{2}, z) - 1\},$$

$$\Sigma_2 = e^z \sum_{k=m}^{\infty} \frac{a_k(\nu)}{(2z)^k} P(k + \nu + \tfrac{1}{2}, z),$$

$$\Sigma_3 = e^{-z} \sum_{k=0}^{\infty} \frac{a_k(\nu)}{(2z)^k} P^*(k + \nu + \tfrac{1}{2}, -z),$$

and $P^*(a, -z)$ is defined in (A.3).

When m is chosen to correspond to optimal truncation, the theory of the Stokes phenomenon shows that $F(\theta)$ should have the approximate form (Berry, 1989; Paris and Kaminski, 2001, p. 248)

$$F(\theta) \simeq \cos \pi a + i \sin \pi a \, \text{erf}\,(\theta |z|^{\frac{1}{2}}) \quad (a = \nu + \tfrac{1}{2})$$

for fixed large $|z|$ near $\theta = 0$; see also (1.7.4). Thus, if we now write the Stokes multiplier in the form $S(\theta) \equiv \tfrac{1}{2} + \{F(\theta) - \cos \pi a\}/(2i \sin \pi a)$, then

$$S(\theta) \simeq \tfrac{1}{2} + \tfrac{1}{2}\text{erf}\,(\theta |z|^{\frac{1}{2}}) \tag{2.2.34}$$

for large $|z|$ in the neighbourhood of $\theta = 0$. The variation of the real and imaginary parts of $S(\theta)$, computed via (2.2.33) when $|z| = 10$ and $\nu = \tfrac{1}{4}$ (so that $m = 20$), is shown in Fig. 2.6. It was found numerically that the sum Σ_2 representing the tail of the first Hadamard series in (2.2.31) made a negligible contribution to $S(\theta)$. The predicted behaviour of the real part of $S(\theta)$ in (2.2.34) is found to be indistinguishable from the calculated value on the scale of the figure.

A discussion of the Stokes multiplier resulting from application of the expansion (2.2.21) is given in Paris (2004b).

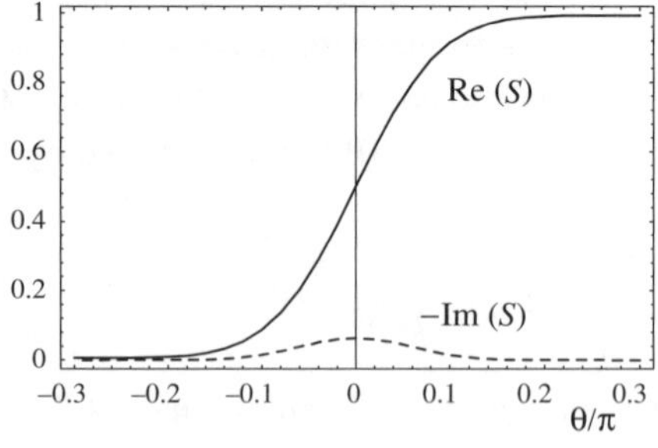

Figure 2.6 The behaviour of the real (solid) and imaginary (dashed) parts of the Stokes multiplier $S(\theta)$ for $I_\nu(z)$ across the Stokes line $\theta = \arg z = 0$, when $|z| = 10$ and $\nu = \tfrac{1}{4}$.

2.2.6 The Hadamard series for $_1F_1(a; a + b; z)$

We present in this section the analogues of the various types of Hadamard series developed in the earlier sections for the confluent hypergeometric function $_1F_1(a; a + b; z)$, of which the Bessel function $I_\nu(z)$ is a special case. These follow by an obvious modification to the previous analysis and accordingly we shall omit the details. We have

$$_1F_1(a; c; z) = \sum_{k=0}^{\infty} \frac{(a)_k z^k}{(c)_k \, k!} \qquad (|z| < \infty)$$

provided $c \neq 0, -1, -2, \ldots$, and from Abramowitz and Stegun (1965, p. 505) the integral representation

$$\frac{\Gamma(a)}{\Gamma(a + b)} \, _1F_1(a; a + b; z) = \frac{e^z}{\Gamma(b)} \int_0^1 e^{-zt} t^{b-1} (1 - t)^{a-1} dt \qquad (2.2.35)$$

valid when $\mathrm{Re}(a) > 0$ and $\mathrm{Re}(b) > 0$. It is sufficient to consider only the sector $|\arg z| \leq \frac{1}{2}\pi$, since evaluation of $_1F_1(a; a+b; z)$ outside this sector can be achieved by means of Kummer's formula

$$_1F_1(a; a + b; z) = e^z \, _1F_1(b; a + b; -z).$$

With the amplitude function $f(t) = t^{b-1}(1 - t)^{a-1}$, the arguments presented in §2.2.1 produce the single-stage expansion

$$\frac{\Gamma(a)}{\Gamma(a + b)} \, _1F_1(a; a + b; z) = z^{-b} e^z \sum_{k=0}^{\infty} \frac{(1 - a)_k (b)_k}{k! \, z^k} P(k + b, z); \qquad (2.2.36)$$

see also Muller (2001). From (2.2.6), the late terms in this series decay like

$$\frac{(1 - a)_k (b)_k}{k! \, \Gamma(k + b + 1)} = O(k^{-a-1}) \qquad (k \to \infty)$$

so that (2.2.36) is absolutely convergent when $\mathrm{Re}(a) > 0$.

The modified form of this expansion (see §2.2.2) is given by

$$\frac{\Gamma(a)}{\Gamma(a + b)} \, _1F_1(a; a + b; z) = z^{-b} e^z \left\{ \sum_{k=0}^{M-1} \frac{(1 - a)_k (b)_k}{k! \, z^k} P(k + b, z) + T_M(z) \right\}$$

$$(2.2.37)$$

for positive integer M, where

$$T_M(z) = z^{b-a} e^z \sum_{r=0}^{\infty} \sigma_r \xi^{r+a}, \qquad \xi = \frac{z}{M}. \qquad (2.2.38)$$

The coefficients σ_r are defined by

$$\sigma_r = M^{r+a} \sum_{k=M}^{\infty} \frac{(1-a)_k (b)_k}{k!\, \Gamma(k+r+b+1)}$$

$$= M^{r+a} \left\{ \frac{\Gamma(r+a)}{r!\, \Gamma(r+a+b)} - \sum_{k=0}^{M-1} \frac{(1-a)_k (b)_k}{k!\, \Gamma(k+r+b+1)} \right\};$$

compare (2.2.13).

The analogue of the expansion (2.2.21) is

$$\frac{\Gamma(a)}{\Gamma(a+b)}\, {}_1F_1(a; a+b; z) = z^{-b} e^z \sum_{k=0}^{\infty} \frac{(1-a)_k (b)_k}{k!\, z^k}\, P(k+b, |z|)$$

$$+ (ze^{\mp \pi i})^{-a} \frac{\Gamma(a)}{\Gamma(b)} \sum_{k=0}^{\infty} \frac{(-)^k (a)_k (1-b)_k}{k!\, z^k}\, P(k+a, |z| - z) \quad (2.2.39)$$

valid in the sector $|\arg z| \le \frac{1}{3}\pi$, where the upper or lower sign is chosen according as $\arg z > 0$ or $\arg \le 0$, respectively (Paris 2001b). The first series involves the incomplete gamma functions of positive argument so that (2.2.39) is suitable for computation in the neighbourhood of the Stokes line $\arg z = 0$.

The analogue of the 2-stage expansion in (2.2.24) for ${}_1F_1(a; a+b; z)$ is obtained by decomposing the integration path in (2.2.35) into the two halves $[0, \frac{1}{2}]$ and $[\frac{1}{2}, 1]$, as in the case of $I_\nu(z)$. Following the procedure in §2.2.4, we then find that

$$\frac{\Gamma(a)}{\Gamma(a+b)}\, {}_1F_1(a; a+b; z) = z^{-b} e^z \sum_{k=0}^{\infty} \frac{(1-a)_k (b)_k}{k!\, z^k}\, P(k+b, \tfrac{1}{2}z)$$

$$+ (ze^{\mp \pi i})^{-a} \frac{\Gamma(a)}{\Gamma(b)} \sum_{k=0}^{\infty} \frac{(-)^k (a)_k (1-b)_k}{k!\, z^k}\, P(k+a, -\tfrac{1}{2}z), \quad (2.2.40)$$

where the signs are chosen as in (2.2.39). From (2.2.6), the late terms in these series decay like $2^{-k} k^{-\alpha-1}$, where $\alpha = a$ for the first series and $\alpha = b$ for the second series. It follows by analytic continuation that (2.2.40) holds for all finite values of the parameters a and b.

The analogue of the 3-stage expansion in (2.2.27) proceeds by decomposition of the integration path in (2.2.35) into the intervals $[0, \frac{1}{3}]$, $[\frac{1}{3}, \frac{2}{3}]$ and $[\frac{2}{3}, 1]$ to yield

$$\frac{\Gamma(a)}{\Gamma(a+b)}\, {}_1F_1(a; a+b; z) =$$

$$z^{-b} e^z \sum_{k=0}^{\infty} \frac{(1-a)_k (b)_k}{k!\, z^k}\, P(k+b, \tfrac{1}{3}z) + \frac{e^{\frac{1}{2}z}}{\Gamma(b)} \sum_{k=0}^{\infty} \frac{d_k}{z^{k+1}}\, \Delta P(k+1, \tfrac{1}{6}z)$$

$$+ (ze^{\mp \pi i})^{-a} \frac{\Gamma(a)}{\Gamma(b)} \sum_{k=0}^{\infty} \frac{(-)^k (a)_k (1-b)_k}{k!\, z^k}\, P(k+a, -\tfrac{1}{3}z), \quad (2.2.41)$$

where ΔP is defined in (2.2.26) and the choice of signs is made as above.

The coefficients d_k are those appearing in the expansion of the amplitude function $f(t)$ about the point $t = \frac{1}{2}$, namely

$$f(\tfrac{1}{2} + u) = (\tfrac{1}{2} - u)^{a-1}(\tfrac{1}{2} + u)^{b-1} = \sum_{k=0}^{\infty} \frac{d_k}{k!}\, u^k \quad (|u| < \tfrac{1}{2}).$$

A recurrence relation satisfied by the coefficients d_k can be obtained by observing that

$$\frac{df}{du} = \left\{ \frac{b-1}{\frac{1}{2} + u} - \frac{a-1}{\frac{1}{2} - u} \right\} f.$$

Substitution of the above expansion into this differential equation then easily leads to the recurrence

$$d_{k+1} = A_k d_k + B_k d_{k-1}, \tag{2.2.42}$$

where

$$A_k = 2(b - a), \quad B_k = 4k(1 - a - b + k) \quad (k \ge 0)$$

with the initial values $d_{-1} = 0$, $d_0 = 2^{2-a-b}$. By dominant balance arguments (Bender and Orszag, 1978, §5.5), the two solutions of (2.2.42) for non-integer values of a and b possess the leading behaviour $2^k k^{-a}$ and $(-2)^k k^{-b}$ as $k \to \infty$. It follows that the large-k behaviour of d_k is controlled [10] by $2^k k^{-\gamma}$, where $\gamma = \min\{a, b\}$. From (2.2.6), the behaviour of the late terms in these three series is then given by $3^{-k} k^{-\alpha-1}$, where $\alpha = a$ for the first series, $\alpha = \min\{a, b\}$ for the second series and $\alpha = b$ for the third series. The expansion (2.2.41) consequently holds for all finite values of the parameters a and b.

When $a = b = \nu + \frac{1}{2}$, the expansions (2.2.37), (2.2.39) and (2.2.40) reduce to those for $I_\nu(z)$ in (2.2.10), (2.2.21) and (2.2.24), since

$$I_\nu(z) = \frac{(\tfrac{1}{2}z)^\nu e^{-z}}{\Gamma(1 + \nu)} \,{}_1F_1(\nu + \tfrac{1}{2}; 2\nu + 1; 2z).$$

Note, however, that the expansions (2.2.41) and (2.2.27) differ in the choice of the intervals used in the decomposition of the integration path and in the resulting dominant decay factor (3^{-k} instead of 4^{-k}) associated with each series. This is a consequence of the fact that the expansion of the amplitude function $f(t)$ about the point $t = \frac{1}{2}$ for the Bessel function $I_\nu(z)$ involves ascending powers of u^2, whereas that for the ${}_1F_1(a; a + b; z)$ function (when $a \neq b$) involves ascending powers of u.

The treatment of the Stokes phenomenon in the neighbourhood of arg $z = 0$ for the confluent hypergeometric function ${}_1F_1(a; a + b; z)$ from the 2-stage expansion

---

[10] A more detailed treatment using Cauchy's integral (cf. §2.4.3) leads to the estimate $d_k/k! \sim \xi_k(a) + (-)^k \xi_k(b)$ as $k \to \infty$, where $\xi_k(x) = 2^{k+1}(2k)^{-x}/\Gamma(1 - x)$.

in (2.2.40) follows in a similar manner to that given in §2.2.5 for $I_\nu(z)$. A derivation of the Hadamard expansion for the confluent hypergeometric function from a differential equation point of view is described in Kaminski and Paris (2004).

We conclude this section by making a remark concerning the more general Hadamard series

$$S(z) = e^z \sum_{k=0}^{\infty} \frac{(a)_k (b)_k}{k! \, z^{k+c}} \, P(k+c, z), \tag{2.2.43}$$

of which (2.2.36) is a particular case. From the convergence condition (B.6) discussed in Appendix B, the parameters a, b and c are subject to the restriction $\mathrm{Re}(c - a - b) > -1$ for absolute convergence. Substitution of (2.1.4) and the series expansion of the confluent hypergeometric function, followed by reversal of the order of summation, shows that

$$\begin{aligned}
S(z) &= \sum_{r=0}^{\infty} \frac{z^r}{\Gamma(r+c+1)} \sum_{k=0}^{\infty} \frac{(a)_k (b)_k}{k!(r+c+1)_k} \\
&= \sum_{r=0}^{\infty} \frac{\Gamma(r+c-a-b-1) z^r}{\Gamma(r+c-a+1)\Gamma(r+c-b+1)} \\
&= \frac{\Gamma(c-a-b+1)}{\Gamma(c-a+1)\Gamma(c-b+1)} \, {}_2F_2(1, c-a-b+1; c-a+1, c-b+1; z),
\end{aligned}$$

where the sum over k has been identified as the hypergeometric function ${}_2F_1(a, b; r+c+1; 1)$ and evaluated by the Gauss summation formula (2.2.14). We see that only when $c = a$ or $c = b$ (when a, $b \neq 0$) does the ${}_2F_2$ function reduce to the simpler ${}_1F_1$ function.

2.3 Rapidly convergent Hadamard series

In §2.2.4 we introduced the notion of maximal integration interval for a Laplace integral, which is determined by the singularity structure of the amplitude function $f(t)$. In the case of the $I_\nu(z)$ Bessel function, the endpoints of the integration interval $[0, 1]$ in (2.2.1) coincide with the branch-point singularities (when ν is non-integer) of $f(t) = \{t(1-t)\}^{\nu-1/2}$. Consequently, when expanding $f(t)$ about $t = 0$ the series will converge in a disc of unit radius. The price to be paid for integrating up to the circle of convergence of $f(t)$ is an algebraic decay of the late terms in the resulting Hadamard series. This slow decay can be transformed by a simple rearrangement of the tail of the series to produce more rapid convergence; see §2.2.3.

We now consider the situation when the integration path is contained within the disc of convergence of the series expansion of $f(t)$ about the lower endpoint, where we would expect rapid convergence of the resulting (single-stage) Hadamard series. The examples we give all involve an amplitude function $f(t)$ with an infinite radius of convergence. The situation when the integration path extends beyond the disc of convergence of $f(t)$ is discussed in §2.4.1.

Example 2.1 Consider the integral

$$I(x) = \int_1^\infty e^{-xt^2} \cos at\, dt = \frac{1}{2}\sqrt{\frac{\pi}{x}} e^{-a^2/(4x)} \operatorname{Re}\left\{\operatorname{erfc}\left(x^{\frac{1}{2}} + \frac{ia}{2x^{\frac{1}{2}}}\right)\right\}$$

for $a > 0$ and $x > 0$. Although $I(x)$ can be evaluated in terms of the complementary error function we shall derive its Hadamard expansion as follows.

From the standard result

$$\int_0^\infty e^{-xt^2} \cos at\, dt = e^{-a^2/(4x)},$$

we can write

$$I(x) = e^{-a^2/(4x)} - \int_0^1 e^{-xt^2} \cos at\, dt.$$

The series expansion of the amplitude function $f(t) = \cos at$ about $t = 0$ has an infinite radius of convergence, and we find

$$I(x) = e^{-a^2/(4x)} - \sum_{k=0}^\infty \frac{(-)^k a^{2k}}{(2k)!} \int_0^1 t^{2k} e^{-xt^2}\, dt$$

$$= e^{-a^2/(4x)} - \frac{1}{2}\sqrt{\frac{\pi}{x}} \sum_{k=0}^\infty \frac{(-)^k}{k!} \left(\frac{a^2}{4x}\right)^k P(k + \tfrac{1}{2}, x) \qquad (2.3.1)$$

upon use of the duplication formula for the gamma function. From (2.2.6), the late terms in the above Hadamard series possess the controlling behaviour

$$\frac{e^{-x}(\tfrac{1}{2}a)^{2k}}{k!\,\Gamma(k + \tfrac{3}{2})} \qquad (k \to \infty).$$

The rapid decay of this series (for finite values of a) is illustrated in Fig. 2.7(a).

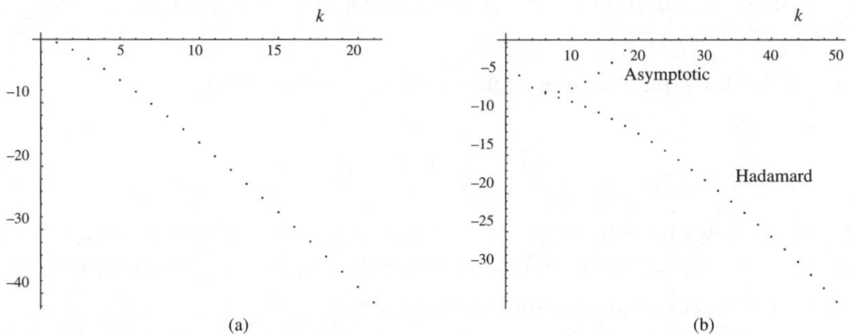

Figure 2.7 The behaviour of the terms (on a $\log_{10}$ scale) of the Hadamard series (a) in (2.3.1) when $x = 40$, $a = 3$ and (b) in (2.3.2) when $x = 20$, $v = \frac{1}{2}$.

Example 2.2 Consider the integral

$$I_{\pm}(x; v) = \int_0^1 t^{v-1} e^{-x(t \pm t^3/3)} dt \qquad (\operatorname{Re}(v) > 0)$$

for large positive values of x. The exponential factor has saddle points at $t = \pm i$ or $t = \pm 1$ in the case of the upper or lower signs, respectively. In the case of the negative sign being chosen the integration path passes along a steepest descent path to the subdominant saddle at $t = 1$.

If we expand the factor $\exp(\mp xt^3/3)$ about $t = 0$ (which has an infinite radius of convergence) we find

$$I_{\pm}(x; v) = \sum_{k=0}^{\infty} \frac{(\pm\frac{1}{3}x)^k}{k!} \int_0^1 t^{3k+v-1} e^{-xt} dt$$

$$= \sum_{k=0}^{\infty} \frac{(\pm)^k \Gamma(3k + v)}{3^k k! x^{2k+v}} P(3k + v, x). \qquad (2.3.2)$$

The asymptotic expansion of $I_{\pm}(x; v)$ is obtained by formally replacing the incomplete gamma function $P(3k + v, x)$ by unity to obtain

$$I_{\pm}(x; v) \sim \sum_{k=0}^{\infty} \frac{(\pm)^k \Gamma(3k + v)}{3^k k! \, x^{2k+v}} \qquad (x \to +\infty).$$

From (2.2.6), the late terms (corresponding to $k \gg k_{\text{opt}}$, where the optimal truncation index of the asymptotic expansion satisfies $k_{\text{opt}} \simeq \frac{1}{3}x$) have the controlling behaviour

$$\frac{e^{-x}(\frac{1}{3}x)^k}{k! \, (3k + v)} \qquad (k \gg k_{\text{opt}}).$$

The decay of the absolute value of the terms in the Hadamard series for $I_{\pm}(x; v)$ in (2.3.2) and its asymptotic expansion is shown in Fig. 2.7(b). It is seen that the terms initially follow those of the asymptotic expansion down to roughly the optimal truncation index k_{opt} and thereafter go over into a more rapid factorial decay.

Example 2.3 Our final example in this section is the integral

$$I(x) = \int_0^1 e^{-xt} J_0(t) \, dt, \qquad (2.3.3)$$

where $J_0(t)$ denotes the zero-order Bessel function. If we substitute the series expansion of $J_0(t)$ about $t = 0$ (which has infinite radius of convergence) we obtain, upon reversal of the order of summation and integration,

$$I(x) = \sum_{k=0}^{\infty} \frac{(-)^k 2^{-2k}}{(k!)^2} \int_0^1 e^{-xt} t^{2k} dt$$

$$= \sum_{k=0}^{\infty} \frac{(-)^k 2^{-2k}}{(k!)^2} \frac{(2k)!}{x^{2k+1}} P(2k+1, x)$$

$$= \frac{1}{\sqrt{\pi}} \sum_{k=0}^{\infty} \frac{(-)^k \Gamma(k+\frac{1}{2})}{k! \, x^{2k+1}} P(2k+1, x). \tag{2.3.4}$$

From (2.2.6), the late terms in this series possess the controlling behaviour

$$\frac{e^{-x} 4^{-k}}{(k!)^2 (2k+1)} \qquad (k \to \infty),$$

which is again an example of very rapid convergence. We have computed $I(x)$ by means of (2.3.4) for large $x > 0$ to high precision and found it to be much faster than the numerical integration routine in *Mathematica*.

The integral in (2.3.3) can easily be generalised to yield the Hadamard expansion of the more general integral

$$\int_0^1 e^{-xt^\mu} J_\nu(at) dt = \frac{1}{\mu} \sum_{k=0}^{\infty} \frac{(-)^k (\frac{1}{2}a)^{\nu+2k} \Gamma(\lambda_k)}{k! \, \Gamma(k+\nu+1) x^{\lambda_k}} P(\lambda_k, x)$$

for $\nu > -1$, $a > 0$ and $\mu > 0$, where $\lambda_k = (2k+\nu+1)/\mu$.

A more standard way of obtaining an expansion for $I(x)$ in (2.3.3) would be to use Watson's lemma;[11] see §1.2.4. Alternatively, if we make use of the standard result (Watson, 1952, p. 388)

$$\int_0^\infty e^{-xt} J_0(t) dt = \frac{1}{\sqrt{1+x^2}},$$

we obtain

$$I(x) = \frac{1}{\sqrt{1+x^2}} - e^{-x} \int_0^\infty e^{-xu} J_0(1+u) \, du$$

$$= \frac{1}{\sqrt{1+x^2}} - e^{-x} \sum_{n=0}^{\infty} x^{-n} J_0^{(n)}(1), \tag{2.3.5}$$

where $J_0^{(n)}(1)$ denotes the nth derivative of $J_0(u)$ evaluated at $u = 1$. However, the difficulty in the use of (2.3.5) for high-precision computation is the evaluation of $J_0^{(n)}(1)$ for large values of n.

The large-n behaviour of these derivatives can be obtained from Cauchy's integral formula

$$J_0^{(n)}(1) = \frac{n!}{2\pi i} \oint \frac{J_0(1+u)}{u^{n+1}} \, du,$$

where the integration path surrounds the origin in the positive sense. We now expand the path into a circle of large radius and employ the asymptotic form

---

[11] We remark that application of Watson's lemma directly to (2.3.3) leads to a convergent expansion with the sum $(1+x^2)^{-1/2}$.

$J_0(1+u) \sim (e^{i(\phi+u)} + e^{-i(\phi+u)})/\sqrt{(2\pi u)}$ for large $|u|$, where $\phi = 1 - \frac{1}{4}\pi$. This results in the integral

$$\frac{n!}{2\pi i} \frac{1}{\sqrt{2\pi}} \oint e^{i(\phi+u)} \frac{du}{u^{n+\frac{3}{2}}} = \frac{n!}{2\pi i} \frac{n^{-n-\frac{1}{2}} e^{i\phi}}{\sqrt{2\pi}} \oint e^{n(iw-\log w)} w^{-\frac{3}{2}} dw,$$

together with the conjugate integral, where we have made the change of variable $u = nw$. The exponential factor in this last integral has a saddle point at $w = -i$ and the steepest descent path through this point is locally horizontal. Upon use of Stirling's formula $n! \sim \sqrt{2\pi}\, e^{-n} n^{n+1/2}$ as $n \to \infty$ and the saddle-point approximation in (1.2.30), we find the large-n contribution

$$\frac{\exp(\frac{1}{2}\pi ni + i)}{\sqrt{2\pi n}}.$$

When combined with the conjugate contribution, this produces the leading behaviour

$$J_0^{(n)}(1) \sim \sqrt{\frac{2}{\pi n}} \cos \pi \left(\frac{n}{2} + \frac{1}{\pi} \right) \qquad (n \to \infty).$$

Thus, $J_0^{(n)}(1) = O(n^{-1/2})$ as $n \to \infty$, so that the expansion (2.3.5) is, in fact, convergent when $x > 1$.

2.4 Hadamard series on an infinite interval

We now consider the situation when the interval of integration extends beyond the circle of convergence of the series expansion of the amplitude function $f(t)$. To illustrate the general procedure we shall consider the confluent hypergeometric function $U(a, a+b, z)$, which has the Laplace integral representation (Abramowitz and Stegun, 1965, p. 505)

$$U(a, a+b, z) = \frac{1}{\Gamma(a)} \int_0^\infty e^{-zt} f(t)\, dt, \qquad f(t) = t^{a-1}(1+t)^{b-1} \qquad (2.4.1)$$

valid for $\mathrm{Re}(z) > 0$ and $\mathrm{Re}(a) > 0$, where we put $z = xe^{i\theta}$, with $x > 0$ and $\theta = \arg z$. If we rotate the path of integration to coincide with the ray $t = \tau e^{-i\theta}$ $(0 \leq \tau < \infty)$, we obtain

$$U(a, a+b, z) = \frac{e^{-i\theta}}{\Gamma(a)} \int_0^\infty e^{-x\tau} f(\tau e^{-i\theta})\, d\tau \qquad (2.4.2)$$

valid in $|\theta| < \pi$ by analytic continuation; see Olver (1997, p. 532).

2.4.1 Subdivision of the integration path

The integration path $[0, \infty)$ in the τ-plane is subdivided into intervals of length ω_n $(n = 0, 1, 2, \ldots)$, where $\omega_0 = \vartheta$, $0 < \vartheta \leq 1$. The midpoints of the intervals are

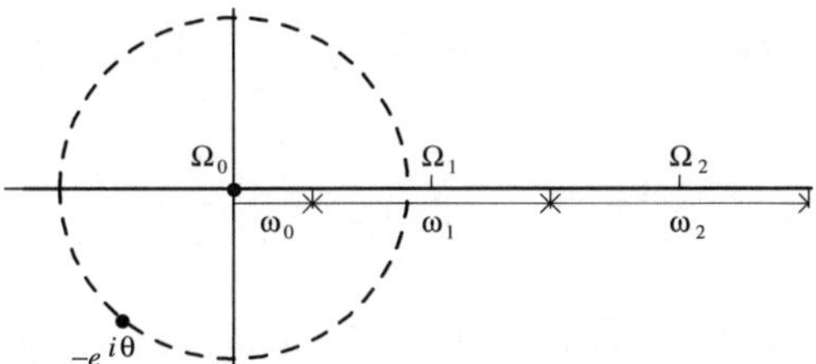

Figure 2.8 The subdivision of the integration path $[0, \infty)$ in (2.4.2). The heavy dots denote the branch points of $f(\tau)$ at $\tau = 0$ and $\tau = -e^{i\theta}$.

denoted by $\tau = \Omega_n$ $(n \geq 1)$ with $\Omega_0 = 0$; see Fig. 2.8. Then we have the simple geometric constraint

$$\Omega_n = \sum_{r=0}^{n-1} \omega_r + \tfrac{1}{2}\omega_n \qquad (n \geq 1). \tag{2.4.3}$$

We expand the amplitude function $f(\tau e^{-i\theta})$ in (2.4.2) about the points $\tau = \Omega_n$. For this reason, the points Ω_n will be called the *expansion points*. For the zeroth interval we use forward expansion, while for the intervals with $n \geq 1$ we use forward-reverse expansion; see §2.2.4. The domain of convergence about $\tau = 0$ is controlled by the singularity at $\tau = -e^{i\theta}$ and so is the unit disc. The intervals of convergence on the real τ-axis about $\tau = \Omega_n$ $(n \geq 1)$ are controlled by the singularity at $\tau = 0$ when $|\theta| \leq \tfrac{1}{2}\pi$; when $\tfrac{1}{2}\pi < |\theta| \leq \pi$, the second singularity at $\tau = -e^{i\theta}$ will play a role and will result in θ-dependent intervals. However, we shall confine our attention here to the sector $|\theta| \leq \tfrac{1}{2}\pi$, since evaluation of $U(a, a+b, z)$ outside this sector can be achieved by use of the continuation formula (Temme, 1996, p. 177)

$$e^{-z}U(a, a+b, z) = \frac{e^{\mp\pi i a}\Gamma(b)}{\Gamma(a+b)} {}_1F_1(b; a+b; ze^{\mp\pi i})$$
$$- e^{\mp\pi i(a+b)}\frac{\Gamma(b)}{\Gamma(a)} U(b, a+b, ze^{\mp\pi i}), \tag{2.4.4}$$

where either upper or lower signs are taken throughout. In this way we can work with intervals ω_n that are independent of θ. For the moment the Ω_n are free to be chosen; their selection will be made in §2.4.3.

2.4.2 The Hadamard expansion

The contribution from the zeroth interval $0 \leq \tau \leq \omega_0$ $(\omega_0 = \vartheta)$ proceeds as in §2.2.6. Upon expanding $f(\tau e^{-i\theta})$ about $\tau = 0$, we have

$$\frac{e^{-i\theta}}{\Gamma(a)}\int_0^{\omega_0} e^{-x\tau} f(\tau e^{-i\theta})\,d\tau = \frac{e^{-ia\theta}}{\Gamma(a)}\sum_{k=0}^{\infty}\binom{b-1}{k} e^{-ik\theta}\int_0^{\vartheta} e^{-x\tau}\tau^{k+a-1}d\tau$$

$$= z^{-a}\sum_{k=0}^{\infty}(-)^k\frac{(a)_k(1-b)_k}{k!\,z^k}\,P(k+a,\vartheta x). \quad (2.4.5)$$

To evaluate the contribution from the nth interval $-\frac{1}{2}\omega_n \le \tau - \Omega_n \le \frac{1}{2}\omega_n$ with $n \ge 1$, we require the expansion of the amplitude function $f(\tau e^{-i\theta})$ about the point $\tau = \Omega_n$ in the form

$$f((\Omega_n + u)e^{-i\theta}) = (\hat{\Omega}_n + ue^{-i\theta})^{a-1}(1 + \hat{\Omega}_n + ue^{-i\theta})^{b-1}, \quad \hat{\Omega}_n \equiv \Omega_n e^{-i\theta}$$

$$= \sum_{k=0}^{\infty}\frac{c_{k,n}(\theta)}{k!}(ue^{-i\theta})^k \qquad (|u| < \Omega_n;\ |\theta| \le \tfrac{1}{2}\pi). \quad (2.4.6)$$

A recurrence relation satisfied by the coefficients $c_{k,n}(\theta)$ is obtained by observing that

$$e^{i\theta}\frac{df}{du} = \left\{\frac{a-1}{\hat{\Omega}_n + ue^{-i\theta}} + \frac{b-1}{1+\hat{\Omega}_n + ue^{-i\theta}}\right\} f.$$

Substitution of the expansion (2.4.6) into this differential equation leads to the recurrence

$$c_{k+1,n}(\theta) = A_k c_{k,n}(\theta) + B_k c_{k-1,n}(\theta) \qquad (k \ge 0), \qquad (2.4.7)$$

where

$$A_k = \frac{a-1+(a+b-2)\hat{\Omega}_n - (1+2\hat{\Omega}_n)k}{\hat{\Omega}_n(1+\hat{\Omega}_n)},$$

$$B_k = \frac{(a+b-k-1)k}{\hat{\Omega}_n(1+\hat{\Omega}_n)},$$

subject to the initial values $c_{-1,n}(\theta) = 0$, $c_{0,n}(\theta) = f(\hat{\Omega}_n)$.

The contribution from the nth interval then takes the form

$$\frac{e^{-i\theta}}{\Gamma(a)}\int_{\Omega_n-\frac{1}{2}\omega_n}^{\Omega_n+\frac{1}{2}\omega_n} e^{-x\tau} f(\tau e^{-i\theta})\,d\tau = \frac{e^{-i\theta-\Omega_n x}}{\Gamma(a)}\int_{-\frac{1}{2}\omega_n}^{\frac{1}{2}\omega_n} e^{-xu} f((\Omega_n+u)e^{-i\theta})\,du$$

$$= \frac{e^{-i\theta-\Omega_n x}}{\Gamma(a)}\sum_{k=0}^{\infty}\frac{c_{k,n}(\theta)}{k!}e^{-ik\theta}\int_{-\frac{1}{2}\omega_n}^{\frac{1}{2}\omega_n} e^{-xu}u^k\,du$$

$$= \frac{e^{-\Omega_n x}}{\Gamma(a)}\,S_n(z;\theta), \qquad (2.4.8)$$

where

$$S_n(z;\theta) = \sum_{k=0}^{\infty}\frac{c_{k,n}(\theta)}{z^{k+1}}\Delta P(k+1,\tfrac{1}{2}\omega_n x)$$

and ΔP is defined in (2.2.26). Then, from (2.4.2), (2.4.5) and (2.4.8), we obtain the Hadamard expansion for $U(a, a + b, z)$ when $\mathrm{Re}(a) > 0$ in the form

$$U(a, a+b, z) = z^{-a} \sum_{k=0}^{\infty} \frac{(-)^k (a)_k (1-b)_k}{k!\, z^k} P(k+a, \vartheta x) + \frac{1}{\Gamma(a)} \sum_{n=1}^{\infty} e^{-\Omega_n x} S_n(z; \theta)$$

(2.4.9)

valid in $|\arg z| < \pi$. We repeat that the intervals ω_n $(n \geq 1)$ and expansion points Ω_n are free to be chosen. An obvious complication with path rotation is that the coefficients $c_{k,n}(\theta)$ depend on θ; a computational scheme that partly avoids this problem is discussed in §2.4.5.

It is clear that, since the integration interval in (2.4.2) is infinite, the resulting Hadamard expansion involves an infinite number of series each multiplied by an exponential factor $\exp(-\Omega_n x)$ of increasing subdominance. If the integration interval had been finite, the resulting Hadamard expansion would involve a finite number of series; the extension in this case is obvious. The derivations of the Hadamard expansions for both types of confluent hypergeometric functions have been obtained from a differential equation approach in Kaminski and Paris (2004).

2.4.3 Choice of the Ω_n

From (2.2.6) and Stirling's formula, the large-k behaviour of the terms in the zeroth contribution in (2.4.9) is

$$k^{-b-1} e^{-\vartheta x} \vartheta^k \qquad (k \to \infty),$$

so that with $\vartheta < 1$ we have a geometric rate of convergence. The large-k behaviour of the coefficients $c_{k,n}(\theta)$ in (2.4.6) can be found by a routine application of Cauchy's integral formula

$$c_{k,n}(\theta) = \frac{k!}{2\pi i} \oint f(\Omega_n + \zeta) \frac{d\zeta}{\zeta^{k+1}},$$

where the integration path is a small loop surrounding $\zeta = 0$. By expanding the contour to be a circle of arbitrarily large radius, it is readily seen that, provided a is not a positive integer, it is the branch point at $\zeta = -\Omega_n$ that dominates the large-k asymptotics. Evaluation of the resulting loop integral around $\zeta = -\Omega_n$ then yields

$$c_{k,n}(\theta) = k!\, O(k^{-a} \Omega_n^{-k}) \qquad (k \to \infty). \tag{2.4.10}$$

From (2.2.6) and (2.2.26), the terms in the Hadamard series $S_n(z; \theta)$ in (2.4.9) consequently possess the large-k behaviour

$$e^{\frac{1}{2}\omega_n x} O(k^{-a-1} \kappa_n^{-k}), \qquad \kappa_n = \frac{\omega_n}{2\Omega_n} \qquad (k \to \infty).$$

When $|\arg z| \leq \frac{1}{2}\pi$, we have seen that the intervals, and hence the geometric factors κ_n, can be chosen independently of $\arg z$. Accordingly, for $n \geq 1$ we set [12] $\kappa_n = \vartheta$, where $0 < \vartheta < 1$, so that the geometric factors corresponding to each interval are the same as that for the zeroth interval; that is, we set

$$\omega_n = 2\vartheta\,\Omega_n \qquad (n \geq 1). \tag{2.4.11}$$

From (2.4.3), the expansion points Ω_n then have the values

$$\Omega_1 = \frac{\vartheta}{1-\vartheta}, \qquad \Omega_n = \left(\frac{1+\vartheta}{1-\vartheta}\right)\Omega_{n-1} \quad (n \geq 2). \tag{2.4.12}$$

With this particular choice of the Ω_n, we are assured (when a is not a positive integer) that each constituent Hadamard series in (2.4.9) is associated with the same geometric convergence rate $\vartheta < 1$ when $|\arg z| \leq \frac{1}{2}\pi$. Furthermore, since convergence is without restriction on the parameters a and b, it follows that (2.4.9) holds for all finite values of a and b by analytic continuation. It is also an obvious consequence of our path decomposition in (2.4.2) that truncation of the sum over n in (2.4.9) after N terms yields an error of $O(e^{-\Omega_N x})$.

An alternative decomposition scheme of the τ-axis was employed in Paris (2001a, 2001b). This used the intervals ω_n and took the expansion point $\tau = \Omega_n$ to be *the left-hand endpoint* of the nth interval, so that

$$\Omega_n = \sum_{r=0}^{n-1} \omega_r \qquad (n \geq 1);$$

see Fig. 2.9. The values of the ω_n were chosen to be the radii of convergence of the series expansion of $f(\tau)$ about $\tau = 0$ and $\tau = \Omega_n$; that is, they were chosen to be the *maximal integration intervals*. In the case of $U(a, a+b, z)$ when $|\arg z| \leq \frac{1}{2}\pi$, this corresponds to the choice

$$\omega_0 = 1, \quad \omega_n = 2^{n-1}, \quad \Omega_n = 2^n \quad (n \geq 1).$$

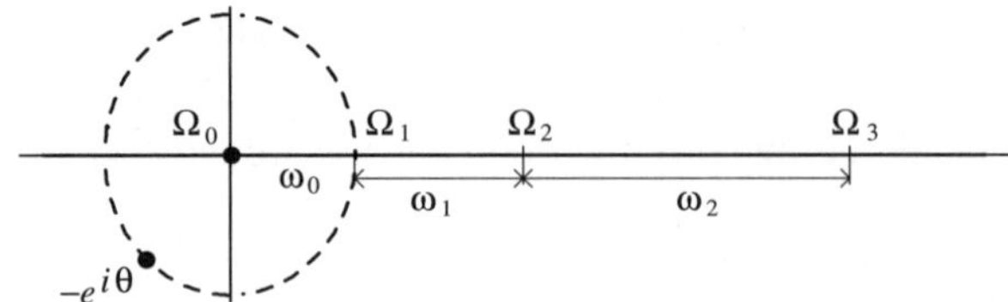

Figure 2.9 An alternative subdivision of the integration path $[0, \infty)$ in (2.4.2). The heavy dots denote the branch points of $f(\tau)$ and $\omega_0 = 1$.

---

[12] Other choices of the intervals ω_n are, of course, possible.

The integration used forward expansion over each interval to produce the expansion

$$U(a, a+b, z) = z^{-a} \sum_{k=0}^{\infty} \frac{(-)^k (a)_k (1-b)_k}{k! \, z^k} P(k+a, x) + \frac{1}{\Gamma(a)} \sum_{n=1}^{\infty} e^{-\Omega_n x} \mathcal{S}_n(z; \theta),$$

$$(2.4.13)$$

where

$$\mathcal{S}_n(z; \theta) = \sum_{k=0}^{\infty} \frac{c_{k,n}(\theta)}{z^{k+1}} P(k+1, \omega_n x).$$

The coefficients $c_{k,n}(\theta)$ in the above expansion are defined as in (2.4.6) and (2.4.7) but, because the points Ω_n are not the same, they will differ (for the same θ and parameters a and b) from those appearing in the expansion (2.4.8). The application of this form of expansion is discussed in Paris (2001b). Since the integration intervals are maximal, it is clear that each series in (2.4.13) has to be used in its modified form as discussed in §2.2.2, with the associated additional computational effort that this entails.

2.4.4 A numerical example

We consider the modified Bessel function $K_\nu(z)$ given by

$$e^z K_\nu(z) = \pi^{\frac{1}{2}} (2z)^\nu \, U(\nu + \tfrac{1}{2}, 2\nu + 1, 2z).$$

From (2.4.9) with $a = b = \nu + \tfrac{1}{2}$ and $z = xe^{i\theta}$, we find

$$e^z K_\nu(z) = \sqrt{\frac{\pi}{2z}} \left\{ \sum_{k=0}^{\infty} \frac{(-)^k a_k(\nu)}{(2z)^k} P(k + \nu + \tfrac{1}{2}, 2\vartheta x) \right.$$

$$\left. + \frac{(2z)^{\nu + \frac{1}{2}}}{\Gamma(\nu + \frac{1}{2})} \sum_{n=1}^{\infty} e^{-\Omega_n x} S_n(2z; \theta) \right\} \qquad (2.4.14)$$

valid for $\mathrm{Re}(\nu) > -\tfrac{1}{2}$, where

$$S_n(2z; \theta) = \sum_{k=0}^{\infty} \frac{c_{k,n}(\theta)}{(2z)^{k+1}} \Delta P(k + 1, \omega_n x)$$

and the coefficients $a_k(\nu)$ are defined in (2.1.2). The coefficients $c_{k,n}(\theta)$ are obtained from the recurrence (2.4.7) and we recall that ΔP is defined in (2.2.26).

For $|\arg z| \leq \tfrac{1}{2}\pi$, we can make the choices $\vartheta = \tfrac{1}{2}$ and $\vartheta = \tfrac{1}{3}$, for example, to find from (2.4.11) and (2.4.12) the expansion points Ω_n and intervals ω_n given by

$$\left. \begin{array}{ll} \vartheta = \tfrac{1}{2}: & \Omega_n = 3^{n-1}, \quad \omega_n = \Omega_n \\[2mm] \vartheta = \tfrac{1}{3}: & \Omega_n = 2^{n-2}, \quad \omega_n = \tfrac{2}{3}\Omega_n \end{array} \right\} \quad (n \geq 1). \qquad (2.4.15)$$

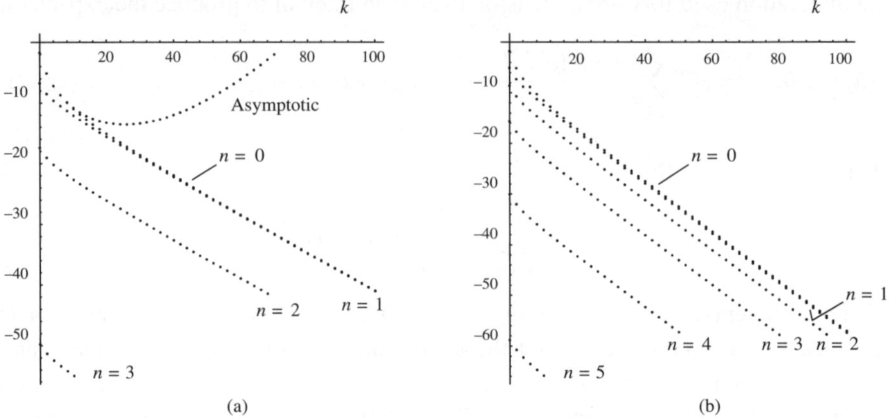

Figure 2.10 The behaviour of the terms (on a $\log_{10}$ scale) in the constituent series (labelled by n) in (2.4.14) for the Bessel function $e^z K_\nu(z)$ when $z = 15$, $\nu = \frac{3}{4}$: (a) $\vartheta = \frac{1}{2}$ and (b) $\vartheta = \frac{1}{3}$. The series labelled by $n = 0$ corresponds to the first series in (2.4.14).

Each Hadamard series in (2.4.14) is then associated with the same geometric convergence rate ϑ.

We present the results of computations for $e^z K_\nu(z)$ using (2.4.14) when $z = 15e^{i\theta}$, $\nu = \frac{3}{4}$ for different levels n. Figure 2.10 shows the decay of the terms in the various series in (2.4.14) against ordinal number k when $\vartheta = \frac{1}{2}$ and $\vartheta = \frac{1}{3}$. The well-known asymptotic expansion (Abramowitz and Stegun, 1965, p. 378)

$$e^z K_\nu(z) \sim \sqrt{\frac{\pi}{2z}} \sum_{k=0}^{\infty} \frac{(-)^k a_k(\nu)}{(2z)^k},$$

which results from the first series in (2.4.14) when the incomplete gamma functions $P(k + \nu + \frac{1}{2}, 2\vartheta x)$ are formally replaced by unity, is also shown for comparison. It is clear that for the smaller value of ϑ the terms decay more rapidly (like 3^{-k} as opposed to 2^{-k}), but that the separation between the different exponential levels is less. In Table 2.4 we show the magnitude of the error as a function of $\theta = \arg z$ in the range $0 \leq \theta \leq \frac{1}{2}\pi$. It is seen that with path rotation the magnitude of the error at fixed truncation indices and number of exponential levels n is roughly independent of θ.

Evaluation of $K_\nu(z)$ in the sectors $\frac{1}{2}\pi < |\theta| \leq \pi$ can be accomplished by means of the continuation formula given in Abramowitz and Stegun (1965, p. 376)

$$K_\nu(z) = e^{\pm\pi i\nu} K_\nu(ze^{\mp\pi i}) \mp \pi i I_\nu(ze^{\mp\pi i});$$

compare (2.4.4). Thus, for example, computation near the Stokes lines $\theta = \pm\pi$ can be reduced to the computation of $K_\nu(z)$ and $I_\nu(z)$ near $\theta = 0$.

Table 2.4 *Values of the absolute error in the computation
of $e^z K_v(z)$ when $z = 15e^{i\theta}$ and $v = \frac{3}{4}$ for different
values of θ using (2.4.14) with $\vartheta = \frac{1}{2}, \frac{1}{3}$*

θ/π	$\vartheta = \frac{1}{2} \ (n \leq 2)^a$	$\vartheta = \frac{1}{3} \ (n \leq 4)$
0	4.335×10^{-43}	2.619×10^{-60}
0.125	8.286×10^{-43}	6.720×10^{-59}
0.250	9.752×10^{-43}	8.164×10^{-59}
0.375	5.250×10^{-43}	2.961×10^{-59}
0.500	6.435×10^{-43}	5.645×10^{-59}

[a] The number of levels employed is indicated and the zeroth series is
truncated after $k = 100$ terms.

2.4.5 A computational simplification

With the above scheme, the coefficients $c_{k,n}(\theta)$ have to be computed for each value
of θ, which clearly represents an inconvenience. An alternative procedure which
partly avoids this and which incurs only a modest loss of precision, is to rotate the
integration path in (2.4.1) by the acute angle ψ to obtain

$$U(a, a + b, z) = \frac{e^{-i\psi}}{\Gamma(a)} \int_0^\infty e^{-z'\tau} f(\tau e^{-i\psi}) \, d\tau, \qquad z' = xe^{i(\theta - \psi)}$$

valid in $|\theta - \psi| \leq \frac{1}{2}\pi - \delta, \delta > 0$. The expansion procedure described in §2.4.3 then
yields, when $\text{Re}(a) > 0$,

$$U(a, a + b, z) = z^{-a} \sum_{k=0}^{\infty} \frac{(-)^k (a)_k (1 - b)_k}{k! \, z^k} P(k + a, \vartheta z')$$

$$+ \frac{1}{\Gamma(a)} \sum_{n=1}^{\infty} e^{-\Omega_n z'} S_n(z'; \psi), \qquad (2.4.16)$$

where

$$S_n(z'; \psi) = \sum_{k=0}^{\infty} \frac{c_{k,n}(\psi)}{z^{k+1}} \Delta P(k + 1, \tfrac{1}{2}\omega_n z').$$

This expansion reduces to (2.4.9) when $\psi = \theta$; when $\psi \neq \theta$, the incomplete gamma
functions now have complex argument z'.

The sector $0 \leq \theta \leq \frac{1}{2}\pi$ can then be divided into two halves. In the sector
$0 \leq \theta \leq \frac{1}{4}\pi$, we can for example set $\psi = \frac{1}{8}\pi$, and compute the corresponding
coefficients $c_{k,n}(\frac{1}{8}\pi)$ by means of the recurrence (2.4.7). Then, for arg z in this sec-
tor, we can employ (2.4.16) with the coefficients $c_{k,n}(\frac{1}{8}\pi)$. For $|\theta - \psi| \leq \frac{1}{8}\pi$, the
loss of precision incurred by this simplification is found to be slight. Similarly, in
$\frac{1}{4}\pi \leq \theta \leq \frac{1}{2}\pi$, we can take $\psi = \frac{3}{8}\pi$ and use (2.4.16) with the coefficients $c_{k,n}(\frac{3}{8}\pi)$.

Table 2.5 *Values of the absolute error in the computation of $e^z K_\nu(z)$ when $z = 15e^{i\theta}$ and $\nu = \frac{3}{4}$ for different values of θ using (2.4.16) with convergence factors $\vartheta = \frac{1}{2}, \frac{1}{3}$*

θ/π	$\psi = \frac{1}{8}\pi$	θ/π	$\psi = \frac{3}{8}\pi$
$\vartheta = \frac{1}{2}$ $(n \le 2)^a$			
0	2.685×10^{-42}	0.250	1.568×10^{-42}
0.125	8.286×10^{-43}	0.375	5.250×10^{-43}
0.250	2.523×10^{-42}	0.500	1.773×10^{-42}
$\vartheta = \frac{1}{3}$ $(n \le 4)$			
0	1.464×10^{-58}	0.250	5.787×10^{-59}
0.125	6.720×10^{-59}	0.375	2.961×10^{-59}
0.250	1.403×10^{-58}	0.500	6.791×10^{-59}

[a] The number of levels employed is indicated and the zeroth series is truncated after $k = 100$ terms.

The results of this modification to the procedure applied to $e^z K_\nu(z)$ are displayed in Table 2.5, where it is seen that the absolute error varies only slightly in the sectors $|\theta - \psi| \le \frac{1}{8}\pi$. If, however, we try to compute beyond these sectors the accuracy begins to deteriorate. For example, with $\psi = \frac{1}{8}\pi$ and $\vartheta = \frac{1}{3}$, the precision in the computation of $e^z K_\nu(z)$ drops 7 orders of magnitude when $\theta = \frac{3}{8}\pi$.

2.5 Examples

In this section we present examples of the Hadamard expansion of functions that are not directly related to either of the two confluent hypergeometric functions discussed in §§2.2.6 and 2.4.

Example 2.4 Consider the Fourier integral

$$I(x) = \int_0^1 e^{ixt^3} dt,$$

where x is a positive variable. The integration path can be deformed to pass down the valley at infinity along the steepest descent path $\arg t = \frac{1}{6}\pi$ from the origin and back along the steepest ascent path through $t = 1$; see Example 1.5 with $n = 3$ and also Erdélyi (1956, p. 45) for details. The integral along the ray $\arg t = \frac{1}{6}\pi$ can be evaluated in terms of a gamma function. Hence

$$I(x) = e^{\pi i/6} \frac{\Gamma(\frac{4}{3})}{x^{1/3}} - \frac{1}{3} i e^{ix} \int_0^\infty e^{-xt}(1 + it)^{-2/3} dt, \qquad (2.5.1)$$

where on the path from $\infty e^{\pi i/6}$ to $t = 1$ we have made the change of variable $t \mapsto (1 + it)^{1/3}$. This transformation has expressed the integral in the form of a Laplace integral in which the amplitude function $f(t) = (1 + it)^{-2/3}$ has a branch-point singularity at $t = i$.

From (2.4.2), the integral appearing in (2.5.1) can be written as a multiple of the confluent hypergeometric function $U(1, \frac{4}{3}, xe^{-\frac{1}{2}\pi i})$. The integration path $[0, \infty)$ is subdivided into intervals of length $\omega_0 = \vartheta \leq 1$ and ω_n ($n \geq 1$), with $\Omega_0 = 0$ and their midpoints at $t = \Omega_n$, as illustrated in Fig. 2.8. Let ρ_n be the length of the line joining the point $t = \Omega_n$ and the singularity $t = i$ and ψ_n be the acute angle this line makes with the positive t-axis; that is

$$1 + i\Omega_n = i\rho_n e^{-i\psi_n} \qquad (\rho_0 = 1, \ \psi_0 = \tfrac{1}{2}\pi).$$

Then

$$(1 + i\Omega_n + iu)^{-2/3} = (1 + i\Omega_n)^{-2/3} \left(1 + \frac{ue^{i\psi_n}}{\rho_n}\right)^{-2/3}$$

$$= (1 + i\Omega_n)^{-2/3} \sum_{k=0}^{\infty} (-)^k (\tfrac{2}{3})_k \left(\frac{ue^{i\psi_n}}{\rho_n}\right)^k,$$

when $|u| < \rho_n$, so that the coefficients $c_{k,n}(-\frac{1}{2}\pi)$ in (2.4.6) are given by

$$c_{k,n}(-\tfrac{1}{2}\pi) = (1 + i\Omega_n)^{-2/3} \frac{d_{k,n}}{\rho_n^k}, \qquad d_{k,n} = (-)^k (\tfrac{2}{3})_k k! e^{ik\psi_n}.$$

From (2.4.9), the Hadamard expansion for the integral $I(x)$ takes the form

$$I(x) = e^{\pi i/6} \frac{\Gamma(\frac{4}{3})}{x^{1/3}} - \frac{1}{3} e^{ix+\pi i/6} \sum_{n=0}^{\infty} e^{-\Omega_n x} S_n(x), \qquad (2.5.2)$$

where

$$S_0(x) = e^{\pi i/3} \sum_{k=0}^{\infty} \frac{d_{k,0}}{x^{k+1}} P(k + 1, \vartheta x)$$

and

$$S_n(x) = \rho_n^{1/3} e^{2i\psi_n/3} \sum_{k=0}^{\infty} \frac{d_{k,n}}{(\rho_n x)^{k+1}} \Delta P(k + 1, \tfrac{1}{2}\omega_n x)$$

for $n \geq 1$, and ΔP is defined in (2.2.26). The asymptotic expansion of $I(x)$ for large x can be recovered by taking the Hadamard series with $n = 0$ and formally replacing the incomplete gamma function $P(k + 1, \vartheta x)$ by unity to yield

$$I(x) \sim e^{\pi i/6} \frac{\Gamma(\frac{4}{3})}{x^{1/3}} + \frac{1}{3} e^{ix} \sum_{k=0}^{\infty} \frac{(\frac{2}{3})_k}{(ix)^{k+1}} \qquad (x \to +\infty),$$

as given in Erdélyi (1956, p. 46); see also (1.3.25).

From (2.2.6), the late terms ($k \gg 1$) in the zeroth Hadamard series $S_0(x)$ are associated with the geometric decay factor ϑ. The values of the expansion points Ω_n are controlled by the singularity at $t = i$ and consequently the radius of convergence about $t = \Omega_n$ is $\rho_n = (1 + \Omega_n^2)^{1/2}$. If we require each Hadamard series $S_n(x)$ with $n \geq 1$ also to have the same geometric decay factor, we must choose

$$\frac{\omega_n}{2\rho_n} = \frac{\omega_n}{2\sqrt{1 + \Omega_n^2}} = \vartheta.$$

If we denote the juncture between the adjacent intervals associated with the expansion points Ω_{n-1} and Ω_n by $t = \lambda_n$, we have from (2.4.3)

$$\lambda_1 = \vartheta, \qquad \lambda_n = \Omega_{n-1} + \tfrac{1}{2}\omega_{n-1} = \Omega_n - \tfrac{1}{2}\omega_n \quad (n \geq 2).$$

Then some straightforward algebra shows that

$$\Omega_n = \frac{\lambda_n + \vartheta(\lambda_n^2 + 1 - \vartheta^2)^{1/2}}{1 - \vartheta^2} \quad (n \geq 1).$$

However, it is not necessary to adhere rigidly to this specification of the expansion points Ω_n. If, for $n \geq 1$, we set instead

$$\frac{\omega_n}{2\Omega_n} = \vartheta,$$

so that the ω_n and Ω_n are as given in (2.4.11) and (2.4.12), then the zeroth series $S_0(x)$ is again associated with geometric decay factor ϑ, while the $S_n(x)$ with $n \geq 1$ are associated with the decay factor

$$\frac{\omega_n}{2\rho_n} = \frac{\vartheta\Omega_n}{\sqrt{1 + \Omega_n^2}} < \vartheta.$$

Consequently, with this second and simpler choice of Ω_n the expansion (2.5.2) can still be said to be associated with the geometric decay factor ϑ.

Example 2.5 The magnetoplasma dispersion function is defined by the integral

$$F(x, \omega) = \frac{1}{\pi} \int_0^\pi e^{-x(1-\cos t)} \cos \omega t \, dt \qquad (x > 0) \tag{2.5.3}$$

which, although closely related to the Bessel function family, has properties that differ significantly from those of any combination of Bessel and associated functions. The function $F(x, \omega)$ is seen to be related to the first integral appearing in the definition of the modified Bessel function $I_\omega(x)$ given in (C.1) in Appendix C, and that when $\omega = m$, an integer, $F(x, m) = e^{-x} I_m(x)$.

Other forms of $F(x, \omega)$ which clearly indicate these differences can be obtained by using the standard result $e^{x \cos t} = \sum_{n=-\infty}^{\infty} I_n(x) \cos nt$, followed by termwise integration, to obtain

$$F(x, \omega) = e^{-x} \frac{\sin \pi\omega}{\pi} \sum_{n=-\infty}^{\infty} \frac{(-)^n I_n(x)}{\omega - n}.$$

Alternatively, a factor e^{-2x} can be extracted from the integrand in (2.5.3) and the resulting exponential can be written as $\exp\{x(1+\cos t)\}$. Expansion of this latter factor then produces

$$
\begin{aligned}
F(x, \omega) &= \frac{e^{-2x}}{\pi} \int_0^\pi e^{x(1+\cos t)} \cos \omega t \, dt \\
&= \frac{e^{-2x}}{\pi} \sum_{n=0}^\infty \frac{(2x)^n}{n!} \int_0^\pi \cos \omega t \, (\cos \tfrac{1}{2}t)^{2n} \, dt \\
&= e^{-2x} \frac{\sin \pi \omega}{\omega} \, {}_2F_2(\tfrac{1}{2}, 1; 1 - \omega, 1 + \omega; 2x)
\end{aligned}
$$

upon use of the integral

$$
\int_0^\pi \cos \omega t \, (\cos \tfrac{1}{2}t)^{2n} \, dt = \frac{\sin \pi \omega}{\omega} \frac{n! \, (\tfrac{1}{2})_n}{(1 - \omega)_n (1 + \omega)_n}
$$

for non-integer ω and $n = 1, 2, \ldots$; see Gradshteyn and Ryzhik (1980, p. 142, 2.538(1)).

To determine the Hadamard series representation of $F(x, \omega)$ we make use of the series expansions (Hobson, 1925, p. 276)

$$
\frac{\cos \omega t}{\cos \tfrac{1}{2}t} = \sqrt{\pi} \sum_{k=0}^\infty \frac{a_k(\omega)}{\Gamma(k + \tfrac{1}{2})} u^{2k}, \qquad \frac{\sin \omega t}{\cos \tfrac{1}{2}t} = \omega \sqrt{\pi} \sum_{k=0}^\infty \frac{b_k(\omega)}{k!} u^{2k+1}, \quad (2.5.4)
$$

valid in $|t| < \pi$, where $u = \sin \tfrac{1}{2}t$ and the coefficients

$$
a_k(\omega) = \frac{(\tfrac{1}{2} - \omega)_k (\tfrac{1}{2} + \omega)_k}{k!}, \qquad b_k(\omega) = \frac{(1 - \omega)_k (1 + \omega)_k}{\Gamma(k + \tfrac{3}{2})}.
$$

Then, with the change of variable $u = \sin \tfrac{1}{2}t$ in (2.5.3), we obtain

$$
\begin{aligned}
F(x, \omega) &= \frac{2}{\pi} \int_0^1 \frac{\cos \omega t}{\cos \tfrac{1}{2}t} e^{-2xu^2} \, du \\
&= \frac{1}{\sqrt{2\pi x}} \sum_{k=0}^\infty \frac{a_k(\omega)}{\Gamma(k + \tfrac{1}{2})(2x)^k} \int_0^{2x} w^{k - \tfrac{1}{2}} e^{-w} \, dw \\
&= \frac{1}{\sqrt{2\pi x}} \sum_{k=0}^\infty \frac{a_k(\omega)}{(2x)^k} P(k + \tfrac{1}{2}, 2x) \qquad (2.5.5)
\end{aligned}
$$

as obtained in Schmitt (1974). This is a single-stage expansion (see §2.2.4) which has late terms controlled by the slow algebraic decay $e^{-2x} k^{-3/2}$; see (2.2.6). Computation of the series in this form would require use of the modified expansion described in §2.2.2.

A 2-stage Hadamard expansion for $F(x, \omega)$, associated with a more rapid geometric decay rate, can be obtained as explained in §2.2.5 for the modified Bessel function. We have

$$F(x, \omega) = \frac{1}{\pi} \int_0^{\pi/2} e^{-x(1-\cos t)} \cos \omega t \, dt + \frac{e^{-2x}}{\pi} \int_0^{\pi/2} e^{x(1-\cos t)} \cos \omega(\pi - t) \, dt.$$

The first integral is evaluated as in (2.5.5), with the argument in the incomplete gamma function changed from $2x$ to x. With the substitution $u = \sin \frac{1}{2} t$ and use of the expansions in (2.5.4), the second integral (without the factor e^{-2x}) becomes

$$\frac{2}{\pi} \int_0^{2^{-1/2}} e^{2xu^2} \frac{\cos \omega(\pi - t)}{\cos \frac{1}{2} t} \, du = \frac{S(x, \omega)}{\sqrt{2\pi x}},$$

where, with the help of (2.2.23),

$$S(x, \omega) = \cos \pi \omega \sum_{k=0}^{\infty} \frac{a_k(\omega)(2x)^{-k}}{\Gamma(k + \frac{1}{2})} \int_0^x w^{k-\frac{1}{2}} e^w \, dw$$

$$+ \omega \sin \pi \omega \sum_{k=0}^{\infty} \frac{b_k(\omega)(2x)^{-k-\frac{1}{2}}}{\Gamma(k + 1)} \int_0^x w^k e^w \, dw$$

$$= \cos \pi \omega \sum_{k=0}^{\infty} \frac{a_k(\omega)}{(2x)^k} P^*(k + \tfrac{1}{2}, -x)$$

$$- \omega \sin \pi \omega \sum_{k=0}^{\infty} \frac{(-)^k b_k(\omega)}{(2x)^{k+\frac{1}{2}}} P(k + 1, -x)$$

and $P^*(a, -x) := e^{-\pi i a} P(a, -x)$ when $x > 0$; see (A.4).

The 2-stage expansion of $F(x, \omega)$ is then given by

$$F(x, \omega) = \frac{1}{\sqrt{2\pi x}} \left\{ \sum_{k=0}^{\infty} \frac{a_k(\omega)}{(2x)^k} P(k + \tfrac{1}{2}, x) + e^{-2x} S(x, \omega) \right\}. \tag{2.5.6}$$

Each Hadamard series in (2.5.6) now has late terms associated with the geometric decay factor 2^{-k}, so that use of their associated modified forms is not necessary; compare §2.2.4. The extension of the expansions (2.5.5) and (2.5.6) to complex values of x is straightforward.

Example 2.6 Consider the integral

$$I(x; \phi) = \int_0^{\infty} \frac{e^{-xt} \, dt}{(t + e^{i\phi})(t + e^{-i\phi})}, \tag{2.5.7}$$

where $x > 0$ and $\phi \geq 0$. The integrand has two simple poles at $t = -e^{\pm i\phi}$ (when $\phi > 0$), which become a double pole at $t = -1$ when $\phi = 0$. The presentation given here is based on Kaminski and Paris (2003).

Straightforward decomposition of the amplitude factor yields

$$I(x; \phi) = \frac{i}{2 \sin \phi} \int_0^{\infty} e^{-xt} \left\{ \frac{1}{t + e^{i\phi}} - \frac{1}{t + e^{-i\phi}} \right\} dt,$$

where each integral can be expressed in terms of the exponential integral $E_1(xe^{\pm i\phi})$ since

$$\int_0^\infty \frac{e^{-xt}}{t+e^{i\phi}}\,dt = e^{xe^{i\phi}}\int_{xe^{i\phi}}^\infty e^{-\tau}\tau^{-1}d\tau = e^{xe^{i\phi}}E_1(xe^{i\phi}).$$

From Abramowitz and Stegun (1965, p. 510), we have $e^z E_1(z) = U(1,1,z)$, where $U(a,b,z)$ is the confluent hypergeometric function of the second kind. It then follows that $I(x;\phi)$ can be written as

$$I(x;\phi) = \frac{i}{2\sin\phi}\{U(1,1,xe^{i\phi}) - U(1,1,xe^{-i\phi})\}. \tag{2.5.8}$$

From (2.4.9) with $z = xe^{i\phi}$, we obtain the Hadamard expansion [13]

$$U(1,1,z) = \sum_{k=0}^\infty \frac{(-)^k k!}{z^{k+1}} P(k+1,\vartheta x) + \sum_{n=1}^\infty e^{-\Omega_n x} S_n(z;\phi),$$

where $0 < \vartheta < 1$ and

$$S_n(z;\phi) = \sum_{k=0}^\infty \frac{c_{k,n}(\phi)}{z^{k+1}} \Delta P(k+1,\tfrac{1}{2}\omega_n x).$$

The coefficients $c_{k,n}(\phi)$ are obtained from (2.4.6) as

$$(1 + (\Omega_n + u)e^{\mp i\phi})^{-1} = \sum_{k=0}^\infty \frac{c_{k,n}(\pm\phi)}{k!}(ue^{\mp i\phi})^k \qquad (|u| < \rho_n)$$

to yield

$$c_{k,n}(\pm\phi) = \frac{(-)^k k!}{\rho_n^{k+1}} e^{\pm i(k+1)(\phi-\psi_n)},$$

where we have set $e^{\pm i\phi} + \Omega_n = \rho_n e^{\pm i\psi_n}$ ($\rho_n > 0$). The quantity ρ_n is the length of the line joining the expansion point $t = \Omega_n$ to the poles at $t = -e^{\pm i\phi}$ and ψ_n is the acute angle this line makes with the real axis; see Fig. 2.11 which illustrates the case of the pole at $t = -e^{-i\phi}$. Since $\Omega_0 = 0$, we must have $\rho_0 = 1$ and $\psi_0 = \phi$.

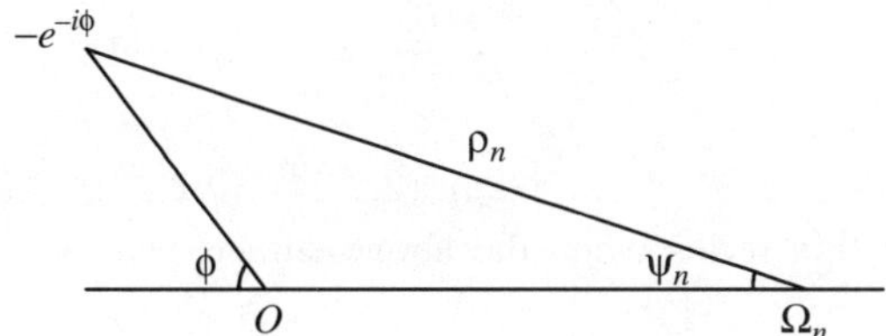

Figure 2.11 The quantities ρ_n and ψ_n for the integral $I(x;\phi)$.

---

[13] We remark that an expansion for $E_1(z) = e^{-z}U(1,1,z)$ in the form $E_1(z) + \gamma + \ln z = \sum_{n=1}^\infty n^{-1} P(n,z)$ has been studied in Gautschi, Harris and Temme (2003).

It then follows from (2.5.8) that

$$I(x; \phi) = \sum_{k=0}^{\infty} \frac{(-)^k k!}{x^{k+1}} \frac{\sin(k+1)\phi}{\sin\phi} P(k+1, \vartheta x) + \sum_{n=1}^{\infty} e^{-\Omega_n x} \mathcal{S}_n(x; \phi),$$

where

$$\mathcal{S}_n(x; \phi) = \frac{S_n(z; -\phi) - S_n(z; \phi)}{2i\sin\phi}$$

$$= \sum_{k=0}^{\infty} \frac{(-)^k k!}{(\rho_n x)^{k+1}} \frac{\sin(k+1)\psi_n}{\sin\phi} \Delta P(k+1, \tfrac{1}{2}\omega_n x) \quad (n \geq 1).$$

The intervals ω_n and expansion points Ω_n can be chosen as in §2.4.3 when $\phi \leq \tfrac{1}{2}\pi$.

The asymptotic expansion of $I(x; \phi)$ is obtained by retaining only the zeroth Hadamard series and formally replacing the incomplete gamma functions by unity to find

$$I(x; \phi) \sim \sum_{k=0}^{\infty} \frac{(-)^k k!}{x^{k+1}} \frac{\sin(k+1)\phi}{\sin\phi} \qquad (x \to +\infty).$$

When $\phi = 0$ the limiting form of the ratio of sines is taken. A discussion of the integral $I(x; \phi)$ when $\phi > \tfrac{1}{2}\pi$ and, in particular, when $\phi \to \pi$, where the poles straddle the integration path, has been given in Kaminski and Paris (2003); see also Paris (2007b).

2.6 Bounds on the tails of Hadamard series

In this final section we derive error bounds on the tails of the Hadamard series encountered in §§2.2 and 2.4 when these series are truncated after M terms, where M is chosen to be greater than the optimal trunction index for the associated Poincaré asymptotic series. Similar bounds have been given in Borwein, Borwein and Chan (2008). Hadamard series of the type

$$S(z) = \sum_{k=0}^{\infty} \frac{(1-\alpha)_k (\beta)_k}{k! \, z^{k+\beta}} P(k+\beta, \pm\vartheta z),$$

with $\mathrm{Re}(z) \geq 0$ and $0 < \vartheta < 1$, arise in (2.2.24), (2.2.27), (2.2.40), (2.2.41) and (2.4.9). The tail of this series when the above sum is truncated after M terms in computations is

$$T_M^{\pm}(z) = \sum_{k=M}^{\infty} \frac{(1-\alpha)_k (\beta)_k}{k! \, z^{k+\beta}} P(k+\beta, \pm\vartheta z). \tag{2.6.1}$$

From (2.1.4), we can write the tails in the form

$$T_M^{\pm}(z) = \frac{1}{\Gamma(\beta)} \sum_{k=M}^{\infty} \frac{(1-\alpha)_k}{k!\, z^{k+\beta}} \int_0^{\pm\vartheta z} e^{-w} w^{k+\beta-1}\, dw$$

$$= \frac{1}{\Gamma(\beta)} \sum_{k=M}^{\infty} \frac{(1-\alpha)_k}{k!}\, I_k^{\pm}(z),$$

where

$$I_k^{\pm}(z) = \int_0^{\pm\vartheta} e^{-uz} u^{k+\beta-1}\, du.$$

Since the truncation index M has to be chosen to scale like $|z|$ for the Hadamard series to yield greater accuracy than the associated asymptotic series (obtained by replacing the incomplete gamma function P by unity), we can assume M chosen sufficiently large such that $k - \text{Re}(\alpha) > 0$ and $k + \text{Re}(\beta) > 0$ in (2.6.1). Then, use of the inequality $|\Gamma(x + iy)| \le \Gamma(x)$ for $x > 0$ and real y shows that

$$|T_M^{\pm}(z)| \le \frac{1}{|\Gamma(1-\alpha)\Gamma(\beta)|} \sum_{k=M}^{\infty} \frac{\Gamma(1-\alpha_r+k)}{k!} |I_k^{\pm}(z)|,$$

where $\alpha_r \equiv \text{Re}(\alpha)$ and similarly for β.

In the integral $I_k^{+}(z)$, the modulus of the integrand $|e^{-uz} u^{k+\beta-1}|$ has a maximum at $u_0 = (k + \beta_r - 1)/\text{Re}(z)$. For M sufficiently large such that $u_0 \ge \vartheta$, the maximum will lie outside the integration interval with the result that the modulus of the integrand is monotonically increasing in $[0, \vartheta]$. A similar argument applied to $I_k^{-}(z)$ (with u replaced by $-u$) shows that the modulus of the integrand is also monotonically increasing in $[0, \vartheta]$. Hence

$$|I_k^{\pm}(z)| \le \int_0^{\vartheta} e^{\mp u\,\text{Re}(z)} u^{k+\beta_r-1}\, du \le e^{\mp\vartheta\,\text{Re}(z)} \vartheta^{k+\beta_r}$$

when $\text{Re}(z) \ge 0$. It then follows that

$$|T_M^{\pm}(z)| \le \frac{\vartheta^{\beta_r} e^{\mp\vartheta\,\text{Re}(z)}}{|\Gamma(1-\alpha)\Gamma(\beta)|} \sum_{k=M}^{\infty} \frac{\Gamma(1-\alpha_r+k)}{k!} \vartheta^k$$

$$= A_M \vartheta^{M+\beta_r} e^{\mp\vartheta\,\text{Re}(z)} \sum_{k=0}^{\infty} \frac{(1-\alpha_r+M)_k}{(M+1)_k} \vartheta^k, \tag{2.6.2}$$

where

$$A_M = \frac{\Gamma(1-\alpha_r+M)}{|\Gamma(1-\alpha)\Gamma(\beta)|\, M!}.$$

The sum in (2.6.2) can be expressed as the Gauss hypergeometric function

$${}_2F_1\left(\begin{array}{c} 1, 1-\alpha_r+M; \\ M+1; \end{array} \vartheta\right) = \frac{1}{1-\vartheta}\, {}_2F_1\left(1, \alpha_r; M+1; \frac{\vartheta}{\vartheta-1}\right)$$

Table 2.6 *Comparison of computed values of $T_M^{\pm}(z)$ with the bounds in (2.6.3) for different z when $\vartheta = \frac{1}{2}$, $\alpha = \beta = \frac{3}{4}$ and $M = 25$*

z	$T_M^+(z)^a$	Bound	z	$\lvert T_M^+(z)\rvert$	Bound
10	2.133(−13)	4.659(−12)	$20e^{\pi i/8}$	5.701(−15)	6.721(−14)
15	1.964(−14)	3.825(−13)	$20e^{\pi i/4}$	2.815(−14)	5.873(−13)
20	1.831(−15)	3.139(−14)	$20e^{3\pi i/8}$	6.150(−13)	1.506(−11)
25	1.735(−16)	2.577(−15)	$20e^{\pi i/2}$	2.448(−11)	6.915(−10)

z	$\lvert T_M^-(z)\rvert$	Bound	z	$\lvert T_M^-(z)\rvert$	Bound
10	1.173(−9)	1.026(−7)	$20e^{\pi i/8}$	7.115(−8)	7.115(−6)
15	1.323(−8)	1.250(−6)	$20e^{\pi i/4}$	8.541(−9)	8.142(−7)
20	1.499(−7)	1.523(−5)	$20e^{3\pi i/8}$	3.612(−10)	3.175(−8)
25	1.707(−6)	1.856(−4)	$20e^{\pi i/2}$	8.832(−12)	6.915(−10)

^{a} We have adopted the notation $x(-y)$ for $x \times 10^{-y}$.

by means of the standard transformation in Abramowitz and Stegun (1965, Eq. (15.3.4)). Consequently we obtain the bound valid for $\mathrm{Re}(z) \geq 0$

$$|T_M^{\pm}(z)| \leq \frac{A_M}{1 - \vartheta}\, \vartheta^{M+\beta_r} e^{\mp \vartheta \mathrm{Re}(z)} \,{}_2F_1\left(1,\ \alpha_r;\ M+1;\ \frac{\vartheta}{\vartheta - 1}\right). \tag{2.6.3}$$

It is evident that, when $\mathrm{Re}(\alpha) \geq 0$, the sum in (2.6.2) is bounded by $(1 - \vartheta)^{-1}$, so that the hypergeometric function in (2.6.3) can be replaced by unity in this case.[14] To demonstrate the effectiveness of the bounds, in Table 2.6 we present numerical results comparing the value of $T_M^{\pm}(z)$ computed from (2.6.1) with the bounds in (2.6.3) for different values of z when $\vartheta = \frac{1}{2}$, $\alpha = \beta = \frac{3}{4}$ and $M = 25$.

The other type of truncated Hadamard series in (2.4.9) has a tail of the form

$$T_{M,n}(z) = \sum_{k=M}^{\infty} \frac{c_{k,n}(\theta)}{x^{k+1}}\, \Delta P(k+1, \tfrac{1}{2}\omega_n x),$$

where $x = |z|$ and the coefficients $c_{k,n}(\theta)$, defined in (2.4.7), depend on the parameters a and b appearing in $U(a, a+b, z)$. Then, when $\mathrm{Re}(z) \geq 0$, we have with $\hat{c}_{k,n}(\theta) = c_{k,n}(\theta)/k!$

$$|T_{M,n}(z)| \leq \sum_{k=M}^{\infty} |\hat{c}_{k,n}(\theta)| \left| \left\{ \int_0^{\frac{1}{2}\omega_n} - \int_0^{-\frac{1}{2}\omega_n} \right\} e^{-ux} u^k \, du \right|$$

$$\leq (1 + e^{\frac{1}{2}\omega_n x}) \sum_{k=M}^{\infty} |\hat{c}_{k,n}(\theta)| (\tfrac{1}{2}\omega_n)^{k+1}.$$

[14] This also readily follows from the Euler integral definition of ${}_2F_1$ given in Abramowitz and Stegun (1965, Eq. (15.3.1)).

From (2.4.10) and M sufficiently large, there exists a computable constant $C_n(\theta)$ such that

$$|\hat{c}_{k,n}(\theta)| \leq C_n(\theta) k^{-a_r} \Omega_n^{-k-1} \qquad (k \geq M).$$

If we choose the intervals ω_n as in (2.4.11) such that $\vartheta = \omega_n/(2\Omega_n)$ for $n \geq 1$, where $0 < \vartheta < 1$, we then obtain the bound

$$\frac{|T_{M,n}(z)|}{1 + e^{\frac{1}{2}\omega_n x}} \leq C_n(\theta) \sum_{k=M}^{\infty} k^{-a_r} \vartheta^{k+1} = C_n(\theta)\, \vartheta^{M+1} \sum_{k=0}^{\infty} \frac{\vartheta^k}{(k+M)^{a_r}}$$

$$= C_n(\theta)\, \vartheta^{M+1}\, \Phi(\vartheta, a_r, M), \qquad (2.6.4)$$

where

$$\Phi(z, s, a) = \sum_{k=0}^{\infty} \frac{z^k}{(k+a)^s}$$

is Lerch's transcendent (Erdélyi, 1953, p. 27).

If $\mathrm{Re}(a) \geq 0$, the second sum over k in (2.6.4) possesses the simple bound $(1-\vartheta)^{-1}M^{-a_r}$, so that

$$\frac{|T_{M,n}(z)|}{1 + e^{\frac{1}{2}\omega_n x}} \leq C_n(\theta)\, \frac{\vartheta^{M+1}}{1-\vartheta}\, M^{-a_r} \qquad (a_r \equiv \mathrm{Re}(a) \geq 0). \qquad (2.6.5)$$

3

Hadamard expansion of Laplace-type integrals

3.1 Introduction

This chapter is concerned with extending the theory of Hadamard series described in Chapter 2 to the hyperasymptotic evaluation [1] of Laplace-type integrals of the form

$$J(z) = \int_C e^{-z\psi(t)} f(t)\, dt \qquad (3.1.1)$$

for large values of $|z|$, where the phase function $\psi(t)$ possesses saddle points (given by the points where $\psi'(t) = 0$). Throughout this chapter we put $z = xe^{i\theta}$, where $x > 0$ and $\theta = \arg z$. The integration path C, which may be finite or infinite in extent, is supposed to commence at a simple saddle point labelled t_s and to pass along the steepest descent path through t_s defined by

$$\mathrm{Im}(e^{i\theta}\{\psi(t) - \psi(t_s)\}) = 0.$$

The amplitude function $f(t)$ is assumed to possess poles or branch-point singularities which either remain static in the complex t-plane, or may coalesce with a saddle point or approach the integration path C as some unspecified parameter in $\psi(t)$ or $f(t)$ varies.

The change of variable $u = e^{i\theta}\{\psi(t) - \psi(t_s)\}$ in (3.1.1) enables $J(z)$ to be written in the form

$$J(z) = e^{-z\psi(t_s)} \int_{C'} e^{-xu} f(t)\frac{dt}{du}\, du, \qquad (3.1.2)$$

where C' is the image of the path C in the u-plane (the so-called Borel plane). If C is infinite in extent then C' is the positive u-axis. On the other hand, if C is infinite in extent on both sides of t_s, passing to infinity down certain valleys, then C' is the negatively orientated loop about $u = 0$ surrounding the positive real axis.

---

[1] By hyperasymptotic evaluation we mean, in the broadest sense, any method of evaluation that achieves an accuracy greater than that obtainable by optimal truncation of the associated asymptotic expansion.

The above transformation has converted the integral (3.1.1) into a Laplace integral with the exponential factor e^{-xu}. We can now apply the Hadamard expansion theory developed for such integrals in Chapter 2.

3.2 Expansion schemes

In this section we review the two different modes of subdivision of the positive u-axis that were employed in §2.4 to deal with Laplace integrals. We obtain the form of the Hadamard expansions for $J(z)$ in (3.1.2) associated with each of these subdivisions, namely forward expansion and forward-reverse expansion. The former procedure uses forward expansion from the saddle point t_s for all the intervals, while the latter uses forward expansion for the zeroth interval but a forward-reverse expansion for the remaining intervals. We also discuss an important modification of this second procedure that allows us to deal with coalescence problems.

3.2.1 Subdivision of the u-axis

The positive real u-axis in (3.1.2) is subdivided into a series of finite intervals of length ω_n; if C is infinite in extent then the number of intervals will be infinite. In the first mode of subdivision, which we shall refer to as Scheme A, the left-hand endpoints of the intervals are denoted by the points Ω_n, with $\Omega_0 = 0$ corresponding to the saddle point t_s. In the second mode of subdivision, which we shall refer to as Scheme B, the points Ω_n ($n \geq 1$) are chosen to denote the midpoints of their respective intervals of length ω_n. The point $\Omega_0 = 0$ and the length of the zeroth interval is ω_0; the right-hand endpoint of the zeroth interval in this case is Ω_0' (see Fig. 3.1). These two schemes satisfy the following simple geometric constraints:

$$\Omega_n = \sum_{r=0}^{n-1} \omega_r \quad \text{(Scheme A)} \tag{3.2.1}$$

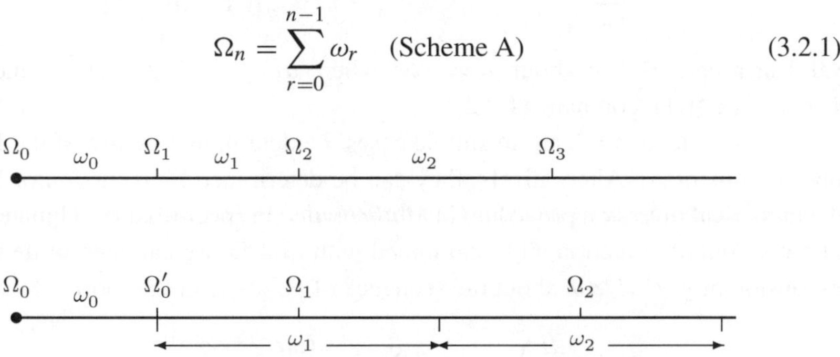

Figure 3.1 Scheme A (upper figure) and Scheme B (lower figure) for the subdivision of the positive u-axis.

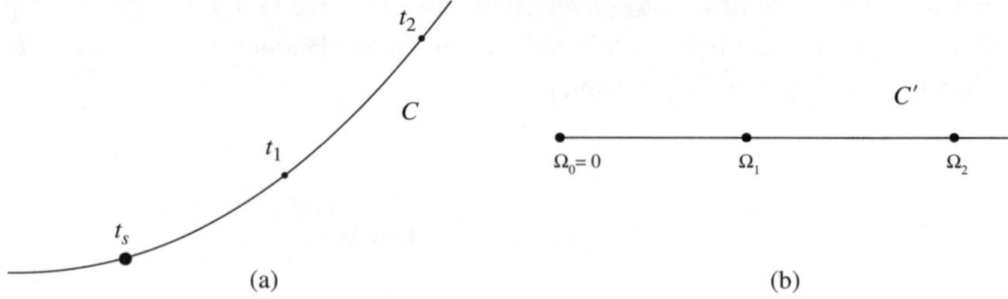

Figure 3.2 The subdivision of (a) the steepest descent path C emanating from a simple saddle t_s and (b) the image path C' in the u-plane. The maps of the points $t_s, t_1, t_2, \ldots$ in the u-plane are denoted by $\Omega_0, \Omega_1, \Omega_2, \ldots$.

and

$$\Omega_n = \sum_{r=0}^{n-1} \omega_r + \tfrac{1}{2}\omega_n \quad \text{(Scheme B)} \tag{3.2.2}$$

for $n \geq 1$.

On the path C in the t-plane the points corresponding to $u = \Omega_n$ are denoted by $t_0 \ (\equiv t_s), t_1, t_2, \ldots$ as indicated in Fig. 3.2. The values of t_n are obtained by straightforward solution of the equation

$$\psi(t_n) - \psi(t_s) = \Omega_n e^{-i\theta} \quad (n \geq 1), \tag{3.2.3}$$

where the roots are chosen to correspond to the integration path under consideration. With $u = \Omega_n + w$, where w is real, the above definitions of u, ω_n and Ω_n show that

$$\psi(t) - \psi(t_n) = (u - \Omega_n)e^{-i\theta} = we^{-i\theta} \quad (n \geq 0). \tag{3.2.4}$$

Inversion of this result (for a simple saddle at $t = t_s$) will yield an expansion[2] of the form

$$t - t_n = \sum_{k=1}^{\infty} a_{k,n}(we^{-i\theta})^{k\mu_n}, \qquad \mu_n = \begin{cases} \tfrac{1}{2} & (n = 0) \\ 1 & (n \geq 1) \end{cases} \tag{3.2.5}$$

valid in a certain disc about $u = \Omega_n$, where $a_{1,0} = \{2/\psi''(t_s)\}^{1/2}$ and $a_{1,n} = 1/\psi'(t_n)$ $(n \geq 1)$; compare §1.2.2.

The coefficients $a_{k,n}$ can, in simple cases, be determined by use of the Lagrange inversion theorem. Alternatively, they can be determined by recursion or by use of the numerical inversion procedure in *Mathematica* in specific cases. Upon expansion of the amplitude function $f(t)$, combined with (3.2.5), we can then write the series expansion of $f(t)dt/dw$ about the sequence of points t_n in the form

$$\left(f(t)\frac{dt}{dw}\right)_{t_n} = e^{-i\theta} \sum_{k=0}^{\infty} c_{k,n} \frac{(we^{-i\theta})^{(k+1)\mu_n-1}}{\Gamma((k+1)\mu_n)}, \tag{3.2.6}$$

---

[2] For a saddle of order $m \geq 3$ we have $\mu_0 = 1/m$; see §1.2.2.

where the $c_{k,n}$ are coefficients with

$$c_{0,n} = a_{1,n}\Gamma(1+\mu_n)f(t_n) \qquad (n=0,1,2,\ldots);$$

see (1.2.14) and (1.2.15) with $m = 1,2$. The $c_{k,n}$ are explicitly independent of $\theta = \arg z$; however, there is an implicit dependence through the variation of the steepest descent path C with θ.

Each of the expansions in (3.2.6) is valid in a disc centred at $u = \Omega_n$ with radius of convergence controlled by the saddle-point distribution of $\psi(t)$ and the singularity structure of $f(t)$. This in turn will control the choice of the values of ω_n and Ω_n to be used in either of the above schemes for the subdivision of the u-axis, since we shall require that the ω_n be chosen so that the expansions (3.2.6) are always used in their respective discs of convergence.

3.2.2 Expansion Scheme A

We start with Scheme A which consists of *forward expansion* about the points $u = \Omega_n$ for all the intervals. The evaluation of the contributions to $J(z)$ from each interval is similar to that described in §2.2.1. The contribution from the nth interval $0 \leq u - \Omega_n \leq \omega_n$ ($n \geq 0$) is, from (3.1.2) and (3.2.6),

$$e^{-z\psi(t_s)-\Omega_n x} \int_0^{\omega_n} e^{-xw} f(t)\frac{dt}{dw}dw$$

$$= e^{-z\psi(t_s)-\Omega_n x} \sum_{k=0}^{\infty} \frac{c_{k,n}e^{-i\theta}}{\Gamma((k+1)\mu_n)} \int_0^{\omega_n} e^{-xw}(we^{-i\theta})^{(k+1)\mu_n-1}dw$$

$$= e^{-z\psi(t_s)-\Omega_n x} \sum_{k=0}^{\infty} \frac{c_{k,n}}{z^{(k+1)\mu_n}} P((k+1)\mu_n, \omega_n x). \tag{3.2.7}$$

The expansion of the integral $J(z)$ in (3.1.1), when C is taken along a semi-infinite steepest descent path commencing at the saddle t_s, then takes the form

$$J(z) = e^{-z\psi(t_s)} \sum_{n=0}^{\infty} e^{-\Omega_n x} S_n(z), \tag{3.2.8}$$

where the Hadamard series $S_n(z)$ for $n \geq 0$ are defined by

$$S_n(z) = \sum_{k=0}^{\infty} \frac{c_{k,n}}{z^{(k+1)\mu_n}} P((k+1)\mu_n, \omega_n x). \tag{3.2.9}$$

Presented in this form, the Hadamard expansion of $J(z)$ in (3.2.8) can be regarded as an 'exactification' of the method of steepest descents.

Each Hadamard series $S_n(z)$ is associated with a decreasing exponential level $\exp(-\Omega_n x)$. An application of Cauchy's integral formula on the lines of that given in Olver (1997, p. 313) shows that the coefficients $c_{k,n}$ possess the large-k growth

$$\frac{c_{k,n}}{\Gamma((k+1)\mu_n)} = O(k^{-1/2}\Lambda_n^{-(k+1)\mu_n}).$$

The quantity Λ_0 denotes the distance in the u-plane from the saddle t_s to the nearest adjacent saddle or singularity of $f(t)$ and, in the simplest situation where the other saddles of $\psi(t)$ are not effective in the determination of the remaining convergence intervals, $\Lambda_n = \Omega_n$ ($n \geq 1$). From the behaviour $P(a, z) \sim z^a e^{-z}/\Gamma(a+1)$ as $a \to +\infty$, the decay of the late terms in (3.2.9) is consequently controlled by

$$e^{-\omega_n x} k^{-3/2} (\omega_n/\Lambda_n)^{(k+1)\mu_n} \qquad (k \to \infty;\ n \geq 0). \tag{3.2.10}$$

This result shows that the Hadamard series $S_n(z)$ in (3.2.9) are absolutely convergent when $\omega_n \leq \Lambda_n$.

The modification of the argument leading to (3.2.8) when the path C in (3.1.1) is infinite in both directions from the saddle point is straightforward. The points corresponding to $u = \Omega_n$ on the path C are now denoted by t_0 ($\equiv t_s$), $t_1^{\pm}$, $t_2^{\pm}, \ldots$, where the '$\pm$' superscripts refer to the halves of C that emanate from and lead up to the saddle t_s, respectively; see §1.2.2. The points $t_n^{\pm}$ will then give rise to the expansions in (3.2.6) with coefficients $c_{k,n}^{\pm}$, where

$$c_{k,0}^{-} = (-)^k c_{k,0}^{+}.$$

The contributions[3] to $J(z)$ from the zeroth intervals between t_s and $t_1^{\pm}$ are, from (3.2.8) and (3.2.9),

$$e^{-z\psi(t_s)} \sum_{k=0}^{\infty} \frac{(\pm 1)^{k-1} c_{k,0}}{z^{(k+1)/2}} P(\tfrac{1}{2}k + \tfrac{1}{2}, \omega_0 x) \tag{3.2.11}$$

respectively, where $c_{k,0} \equiv c_{k,0}^{+}$.

The Hadamard expansion for $J(z)$ in the case when C is infinite in both directions is then obtained by adding the contributions (3.2.8) corresponding to each half. For the interval between t_1^{-} and t_1^{+} containing the saddle point t_s, there is a cancellation of the odd terms in (3.2.11) to produce the zeroth-interval contribution given by

$$2e^{-z\psi(t_s)} \sum_{k=0}^{\infty} \frac{c_{2k,0}}{z^{k+\frac{1}{2}}} P(k + \tfrac{1}{2}, \omega_0 x). \tag{3.2.12}$$

The standard Poincaré asymptotic expansion for $J(z)$ is then obtained from this zeroth-interval contribution by replacing the incomplete gamma functions in (3.2.12) by unity[4] to find

$$J(z) \sim 2e^{-z\psi(t_s)} \sum_{k=0}^{\infty} \frac{c_{2k,0}}{z^{k+\frac{1}{2}}} \qquad (z \to \infty) \tag{3.2.13}$$

valid in some sector; compare (1.2.19).

---

[3] The additional factor of -1 for the contribution between t_s and t_1^{-} appears in (3.2.11) since integration is in the opposite direction.

[4] This is equivalent to replacing the upper limit of integration in the integral in (3.2.7) corresponding to $n = 0$ by $+\infty$.

In Paris (2004b, 2007a), the intervals ω_n were chosen to be the radii of convergence in (3.2.6); that is, the value of ω_0 is dictated by the presence of a neighbouring saddle point of $\psi(t)$ or singularity of $f(t)$, while (in the simplest situation) the remaining intervals are controlled by t_s, so that $\omega_n = \Omega_n$ ($n \geq 1$). This choice uses maximal integration intervals (see §2.2.4) and so maximises the exponential separation between the different levels in (3.2.8), but at the cost of a slower rate of convergence (controlled by $k^{-3/2}$) of each Hadamard series $S_n(z)$. It is, of course, possible to use forward expansion with the interval lengths chosen to satisfy $\omega_n < \Lambda_n$, in order to benefit from the geometric rate of decay of the terms given in (3.2.10).

As described in §2.2.2, the slow algebraic decay of the tail of $S_n(z)$ when using maximal integration intervals can be transformed into a rapid decay comparable with the asymptotic-like phase of the early terms by a simple rearrangement of the series. This rearranged series is called the *modified Hadamard expansion*. We write, for a suitably chosen truncation index $M \equiv M_n$,

$$S_n(z) = \sum_{k=0}^{M-1} \frac{c_{k,n}}{z^{(k+1)\mu_n}} P((k+1)\mu_n, \omega_n x) + T_{M,n}(z), \tag{3.2.14}$$

where, from (2.1.4) and the series expansion of the $_1F_1$ function,

$$\begin{aligned}
T_{M,n}(z) &= \sum_{k=M}^{\infty} \frac{c_{k,n}}{z^{(k+1)\mu_n}} P((k+1)\mu_n, \omega_n x) \\
&= e^{-\omega_n x} \sum_{k=M}^{\infty} c_{k,n}(\omega_n e^{-i\theta})^{(k+1)\mu_n} \sum_{r=0}^{\infty} \frac{(\omega_n x)^r}{((k+1)\mu_n + r)!} \\
&= e^{-\omega_n x} \sum_{r=0}^{\infty} \sigma_{r,n}\, \xi_n^r, \qquad \xi_n = \frac{\omega_n x}{M},
\end{aligned} \tag{3.2.15}$$

upon reversal of the order of summation, with

$$\sigma_{r,n} = M^r \sum_{k=M}^{\infty} \frac{c_{k,n}(\omega_n e^{-i\theta})^{(k+1)\mu_n}}{((k+1)\mu_n + r)!}; \tag{3.2.16}$$

compare (2.2.12). The coefficients $\sigma_{r,n}$ can be computed in the form

$$\sigma_{r,n} = M^r \left\{ s_n - \sum_{k=0}^{M-1} \frac{c_{k,n}(\omega_n e^{-i\theta})^{(k+1)\mu_n}}{((k+1)\mu_n + r)!} \right\}, \tag{3.2.17}$$

where

$$s_n = \sum_{k=0}^{\infty} \frac{c_{k,n}(\omega_n e^{-i\theta})^{(k+1)\mu_n}}{((k+1)\mu_n + r)!} = \frac{1}{r!} \int_0^{\omega_n} (1-v)^r f(t) \frac{dt}{dw} dw$$

with $w = \omega_n v$ ($0 \leq v \leq 1$). This representation for the infinite sum s_n as an integral follows by substitution of the series expansion for $f(t)dt/dw$ in (3.2.6) and evaluation of the resulting integral as a beta function; see (2.2.17). On making an obvious change of integration variable, we finally obtain s_n in the form

$$s_n = \frac{1}{r!} \int_{t_n}^{t_{n+1}} (1 - v)^r f(t)\, dt, \tag{3.2.18}$$

where, from (3.2.4),

$$v \equiv v_n(t) = \frac{\psi(t) - \psi(t_n)}{\omega_n e^{-i\theta}} \qquad (n = 0, 1, 2, \ldots). \tag{3.2.19}$$

In the case when the saddle point is contained in the zeroth interval (with the endpoints $t_1^{\pm}$ in the t-plane) the modified form of (3.2.12) is given by (3.2.14), (3.2.15) and (3.2.17) with $n = 0$ and k replaced by $2k$, and with s_0 in (3.2.18) replaced by

$$s_0 = \frac{1}{r!} \int_{t_1^{-}}^{t_1^{+}} (1 - v)^r f(t)\, dt.$$

3.2.3 Expansion Scheme B

With the subdivision Scheme B we use *forward expansion* for the zeroth interval (of length ω_0) as in Scheme A, but *forward-reverse expansion* about the midpoints Ω_n for the intervals corresponding to $n \geq 1$; see §2.2.4. The contribution from the zeroth interval in the case of a simple saddle at t_s is given by (3.2.9) with $\mu_0 = \frac{1}{2}$ to yield

$$S_0(z) = \sum_{k=0}^{\infty} \frac{c_{k,0}}{z^{(k+1)/2}}\, P(\tfrac{1}{2}k + \tfrac{1}{2}, \omega_0 x). \tag{3.2.20}$$

The contribution from the nth interval $-\frac{1}{2}\omega_n \leq u - \Omega_n \leq \frac{1}{2}\omega_n$ $(n \geq 1)$ is, from (3.2.6) and (2.2.23),

$$e^{-z\psi(t_s)-\Omega_n x} \int_{-\frac{1}{2}\omega_n}^{\frac{1}{2}\omega_n} e^{-xw} f(t)\frac{dt}{dw}\, dw$$

$$= e^{-z\psi(t_s)-\Omega_n x} \sum_{k=0}^{\infty} \frac{c_{k,n}}{k!} e^{-i(k+1)\theta} \int_{-\frac{1}{2}\omega_n}^{\frac{1}{2}\omega_n} e^{-xw} w^k\, dw$$

$$= e^{-z\psi(t_s)-\Omega_n x} \sum_{k=0}^{\infty} \frac{c_{k,n}}{z^{k+1}} \Delta P(k + 1, \tfrac{1}{2}\omega_n x), \tag{3.2.21}$$

where ΔP is defined by

$$\Delta P(k + 1, \tfrac{1}{2}\omega_n x) = P(k + 1, \tfrac{1}{2}\omega_n x) - P(k + 1, -\tfrac{1}{2}\omega_n x); \tag{3.2.22}$$

compare (2.4.8).

Then, with $S_0(z)$ as defined in (3.2.20) and

$$S_n(z) = \sum_{k=0}^{\infty} \frac{c_{k,n}}{z^{k+1}} \Delta P(k + 1, \tfrac{1}{2}\omega_n x) \qquad (n \geq 1), \tag{3.2.23}$$

the Hadamard expansion of the integral $J(z)$ in (3.1.1) is given by

$$J(z) = e^{-z\psi(t_s)} \sum_{n=0}^{\infty} e^{-\Omega_n x} S_n(z) \qquad (3.2.24)$$

when C is taken along a semi-infinite steepest descent path commencing at the simple saddle t_s. This expansion has the same form as that resulting from Scheme A in (3.2.8) but, of course, the individual Hadamard series $S_n(z)$ for $n \geq 1$ are defined differently for Scheme B.

We are now free to choose the length of the zeroth interval ω_0 to be a multiple ϑ $(0 < \vartheta \leq 1)$ of the distance in the u-plane to the nearest saddle point of $\psi(t)$ or singularity of $f(t)$, which we have called Λ_0 in (3.2.10); that is, we set $\omega_0 = \Lambda_0 \vartheta$. For the remaining intervals with $n \geq 1$, the values of ω_n and Ω_n can be chosen so that $\omega_n/(2\Omega_n) = \vartheta$; from (2.4.12), the Ω_n are then given by

$$\Omega_n = \omega_0 \frac{(1+\vartheta)^{n-1}}{(1-\vartheta)^n}, \qquad \omega_0 = \Lambda_0 \vartheta \qquad (n \geq 1). \qquad (3.2.25)$$

The decay of the terms in the Hadamard series in (3.2.20) and (3.2.21) will then be controlled by the geometric factor ϑ^k; see §2.4.3. If ϑ is suitably chosen, the decay of these series can be sufficiently rapid to avoid the separate evaluation of the modified tails, as in (3.2.14) and (3.2.15) in Scheme A. This represents a considerable saving in effort since it is the determination of the coefficients $\sigma_{r,n}$ that is the computationally most expensive part of the process. The disadvantages of the approach in Scheme B are twofold: the reduction in the exponential separation between the different levels in the expansion and the requirement of a greater number of coefficients $c_{k,n}$ than that in (3.2.14). In situations where the calculation of these coefficients is not difficult, however, this latter consideration is not really an issue.

3.3 Examples

In this section we present some examples of the Hadamard expansion process applied to Laplace-type integrals of the form (3.1.1). To explain the procedure we start with some straightforward examples in which z is positive real. The final example in this section considers the case when z is complex, where we encounter the effect on the Hadamard expansion of a Stokes line. In all cases the saddle points of the phase function are well separated so that problems associated with coalescence are avoided. Examples illustrating the modification required in the Hadamard expansion process in the presence of coalescence will be given in §3.5.

Example 3.1 Consider the integral

$$I(x) = \int_{-1}^{1} e^{-x(t-t^3/3)}\,dt$$

for large $x > 0$. Here $f(t) = 1$ and $\psi(t) = t - \frac{1}{3}t^3$ has two saddle points at $t = \pm 1$. Our integration path then runs from the saddle at $t = -1$ (of height $e^{2x/3}$) to the saddle at $t = 1$ (of height $e^{-2x/3}$) along a path of steepest descent.[5] We set

$$u = \psi(t) - \psi(t_s) = t - \tfrac{1}{3}t^3 + \tfrac{2}{3} = (t+1)^2(\tfrac{2}{3} - \tfrac{1}{3}t), \qquad (3.3.1)$$

so that $u = 0$ at $t = -1$ and $u = \frac{4}{3}$ at $t = 1$. Thus

$$(t+1)(\tfrac{2}{3} - \tfrac{1}{3}t)^{1/2} = \pm u^{1/2},$$

where the square root $(\frac{2}{3} - \frac{1}{3}t)^{1/2}$ takes the value 1 at $t = -1$.

Application of the Lagrange inversion theorem in (1.2.12) yields (see also Example 1.4)

$$t^{\pm}(u) + 1 = \sum_{k=1}^{\infty} \frac{(\pm u^{\frac{1}{2}})^k}{k!} \frac{d^{k-1}}{dt^{k-1}}(\tfrac{2}{3} - \tfrac{1}{3}t)^{-k/2}\bigg|_{t=-1}$$

$$= \sum_{k=1}^{\infty} \frac{(\pm u^{\frac{1}{2}})^k}{k!} \frac{\Gamma(\frac{3}{2}k - 1)}{3^{k-1}\Gamma(\frac{1}{2}k)}. \qquad (3.3.2)$$

Hence

$$\frac{dt^{\pm}}{du} = \sum_{k=0}^{\infty} \frac{(\pm)^{k+1}c_k}{\Gamma(\frac{1}{2}k + \frac{1}{2})} u^{(k-1)/2}, \qquad c_k = \frac{\Gamma(\frac{3}{2}k + \frac{1}{2})}{2 \cdot 3^k k!}, \qquad (3.3.3)$$

where $t^{\pm}(u)$ denote the right and left halves of the steepest descent path through $t_s = -1$. The circle of convergence of the expansions (3.3.2) and (3.3.3) is determined by the nearest point in the mapping $t \mapsto u$ where dt/du is singular; that is, at the saddle point $t = 1$. Since the value of u at $t = 1$ is $\frac{4}{3}$, the expansion (3.3.3) converges[6] in the disc $|u| < \frac{4}{3}$.

We now apply forward expansion from $t = -1$ to the saddle at $t = 1$, using dt^{+}/du with the maximum integration interval $\omega_0 = \frac{4}{3}$. From (3.2.8) and (3.2.9), we therefore obtain the single-stage Hadamard expansion

$$I(x) = e^{2x/3} \sum_{k=0}^{\infty} \frac{c_k}{x^{(k+1)/2}} P(\tfrac{1}{2}k + \tfrac{1}{2}, \tfrac{4}{3}x) \qquad (3.3.4)$$

for the inter-saddle contribution. Since this expansion has involved a maximal integration interval, it is expected that the decay of its terms will be algebraic, rather than geometric. This can be confirmed by use of Appendix B, with the parameters defined there $\kappa = 0$, $h = \frac{3}{2}\sqrt{3}$, $\lambda = -1$ and $\xi = \frac{1}{3}$, to obtain from (B.5) the slow algebraic decay of the late terms (excluding the prefactor $\exp(2x/3)$) described by

$$e^{-4x/3}k^{-3/2} \qquad (k \to +\infty).$$

---

[5] This is relative to the saddle at $t = -1$; relative to the saddle at $t = 1$, the path $[-1, 1]$ is a path of steepest ascent.

[6] This can also be established by application of Stirling's formula to the coefficients in (3.3.3) to determine their large-k behaviour; see Example 1.4.

In computations with (3.3.4) we therefore need the modified form of this expansion which, for a suitably chosen truncation index M_0, is from (3.2.14) and (3.2.15) given by

$$I(x) = e^{2x/3} \left\{ \sum_{k=0}^{M_0-1} \frac{c_k}{x^{(k+1)/2}} \, P(\tfrac{1}{2}k + \tfrac{1}{2}, \tfrac{4}{3}x) + T_{M_0}(x) \right\}, \tag{3.3.5}$$

where

$$T_{M_0}(x) = e^{-4x/3} \sum_{r=0}^{\infty} \sigma_r \left(\frac{4x}{3M_0} \right)^r. \tag{3.3.6}$$

From (3.2.17)–(3.2.19), the coefficients σ_r can be written in the form

$$\sigma_r = M_0^r \left\{ \frac{1}{r!} \int_{-1}^{1} (1-v)^r \, dt - \sum_{k=0}^{M_0-1} \frac{c_k (\tfrac{4}{3})^{(k+1)/2}}{\Gamma(\tfrac{1}{2}k + r + \tfrac{3}{2})} \right\},$$

where

$$v = \tfrac{3}{4}u = \tfrac{3}{4}\tau^2(1 - \tfrac{1}{3}\tau), \quad \tau = 1 + t.$$

We remark that in this case the integral appearing in the coefficients σ_r can be evaluated *explicitly* as

$$\frac{1}{r!} \int_0^2 (1-v)^r \, d\tau = \frac{1}{r!} \sum_{k=0}^{r} \sum_{n=0}^{k} \binom{r}{k} \binom{k}{n} (-\tfrac{3}{4})^r (-\tfrac{1}{3})^n \frac{2^{2k+n+1}}{2k+n+1}.$$

An alternative mode of expansion is to employ forward expansion from $t = -1$ to 0 and reverse expansion from $t = 1$ to 0. Then

$$I(x) = \left\{ \int_{-1}^{0} + \int_{0}^{1} \right\} e^{-x(t-t^3/3)} \, dt \equiv I_1 + I_2.$$

The integral I_1 is given by (3.3.4) with the argument of the P function replaced by $\tfrac{2}{3}x$. To deal with the integral I_2, we now expand about the saddle at $t = 1$ and set

$$u = \tfrac{1}{3}t^3 - t + \tfrac{2}{3} = (t-1)^2(\tfrac{2}{3} + \tfrac{1}{3}t).$$

Then we find

$$I_2 = \int_0^1 e^{x(t^3/3-t)} \, dt = -e^{-2x/3} \int_0^{2/3} e^{xu} \frac{dt}{du} \, du. \tag{3.3.7}$$

By Lagrange inversion, we have

$$\frac{dt^{\pm}}{du} = \pm \sum_{k=0}^{\infty} \frac{(\mp)^k c_k}{\Gamma(\tfrac{1}{2}k + \tfrac{1}{2})} u^{(k-1)/2} \quad (|u| < \tfrac{4}{3}),$$

where the c_k are defined in (3.3.3) and $t^{\pm}(u)$ now denote the right and left halves of the steepest ascent path through $t = 1$. Substitution of the expansion for dt^-/du into (3.3.7) and use of the result (2.2.23) then shows that

$$I_2 = e^{-2x/3} \sum_{k=0}^{\infty} \frac{c_k}{x^{(k+1)/2}} \, P^*(\tfrac{1}{2}k + \tfrac{1}{2}, -\tfrac{2}{3}x),$$

where[7] $P^*(a, -x) \equiv e^{-\pi i a} P(a, -x)$ when $x > 0$.

Hence, we obtain the 2-stage Hadamard expansion in the form

$$I(x) = e^{2x/3} \sum_{k=0}^{\infty} \frac{c_k}{x^{(k+1)/2}} \, P(\tfrac{1}{2}k + \tfrac{1}{2}, \tfrac{2}{3}x)$$

$$+ e^{-2x/3} \sum_{k=0}^{\infty} \frac{c_k}{x^{(k+1)/2}} \, P^*(\tfrac{1}{2}k + \tfrac{1}{2}, -\tfrac{2}{3}x); \qquad (3.3.8)$$

compare §2.2.4 for a similar expansion for the modified Bessel function $I_\nu(z)$, but note that the summation index in (3.3.8) is $\tfrac{1}{2}k$ because the endpoints of the integration interval are simple saddle points. By (A.6), each of these Hadamard series has late terms that are now controlled by the geometric factor $2^{-k/2}$, so that their convergence is more rapid than that of the single-stage expansion in (3.3.4).

In Fig. 3.3(a), we show the decay of the terms (on a $\log_{10}$ scale) in the modified expansion in (3.3.5) and (3.3.6) for $e^{-2x/3} I(x)$, with $x = 10$ and the truncation index $M_0 = 36$. These terms are also compared with those of the Poincaré asymptotic expansion

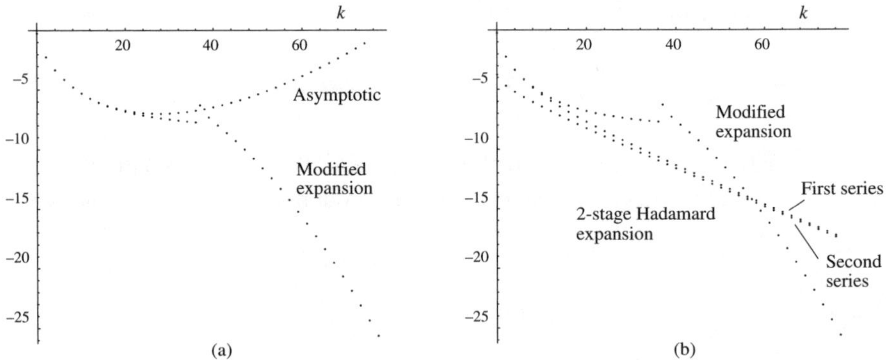

Figure 3.3 The behaviour of the terms (on a $\log_{10}$ scale) in the Hadamard expansions for $e^{-2x/3} I(x)$ when $x = 10$ against ordinal number k: (a) the modified expansion (3.3.5) and the asymptotic series and (b) the 2-stage expansion (3.3.8) compared with the modified expansion (3.3.5).

---

[7] We remark that $P^*(a, -x)$ is real for $x > 0$ and $a > 0$; see Appendix A.

Table 3.1 *The absolute error in $e^{-2x/3}I(x)$ for different x and truncation indices*

x	Modified expansion		2-stage expansion	
	$N_0 = 20$	$N_0 = 40$	$M_0 = 60$	$M_0 = 100$
2	1.915×10^{-22}	1.987×10^{-46}	7.562×10^{-13}	3.474×10^{-19}
5	9.166×10^{-16}	8.175×10^{-32}	1.032×10^{-13}	4.717×10^{-20}
10	2.904×10^{-16}	3.117×10^{-28}	3.794×10^{-15}	1.702×10^{-21}
15	7.257×10^{-17}	3.731×10^{-26}	1.426×10^{-16}	6.194×10^{-23}
20	2.436×10^{-18}	6.750×10^{-26}	5.490×10^{-18}	2.273×10^{-24}

$$e^{-2x/3}I(x) \sim \sum_{k=0}^{\infty} c_k \, x^{-(k+1)/2} \qquad (x \to +\infty),$$

obtained from (3.3.4) by replacing the incomplete gamma functions P by unity. The value of the optimal truncation index $k = k_0$ of this asymptotic series occurs at the smallest term given by

$$\frac{c_{k_0}}{c_{k_0-1}} = \frac{\Gamma(\frac{3}{2}k_0 + \frac{1}{2})}{\Gamma(\frac{3}{2}k_0 - 1)}(3k_0)^{-1} \simeq x^{1/2}.$$

Assuming x to be large, so that $k_0 \gg 1$, we can approximate the ratio of gamma functions by Stirling's formula as $(\frac{3}{2}k_0)^{3/2}$ to obtain $\frac{1}{2}k_0 \simeq \frac{4}{3}x$. Then, for the value $x = 10$ in the figure, the asymptotic series has terms that decrease to a minimum of order 10^{-7} corresponding to an optimal truncation index of $k_0 \simeq 26$. We note the fact that the leading terms in the tail $T_{M_0}(x)$ are somewhat larger than the late terms in the associated finite main sum. In Fig. 3.3(b) the terms in the modified expansion are compared with both those of the 2-stage expansion (3.3.8). It is apparent that the latter series possess terms with a more rapid decay than the series (3.3.4) but that, for the number of terms computed, the modified single-stage expansion produces greater accuracy. This increased precision comes, of course, at the cost of the computation of the modified tail in (3.3.6).

In Table 3.1 we show the absolute error in the computation of $e^{-2x/3}I(x)$ using the expansions (3.3.5) and (3.3.8) for different values of x. For the modified expansion (3.3.5), we chose the truncation index of the finite main sum to be $M_0 = 20$ for $x \leq 5$ and $M_0 = 20 + 2[x]$ for the remaining x-values; the tail $T_{M_0}(x)$ in (3.3.6) is truncated after N_0 terms. For the expansion (3.3.8) each series is truncated after M_0 terms. It is worth remarking that, because the series involve integration from and up to a saddle point, the index is $\frac{1}{2}k$ rather than k (as would result from an integration interval that encloses a saddle point). This requires larger k-values and hence larger truncation indices.

Finally, we observe that it is possible to subdivide the integration path $[-1, 1]$ further to obtain more rapidly convergent series. For example, we could expand about the saddles $t = \pm 1$, over the interval $0 \leq u \leq \frac{1}{3}$, and also about $t = 0$. This would result in series whose terms are controlled by the decay factor 2^{-k}, but which would necessitate the evaluation of the inversion coefficients about $t = 0$.

Example 3.2 Our second example is the gamma function $\Gamma(x)$ for $x > 0$. This function can be expressed in terms of an integral of the form (3.1.1) with $\psi(t) = t - \log(1 + t)$, $f(t) = 1$ by

$$\Gamma(x) = \frac{1}{x} \int_0^\infty e^{-w} w^x dw = x^x e^{-x} \int_{-1}^\infty e^{-x\{t - \log(1+t)\}} dt,$$

where we have made the change of variable $w = x(1 + t)$. The exponential has a saddle point at $t = 0$; accordingly, we divide the integration path into the intervals $[-1, 0]$ and $[0, \infty)$, where the function

$$u = t - \log(1 + t)$$

is montonically decreasing and increasing, respectively. Hence we may write the above integral as (Copson, 1965, p. 54)

$$\frac{\Gamma(x)}{x^x e^{-x}} = \int_0^\infty e^{-xu} \left(\frac{dt^+}{du} - \frac{dt^-}{du} \right) du, \tag{3.3.9}$$

where $t^\pm(u)$ denote the two branches of $t = t(u)$ corresponding to $t^+ \in [0, \infty)$ and $t^- \in [-1, 0]$. Singular points of the above mapping arise where $dt/du = (1+t)/t$ is singular, that is at $t = 0$; this corresponds to the points $u = \pm 2\pi ki$, $k = 0, 1, 2, \ldots$ in the u-plane.

We adopt Scheme B for the subdivision of the u-axis, with the zeroth interval chosen to be $0 \leq u \leq 2\pi\vartheta$, $0 < \vartheta < 1$, together with a sequence of points $u = \Omega_n$ ($n \geq 1$) and associated interval lengths ω_n to be specified presently; see §3.2.3. The inversion of the mapping $t \mapsto u(t)$ about $u = 0$ $(t = 0)$ is given by

$$t^\pm(u) = \sum_{k=1}^\infty (\pm)^k \alpha_k (2u)^{k/2} \qquad (|u| < 2\pi),$$

where a recursion for the coefficients α_k is given in (1.3.14). It then follows that

$$\frac{dt^+}{du} - \frac{dt^-}{du} = \sqrt{2\pi} \sum_{k=0}^\infty \frac{(-)^k \gamma_k}{\Gamma(k + \frac{1}{2})} u^{k-1/2}$$

in $|u| < 2\pi$, where the Stirling coefficients γ_k are defined in terms of α_{2k+1} in (1.3.13).

A similar inversion applies about the points $u = \Omega_n$, which correspond in the t-plane to the points $t_n^\pm$ given by the two solutions of the equation

$$t_n^\pm - \log(1 + t_n^\pm) = \Omega_n \quad (n \geq 1).$$

Then we have the inversion

$$t^{\pm}(u) = t_n^{\pm} + \sum_{k=1}^{\infty} d_{k,n}^{\pm} w^k, \tag{3.3.10}$$

where $w = u - \Omega_n$, to yield

$$\frac{dt^+}{dw} - \frac{dt^-}{dw} = \sum_{k=0}^{\infty} \frac{c_{k,n}}{k!} w^k, \qquad c_{k,n} = (k+1)!\,(d_{k+1,n}^+ - d_{k+1,n}^-).$$

The domain of convergence of these last expansions is $|w| < \Omega_n$ since the nearest singularity to the expansion point $u = \Omega_n$ is that at the origin.

A recursion for the coefficients $d_{k,n}$ can be obtained in a similar manner to that described for the coefficients α_k in (1.3.14). We substitute the expansion (3.3.10) into the expression $t\,dt/dw = 1 + t$ and equate coefficients of powers of w^{k-1} to find (omitting the $\pm$ superscripts and setting $t_n^{\pm} \equiv A$)

$$d_{k-1,n} = kAd_{k,n} + \sum_{r+s=k} r d_{r,n}\, d_{s,n} = kAd_{k,n} + \tfrac{1}{2}k \sum_{r+s=k} d_{r,n}\, d_{s,n}\,,$$

whence

$$d_{k,n} = \frac{d_{k-1,n}}{kA} - \frac{1}{2A} \sum_{r=1}^{k-1} d_{r,n}\, d_{k-r,n} \quad (n \geq 1,\ k \geq 2)$$

with $d_{1,n} = (1 + A)/A$.

From (3.3.9), the contribution from the zeroth interval is

$$\int_0^{2\pi\vartheta} e^{-xu} \left(\frac{dt^+}{du} - \frac{dt^-}{du}\right) du = \sqrt{2\pi} \sum_{k=0}^{\infty} \frac{(-)^k \gamma_k}{x^{k+\frac{1}{2}}} P(k + \tfrac{1}{2}, 2\pi\vartheta),$$

and that from the nth interval (of length ω_n) $-\tfrac{1}{2}\omega_n \leq u - \Omega_n \leq \tfrac{1}{2}\omega_n$ is

$$e^{-\Omega_n x} \int_{-\frac{1}{2}\omega_n}^{\frac{1}{2}\omega_n} e^{-xu} \left(\frac{dt^+}{dw} - \frac{dt^-}{dw}\right) dw = e^{-\Omega_n x} S_n(x),$$

where

$$S_n(x) = \sum_{k=0}^{\infty} \frac{c_{k,n}}{x^{k+1}} \Delta P(k + 1, \tfrac{1}{2}\omega_n x);$$

see (3.2.21) and (3.2.22). The Hadamard expansion for $\Gamma(x)$ then takes the form

$$\frac{\Gamma(x)}{x^x e^{-x}} = \sqrt{2\pi} \sum_{k=0}^{\infty} \frac{(-)^k \gamma_k}{x^{k+\frac{1}{2}}} P(k + \tfrac{1}{2}, 2\pi\vartheta) + \sum_{n=1}^{\infty} e^{-\Omega_n x} S_n(x). \tag{3.3.11}$$

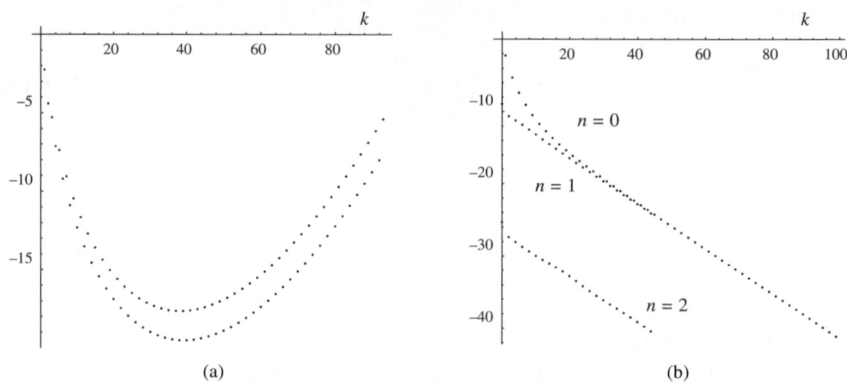

Figure 3.4 The behaviour of the terms (on a $\log_{10}$ scale) against ordinal number k in the expansion of $\Gamma(x)/(x^x e^{-x})$ when $x = 6$: (a) the asymptotic series and (b) the first three Hadamard series in (3.3.11) with $\vartheta = \frac{1}{2}$.

We remark that if only the zeroth level is retained, with the incomplete gamma functions replaced by unity, then we recover the well-known asymptotic expansion

$$\Gamma(x) \sim \sqrt{2\pi}\, e^{-x} x^{x-1/2} \sum_{k=0}^{\infty} (-)^k \gamma_k x^{-k} \qquad (x \to +\infty); \qquad (3.3.12)$$

see (1.3.15).

In Fig. 3.4(a) we display the magnitude of the terms in the asymptotic series (3.3.12) for $\Gamma(x)/(x^x e^{-x})$ when $x = 6$. From the asymptotic behaviour of the Stirling coefficients γ_k for $k \to \infty$ (Dingle, 1973, p. 159)

$$\gamma_k = O\left((2\pi)^{-k}\Gamma(k)\epsilon(k)\right), \qquad \epsilon(k) = \begin{cases} 1 & (k\ \text{odd}) \\ k^{-1} & (k\ \text{even}), \end{cases}$$

the optimal truncation point is given approximately by $k = [2\pi x]$. In addition, we see that the terms exhibit a saw-tooth type behaviour caused by the different rates of growth of γ_k for k of different parity. In Fig. 3.4(b) we show the first three levels in the Hadamard expansion (3.3.11). The expansion points Ω_n have been chosen to satisfy $\omega_n/(2\Omega_n) = \vartheta$, so that the terms in each series are controlled by the geometric decay factor ϑ^k. The values displayed correspond to the particular choice $\vartheta = \frac{1}{2}$ and, from (3.2.25) with $\Lambda_0 = 2\pi$, we have $\Omega_1 = 2\pi$, $\Omega_2 = 6\pi$, $\Omega_3 = 18\pi, \ldots$. The leading terms in the series with $n = 3$ are of order 10^{-74}; computation with the terms illustrated leads to an absolute error in the evaluation of $\Gamma(x)/(x^x e^{-x})$ when $x = 6$ of order 10^{-40}, compared to that resulting from optimal truncation of the asymptotic series of order 10^{-17}. It should be remarked that the terms in the zeroth series ($n = 0$) present the same saw-tooth behaviour as the asymptotic series; for clarity, we have shown only the terms corresponding to odd k. The terms in the series with $n \geq 1$ do not exhibit this behaviour.

Subdivision of the u-axis according to Scheme A is, of course, also possible. An example applied to $\Gamma(x)$ using forward expansion (Scheme A) with maximal integration intervals and where the modified form of the series must be employed, is given in Paris (2004a). But forward expansion using non-maximal integration intervals, at the cost of a reduced exponential separation between the different levels, can also be used and thereby obviate the need to employ the computationally more expensive modified Hadamard series.

Example 3.3 From (1.3.16), the Airy function $\mathrm{Ai}(x^{2/3})$ can be written in the form

$$\mathrm{Ai}(x^{2/3}) = \frac{x^{1/3}}{2\pi i} \int_{\infty e^{-\pi i/3}}^{\infty e^{\pi i/3}} e^{-x(t - t^3/3)} dt \tag{3.3.13}$$

when $x > 0$. The integrand possesses two saddle points situated at $t = \pm 1$ and the associated paths of steepest descent are illustrated in Fig. 1.5(a). By Cauchy's theorem, the integration path can be made to coincide with the steepest descent path through the saddle point at $t_s = 1$.

We set $\psi(t) = t - \frac{1}{3}t^3$, so that

$$\psi(t) - \psi(1) = t - \tfrac{1}{3}t^3 - \tfrac{2}{3} = -\tfrac{1}{3}(t-1)^3 - (t-1)^2 = u$$

which maps the steepest descent path through $t_s = 1$ onto the positive real u-axis. Then

$$e^{2x/3}\mathrm{Ai}(x^{2/3}) = \frac{x^{1/3}}{2\pi i} \int_0^\infty e^{-xu} \left(\frac{dt^+}{du} - \frac{dt^-}{du} \right) du,$$

where $t^\pm(u)$ denote the upper and lower halves of the steepest descent path through $t_s = 1$. The inversion of the above transformation by Lagrange's theorem is given in (1.3.19) and yields

$$\frac{dt^\pm}{du} = \pm i \sum_{k=0}^\infty \frac{(\mp i)^k c_{k,0}}{\Gamma(\frac{1}{2}k + \frac{1}{2})} u^{(k-1)/2}, \qquad c_{k,0} = \frac{\Gamma(\frac{3}{2}k + \frac{1}{2})}{2 \cdot 3^k k!}. \tag{3.3.14}$$

The circle of convergence of this expansion is determined by the nearest point in the mapping $t \mapsto u$ where dt/du is singular; that is, at the saddle point at $t = -1$. Since the value of u at $t = -1$ is $-\frac{4}{3}$, the above expansion converges in the disc $|u| < \Lambda_0$, where $\Lambda_0 = \frac{4}{3}$.

We adopt the subdivision Scheme B in §3.2.3. The contribution from the zeroth interval $0 \leq u \leq \omega_0$ containing the saddle point $t_s = 1$ is then

$$\frac{x^{1/3}}{2\pi i} \int_0^{\omega_0} e^{-xu} \left(\frac{dt^+}{du} - \frac{dt^-}{du} \right) du = \frac{x^{1/3}}{\pi} \sum_{k=0}^\infty \frac{(-)^k c_{2k,0}}{x^{k+\frac{1}{2}}} P(k + \tfrac{1}{2}, \omega_0 x), \tag{3.3.15}$$

where we set $\omega_0 = \Lambda_0 \vartheta = \frac{4}{3}\vartheta$, with $0 < \vartheta < 1$. The points t_n in the t-plane that correspond to the points $u = \Omega_n$ are given by the solution of the cubic equation $u(t_n) = \Omega_n$, with the root t_n being chosen to correspond to the integration path; since

Table 3.2 *Corresponding values of Ω_n and*
t_n ($t_n \equiv t_n^+$) in the subdivision of the
integration path when $\vartheta = \frac{1}{2}$

n	Ω_n	t_n
0	0	1
1	4/3	$1.17765 + 1.07730i$
2	4	$1.41082 + 1.72373i$
3	12	$1.82936 + 2.65323i$
4	36	$2.50000 + 3.96863i$

$x > 0$, we have $t_n^- = (t_n^+)^*$, where the asterisk denotes the complex conjugate. With the points Ω_n chosen according to (3.2.25) and $\Lambda_0 = \frac{4}{3}$, the first few values of t_n when $\vartheta = \frac{1}{2}$ are displayed in Table 3.2 to 5 decimal places.

For the nth interval $-\frac{1}{2}\omega_n \leq u - \Omega_n \leq \frac{1}{2}\omega_n$, we determine the expansion for dt/du about $t = t_n$ as follows. With $t = t_n + \tau$, we have

$$\psi(t_n + \tau) - \psi(t_n) = u - \Omega_n = w \tag{3.3.16}$$

which yields

$$\tau(1 - t_n^2) - \tfrac{1}{3}\tau^3 - t_n\tau^2 = w; \tag{3.3.17}$$

compare (3.2.4). Inversion of this expression to obtain the expansion

$$\tau = t - t_n = \sum_{k=1}^{\infty} d_{k,n} w^k$$

can be carried out in specific cases (i.e., with numerical values) using *Mathematica*. Alternatively, we can substitute this expansion for τ into (3.3.17) and form the Cauchy products to obtain the recursion for the coefficients $d_{k,n}$ (for $k \geq 2$)

$$d_{k,n} = \frac{t_n}{1 - t_n^2} \sum_{m=1}^{k-1} d_{m,n} d_{k-m,n} + \frac{1}{3(1 - t_n^2)} \sum_{r=1}^{k-2} d_{r,n} \sum_{m=1}^{k-r-1} d_{m,n} d_{k-r-m,n}, \tag{3.3.18}$$

where $d_{1,n} = 1/(1 - t_n^2)$. Differentiation of the expansion for τ then produces

$$\frac{dt^+}{dw} = i \sum_{k=0}^{\infty} \frac{c_{k,n}}{k!} w^k \quad (|w| < \Omega_n), \qquad c_{k,n} = -i(k+1)! \, d_{k+1,n},$$

with a conjugate expansion for dt^-/dw.

The contribution from the nth interval (on the upper and lower halves of the integration path) then becomes

$$\frac{x^{1/3}}{2\pi i} e^{-\Omega_n x} \int_{-\frac{1}{2}\omega_n}^{\frac{1}{2}\omega_n} e^{-xw} \left(\frac{dt^+}{dw} - \frac{dt^-}{dw} \right) dw = \frac{x^{1/3}}{\pi} e^{-\Omega_n x} \, \mathrm{Re}\{S_n(x)\},$$

where, from (3.2.23),

$$S_n(x) = \sum_{k=0}^{\infty} \frac{c_{k,n}}{x^{k+1}} \, \Delta P(k+1, \tfrac{1}{2}\omega_n x)$$

and ΔP is defined in (3.2.22). The Hadamard expansion for $\mathrm{Ai}(x^{2/3})$ when $x > 0$ then becomes

$$e^{2x/3}\mathrm{Ai}(x^{2/3}) = \frac{x^{1/3}}{\pi} \left\{ \sum_{k=0}^{\infty} \frac{c_{2k,0}}{x^{k+\frac{1}{2}}} \, P(k+\tfrac{1}{2}, \tfrac{4}{3}\vartheta x) + \sum_{n=1}^{\infty} e^{-\Omega_n x} \mathrm{Re}\{S_n(x)\} \right\},$$

$$(3.3.19)$$

where the quantities Ω_n are specified in (3.2.25) with $\Lambda_0 = \tfrac{4}{3}$ and the ratios $\omega_n/(2\Omega_n) = \vartheta$.

In Fig. 3.5 we present the behaviour of the terms (on a $\log_{10}$ scale) in the expansion (3.3.19) against ordinal number k when $x = 15$. The value of the parameter $\vartheta = \tfrac{1}{2}$, so that $\Omega_1 = \tfrac{4}{3}$, $\Omega_2 = 4$, $\Omega_3 = 12, \ldots$. The first figure compares the terms in the zeroth series (involving the coefficients $c_{2k,0}$) with those of the asymptotic series obtained by replacing $P(k+\tfrac{1}{2}, \tfrac{4}{3}\vartheta x)$ by unity, namely

$$e^{2x/3}\mathrm{Ai}(x^{2/3}) \sim \frac{x^{1/3}}{\pi} \sum_{k=0}^{\infty} \frac{(-)^k c_{2k,0}}{x^{k+\frac{1}{2}}} = \frac{x^{-1/6}}{2\pi} \sum_{k=0}^{\infty} \frac{(-)^k \Gamma(3k+\tfrac{1}{2})}{3^{2k}\,(2k)!\,x^k}$$

valid for $x \to +\infty$.

The late terms in the zeroth series possess the absolute behaviour

$$e^{-\frac{4}{3}\vartheta x} \frac{c_{2k,0}}{\Gamma(k+\tfrac{3}{2})} \, (\tfrac{4}{3}\vartheta)^k = e^{-\frac{4}{3}\vartheta x} O(k^{-3/2}\vartheta^k) \qquad (k \to +\infty),$$

which follows from (A.6) and use of Stirling's formula. The second figure shows the terms in the first three series ($n \le 2$) in the expansion (3.3.19), where the series

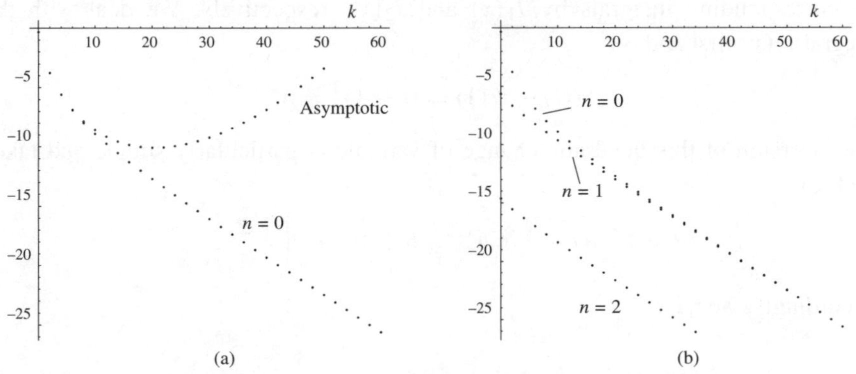

Figure 3.5 The behaviour of the terms (on a $\log_{10}$ scale) in the Hadamard expansion for $e^{2x/3}\mathrm{Ai}(x^{2/3})$ when $x = 15$ and $\vartheta = \tfrac{1}{2}$ against ordinal number k: (a) the zeroth and the asymptotic series and (b) the series in (3.3.19) with $n \le 2$.

Table 3.3 *The absolute error in $e^{2x/3}\mathrm{Ai}(x^{2/3})$ for different x and exponential level n when $\vartheta = \frac{1}{2}$ with the truncation indices cited in the text*

n	$x = 10$	$x = 15$	$x = 20$
0	4.440×10^{-5}	1.237×10^{-6}	3.688×10^{-8}
1	3.820×10^{-11}	1.337×10^{-15}	5.034×10^{-20}
2	1.264×10^{-21}	6.109×10^{-26}	2.080×10^{-28}

are truncated after M_n terms; each of these series has late terms controlled by the geometric decay factor ϑ^k. The leading terms in the series with $n = 3$ are found to be of order 10^{-41} with this value of x, and so are not displayed in the figure. The absolute error in the computation of $e^{2x/3}\mathrm{Ai}(x^{2/3})$ from (3.3.19) is displayed in Table 3.3 for different x as a function of the level n using the terms indicated in Fig. 3.5(b) (i.e., $M_0 = M_1 = 60$, $M_2 = 30$).

Example 3.4 In our next example we make the choice of the simple quadratic expression $\psi(t) = t^2 - 2t$ for the phase function in (3.1.1) but include an amplitude function $f(t)$. We consider the integral

$$I(x) = \frac{1}{\sqrt{\pi}} \int_0^\infty e^{-x(t^2 - 2t)} t^\nu \, dt \qquad (\mathrm{Re}\,(\nu) > -1)$$

for positive values of x. The exponential factor has a simple saddle point at $t = 1$ and $f(t) = t^\nu$ has, for non-integer ν, a branch-point singularity at $t = 0$. For $x > 0$, the integration path is easily verified to be a steepest descent path through the saddle.

We divide the integration path into the two intervals $[0, 2]$ and $[2, \infty)$, and denote the corresponding integrals by $I_1(x)$ and $I_2(x)$, respectively. We deal with the integral $I_1(x)$ first and set

$$\psi(t) - \psi(1) = (t - 1)^2 = u.$$

The inversion of this quadratic change of variable is particularly simple and takes the form

$$t \equiv t^\pm(u) = 1 \pm u^{1/2}, \quad u \geq 0 \qquad \begin{cases} t \geq 1 \\ t \leq 1. \end{cases}$$

Accordingly, we have

$$I_1(x) = \frac{1}{\sqrt{\pi}} \int_0^2 e^{-x\psi(t)} f(t) \, dt$$

$$= \frac{e^x}{\sqrt{\pi}} \int_0^1 e^{-xu} \left\{ f(t^+) \frac{dt^+}{du} - f(t^-) \frac{dt^-}{du} \right\} du,$$

where the Maclaurin expansions of $f(t^\pm)dt^\pm/du$ are given by

$$f(t^\pm)\frac{dt^\pm}{du} = \pm\frac{1}{2}u^{-1/2}(1\pm u^{1/2})^v = \pm\frac{1}{2}\sum_{k=0}^{\infty}(\pm)^k\binom{v}{k}u^{(k-1)/2}$$

valid in $|u| < 1$. Then, on inverting the order of summation and integration we obtain

$$e^{-x}I_1(x) = \frac{1}{\sqrt{\pi}}\sum_{k=0}^{\infty}\binom{v}{2k}\int_0^1 e^{-xu}u^{k-\frac{1}{2}}du$$

$$= \frac{1}{\sqrt{\pi}}\sum_{k=0}^{\infty}\binom{v}{2k}\frac{\Gamma(k+\frac{1}{2})}{x^{k+\frac{1}{2}}}P(k+\tfrac{1}{2},x)$$

$$= \sum_{k=0}^{\infty}\frac{(-\frac{1}{2}v)_k\,(\frac{1}{2}-\frac{1}{2}v)_k}{k!\,x^{k+\frac{1}{2}}}P(k+\tfrac{1}{2},x) \equiv S_0(x), \qquad (3.3.20)$$

where we have used the reflection and duplication formulas for the gamma function. Since we have used the maximal interval of integration $[0, 1]$ in the u-plane in the derivation of (3.3.20), there will be the associated slow algebraic decay of the late terms in $S_0(x)$. From (A.6) and Stirling's formula for the gamma function, the late terms in (3.3.20) are $e^{-x}O(k^{-v-2})$ as $k \to \infty$ and the series is consequently absolutely convergent when $\mathrm{Re}\,(v) > -1$.

The integral $I_2(x)$, with $t = 1 + u^{1/2}$ $(u \geq 1)$, is

$$e^{-x}I_2(x) = \frac{1}{\sqrt{\pi}}\int_2^{\infty} e^{-x(t-1)^2}t^v dt = \frac{1}{\sqrt{\pi}}\int_1^{\infty} e^{-xu}f(t)\frac{dt}{du}du.$$

We shall adopt the subdivision of the u-axis following Scheme B in §3.2.3 and accordingly expand the integrand about the points $u = \Omega_n$ $(n \geq 1)$, where Ω_n will be suitably chosen. With $u = \Omega_n + w$, we have

$$f(t)\frac{dt}{dw} = \frac{(1+\Omega_n^{1/2}\sqrt{1+w/\Omega_n})^v}{2\Omega_n^{1/2}\sqrt{1+w/\Omega_n}} = A_n\sum_{k=0}^{\infty}\frac{c_{k,n}}{k!}w^k$$

valid in $|w| < \Omega_n$, where

$$A_n = \frac{(1+\Omega_n^{1/2})^v}{2\Omega_n^{1/2}}, \qquad c_{0,n} = 1.$$

The contribution from the nth interval $-\frac{1}{2}\omega_n \leq u - \Omega_n \leq \frac{1}{2}\omega_n$ becomes

$$\frac{e^{-\Omega_n x}}{\sqrt{\pi}}\int_{-\frac{1}{2}\omega_n}^{\frac{1}{2}\omega_n} e^{-xw}f(t)\frac{dt}{dw}dw = e^{-\Omega_n x}S_n(x),$$

where, from (3.2.21) and (3.2.23),

$$S_n(x) = \frac{A_n}{\sqrt{\pi}}\sum_{k=0}^{\infty}\frac{c_{k,n}}{x^{k+1}}\Delta P(k+1,\tfrac{1}{2}\omega_n x).$$

This yields the expansion

$$e^{-x} I_2(x) = \sum_{n=1}^{\infty} e^{-\Omega_n x} S_n(x); \qquad (3.3.21)$$

the Hadamard expansion of $I(x) = I_1(x) + I_2(x)$ is then given by the sum of (3.3.20) and (3.3.21).

The modified form of the zeroth series $S_0(x)$ in (3.3.20) is given by

$$S_0(x) = \sum_{k=0}^{M_0-1} \frac{c_{k,0}}{x^{k+\frac{1}{2}}} P(k + \tfrac{1}{2}, x) + T_{M_0,0}(x) \qquad (3.3.22)$$

for a suitably chosen index M_0, where we have put $c_{k,0} \equiv (-\tfrac{1}{2}\nu)_k(\tfrac{1}{2} - \tfrac{1}{2}\nu)_k/k!$ and, from (3.2.15) with $\omega_0 = 1$,

$$T_{M_0,0}(x) = e^{-x} \sum_{r=0}^{\infty} \sigma_{r,0}\, (x/M_0)^r. \qquad (3.3.23)$$

From (3.2.16), the coefficients in this case can be expressed in terms of a truncated Gauss hypergeometric series of unit argument as

$$\sigma_{r,0} = M_0^r \sum_{k=M_0}^{\infty} \frac{c_{k,0}}{\Gamma(k + r + \frac{3}{2})} = \frac{M_0^r}{\Gamma(r + \frac{3}{2})}\, {}_2F_1(-\tfrac{1}{2}\nu, \tfrac{1}{2} - \tfrac{1}{2}\nu; r + \tfrac{3}{2}; 1)_{\downarrow M_0},$$

where the symbol $\downarrow M_0$ denotes that the first M_0 terms of the hypergeometric series are to be deleted. Application of the Gauss summation formula in (2.2.14) then shows that

$$\frac{\sigma_{r,0}}{M_0^r} = \frac{\Gamma(r + \nu + 1)}{\Gamma(r + \frac{1}{2}\nu + \frac{3}{2})\,\Gamma(r + \frac{1}{2}\nu + 1)} - \sum_{k=0}^{M_0-1} \frac{c_{k,0}}{\Gamma(k + r + \frac{3}{2})}.$$

In computations the tail $T_{M_0,0}(x)$ is truncated at the N_0th term.

The behaviour of the terms (on a $\log_{10}$ scale) of the modified zeroth series (3.3.22) and (3.3.23) and the first two series in (3.3.21) against ordinal number k is presented in Fig. 3.6. The graph shows the particular case $x = 15$ and $\nu = \tfrac{1}{4}$, with the intervals for $I_2(x)$ chosen to satisfy $\omega_n/(2\Omega_n) = \vartheta = \tfrac{1}{2}\ (n \geq 1)$; from (3.2.25), the expansion points are therefore $\Omega_1 = 2, \Omega_2 = 6, \Omega_3 = 18, \ldots$. The coefficients $c_{k,n}$ appearing in $S_n(x)$ in (3.3.21) are easily computed using the series inversion facility in *Mathematica*. In Table 3.4 we present, for different values of x with $\nu = \tfrac{1}{4}$, the absolute error in the computation of $e^{-x} I_1(x)$ and $e^{-x} I_2(x)$ from (3.3.20) and the first two levels of (3.3.21). The truncation indices employed are fixed at $M_0 = N_0 = 30$ for $S_0(x)$ and $M_1 = 50, M_2 = 40$ for $S_n(x)$, $n = 1, 2$. The resulting error is determined by comparison with the values obtained by high-precision numerical quadrature. Also shown for comparison is the error associated with the optimally

Table 3.4 *The absolute error in $e^{-x}I_1(x)$ and $e^{-x}I_2(x)$ for different x when $v = \frac{1}{4}$ and $\vartheta = \frac{1}{2}$, with the truncation indices cited in the text; the error in the optimally truncated asymptotic series for $e^{-x}I(x)$ is also shown for comparison*

x	Asymptotic error	|Error|	
		$e^{-x}I_1(x)$	$e^{-x}I_2(x)$
5	1.154×10^{-5}	5.431×10^{-37}	2.514×10^{-21}
10	1.049×10^{-7}	8.590×10^{-30}	1.002×10^{-23}
15	2.215×10^{-10}	1.835×10^{-26}	6.258×10^{-26}
20	3.902×10^{-13}	1.028×10^{-24}	3.929×10^{-28}

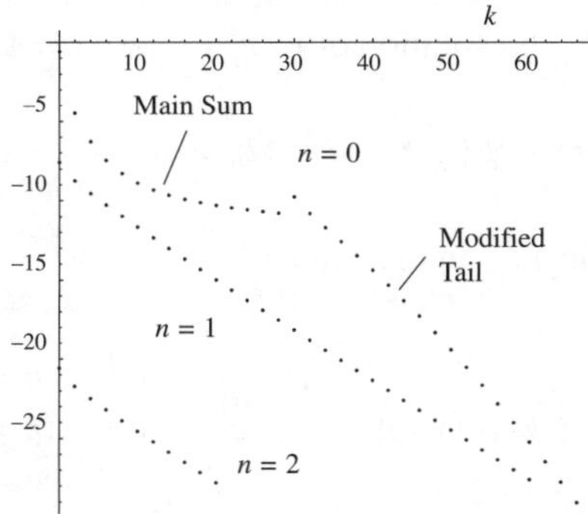

Figure 3.6 The behaviour of the terms (on a $\log_{10}$ scale) in the Hadamard expansion for the integral $e^{-x}I(x)$ when $x = 15$ and $v = \frac{1}{4}$ against ordinal number k.

truncated asymptotic series for $I(x)$, obtained by formally replacing $P(k + \frac{1}{2}, x)$ in (3.3.20) by unity, namely

$$e^{-x}I(x) \sim \sum_{k=0}^{\infty} \frac{(-\frac{1}{2}v)_k(\frac{1}{2} - \frac{1}{2}v)_k}{k!\,x^{k+\frac{1}{2}}} \qquad (x \to +\infty);$$

see also Example 1.1. The integral $I(x)$ has been considered for complex values of x in Paris (2004b).

Example 3.5 We consider the Laplace-type integral

$$J(x) = \int_{-\infty+ic}^{\infty+ic} e^{-x\psi(t)}dt \qquad (c > 0), \tag{3.3.24}$$

where the phase function is given by

$$\psi(t) = t^2 + \frac{2}{t} - 3e^{-2\pi i/3}, \tag{3.3.25}$$

$x > 0$, and the integration path is taken parallel to the real axis; see Lauwerier (1966, p. 52) and also Example 1.7. The phase function has a pole at $t = 0$ and three saddles on the unit circle situated at $t = 1$, $e^{\pm 2\pi i/3}$. The saddles at $t = e^{\pm 2\pi i/3}$ are of unit height, whereas the saddle at $t = 1$ has height $\exp\{-\frac{3}{2}x(1 + i\sqrt{3})\}$ and so is subdominant for $x > 0$. The steepest descent paths through the saddles are illustrated in Fig. 3.7; with $c > 0$, the integration path can be deformed to pass over the saddles at $t = e^{2\pi i/3}$ and $t = 1$, approaching the singularity at $t = 0$ in a direction situated in $\mathrm{Re}(t) > 0$.

We introduce the new variable given by $u = \psi(t) - \psi(t_s)$, where t_s denotes either of the saddles at $e^{2\pi i/3}$ or 1. The presence of the pole in $\psi(t)$ results in a division of the steepest descent path into two separate halves, on each of which u runs from 0 to ∞. We employ the subdivision of the u-axis according to Scheme A illustrated in Fig. 3.1(a). For each steepest descent path, we then select a suitable sequence of values of ω_n – and hence, by (3.2.1), the expansion points Ω_n – and determine the corresponding values of $t_n^\pm$ by solution of $\psi(t_n^\pm) - \psi(t_s) = \Omega_n$ $(n \geq 1)$. Inversion of

$$\psi(t) - \psi(t_n^\pm) = u - \Omega_n = w \quad (n \geq 0)$$

then yields the coefficients $\hat{c}_{k,n}^\pm$ in the expansion

$$\frac{dt^\pm}{dw} = \sum_{k=0}^{\infty} \frac{\hat{c}_{k,n}^\pm \, w^{(k+1)\mu_n - 1}}{\Gamma((k+1)\mu_n)}, \qquad \mu_n = \begin{cases} \frac{1}{2} & (n = 0) \\ 1 & (n \geq 1) \end{cases} \tag{3.3.26}$$

associated with each path; compare (3.2.6). Then, from (3.2.8) and (3.2.9), it follows that $J(x)$ can be written as the sum of two expansions about the saddles $t_s = e^{2\pi i/3}$, 1 of the form

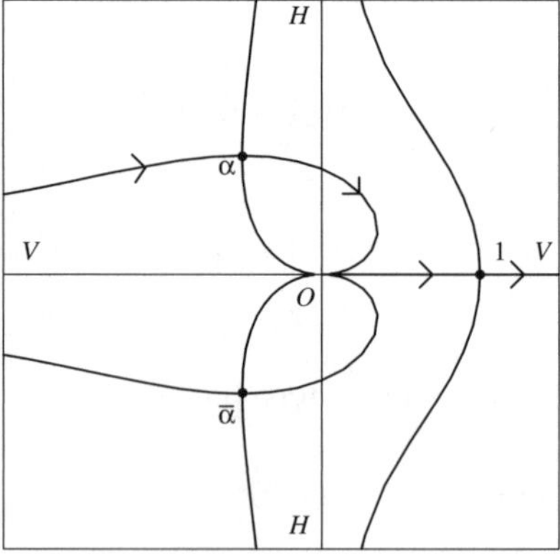

Figure 3.7 The steepest descent and ascent paths for the phase function in (3.3.25). The saddle points are denoted by heavy dots with $\alpha = \exp(2\pi i/3)$. The valleys (V) and hills (H) at infinity are indicated and arrows denote the direction of integration.

$$e^{-x\psi(t_s)} \sum_{k=0}^{\infty} e^{-\Omega_n x} S_n(x),$$

where

$$S_n(x) = \sum_{k=0}^{\infty} \frac{c_{k,n}}{x^{k+\mu_n}} P(k+\mu_n, \omega_n x)$$

with $c_{k,0} = 2\hat{c}_{2k,0}$, $c_{k,n} = \hat{c}^+_{k,n} - \hat{c}^-_{k,n}$ ($n \geq 1$).

For the dominant saddle $t_s = e^{2\pi i/3}$, we choose levels with $n \leq 3$ and the expansion points $\Omega_0 = 0$, $\Omega_1 = 2$, $\Omega_2 = 3$, $\Omega_3 = 4.5$ and $\Omega_4 = 6.75$, so that by (3.2.1) $\omega_0 = 2$, $\omega_1 = 1$, $\omega_2 = 1.5$ and $\omega_3 = 2.25$. Each subdivision with $n \geq 1$ corresponds to the ratio $\omega_n/\Lambda_n = \frac{1}{2}$; see (3.2.10). For the subdominant saddle $t = 1$ we use only the zeroth interval and choose $\omega_0 = 2$. Note that the choice $\omega_0 = 2$ for the zeroth interval about each saddle corresponds to the value $\vartheta = 2/(3\sqrt{3}) \doteq 0.385$, as the radius of convergence of the inversion expansion (3.3.26) about each saddle is $3\sqrt{3}$; see Example 1.7.

The results of computations with $x = 10$ are shown in Fig. 3.8. In the first figure we display the behaviour of the terms in each of the zeroth Hadamard series $\exp\{-x\psi(t_s)\}S_0(x)$ containing the dominant and subdominant saddles. In the second figure we display the terms in the Hadamard series at levels $1 \leq n \leq 3$ on the positive side of the dominant saddle; the terms on the negative side possess a similar behaviour. It is seen that with our choice of intervals ω_n, the terms at each level exhibit a rapid decay, so that high-precision evaluation of $J(10)$ can be achieved without the need for employing the modified form of expansions given in (3.2.14) and (3.2.15). In the computations, we truncated the series $S_n(10)$ after M_n terms, where $M_0 = M_1 = 40$, $M_2 = 20$ and $M_3 = 5$ for the dominant saddle, and $M_0 = 5$

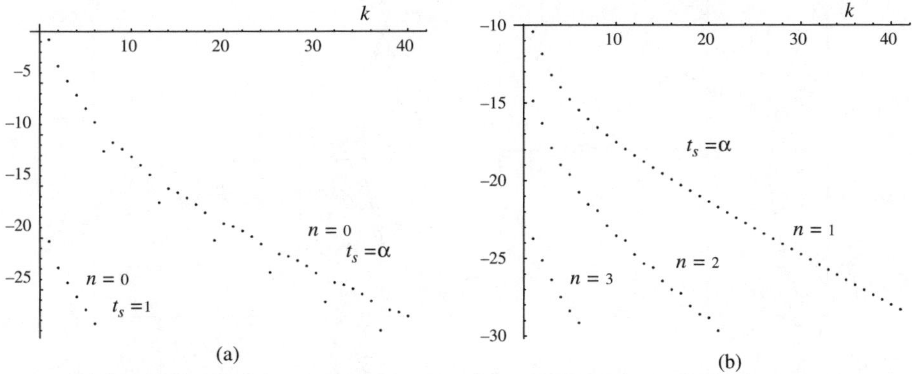

Figure 3.8 Magnitude of the terms (on a $\log_{10}$ scale) against ordinal number k in the Hadamard series for $J(x)$ at different levels when $x = 10$: (a) the terms in the zeroth series containing the saddles at $t_s = \alpha$ ($\alpha = e^{2\pi i/3}$) and $t_s = 1$ and (b) the terms in levels $1 \leq n \leq 3$ on the positive side of the saddle at $t_s = \alpha$.

for the subdominant saddle. The value of $J(10)$ from the Hadamard expansion was compared with that obtained from high-precision numerical quadrature of (3.3.24) with $c = \frac{1}{2}$ and the absolute error was found to be of order 10^{-28}.

Example 3.6 As our final example of this section, we return to the Airy function $\mathrm{Ai}(z^{2/3})$, but now of complex argument. We have

$$\mathrm{Ai}(z^{2/3}) = \frac{z^{1/3}}{2\pi i} \int_{\infty e^{-i(\pi+\theta)/3}}^{\infty e^{i(\pi-\theta)/3}} e^{-z(t-t^3/3)}\,dt, \tag{3.3.27}$$

where[8] $z = xe^{i\theta}$ and $\theta = \arg z$. There are two saddle points situated at $t = \pm 1$; the integration path is taken to be the path of steepest descent through $t = 1$ which passes to infinity in the directions $\arg t = \frac{1}{3}(\pm\pi - \theta)$. The variation in this path for θ in the range $0 \leq \theta \leq \pi + \epsilon$, where ϵ is a small positive angle, is shown in Fig. 3.9. When $\theta = \pi$ (that is $\arg \hat{z} = \frac{2}{3}\pi$, where $\hat{z} = z^{2/3}$), the steepest descent path

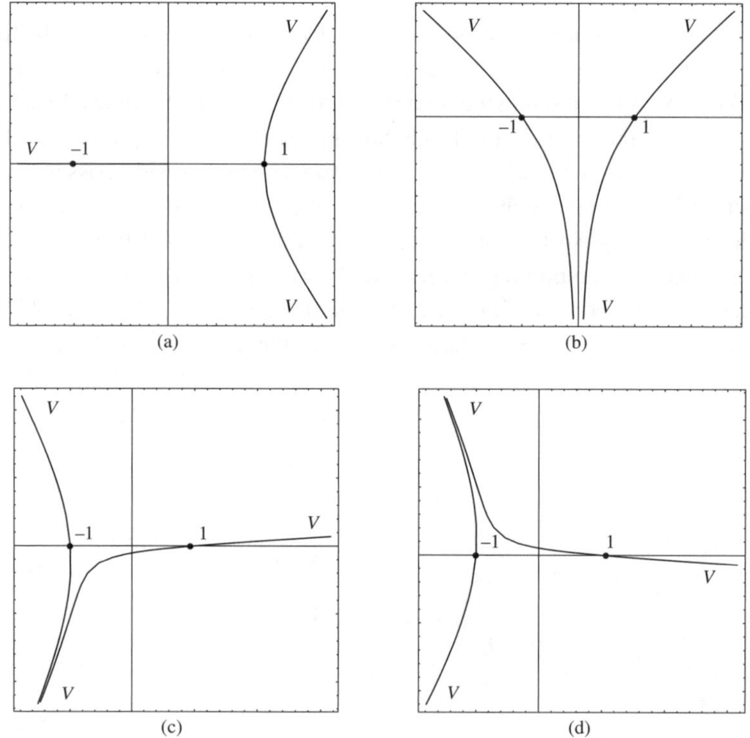

Figure 3.9 The change in the topology of the steepest descent paths for the Airy function in (3.3.27) for (a) $\theta = 0$, (b) $\theta = 0.5\pi$, (c) $\theta = 0.95\pi$ and (d) $\theta = 1.05\pi$. The asymptotic valleys (V) at infinity are indicated.

[8] The standard representation $\mathrm{Ai}(\hat{z})$ is obtained by putting $\hat{z} = z^{2/3}$, so that $\arg \hat{z} = \frac{2}{3}\theta$.

through $t = 1$ connects with the saddle at $t = -1$ to produce a Stokes phenomenon; see §1.7.

We shall pay particular attention in this example to the manner in which the Stokes phenomenon manifests itself in the Hadamard expansion process. We shall consider only the case $\theta \geq 0$; the treatment when $\theta \leq 0$ is similar or can be obtained by taking the conjugate expressions. We shall also limit our consideration to $\theta \leq \pi$ (that is, $\arg \hat{z} \leq \frac{2}{3}\pi$), since computation beyond $\theta = \pi$ can be achieved by use of the connection formula (Abramowitz and Stegun, 1965, p. 446)

$$\mathrm{Ai}(\hat{z}) + e^{2\pi i/3}\mathrm{Ai}(\hat{z}e^{2\pi i/3}) + e^{-2\pi i/3}\mathrm{Ai}(\hat{z}e^{-2\pi i/3}) = 0. \tag{3.3.28}$$

Proceeding in a similar manner to that for the Airy function of positive argument in Example 3.3, we set $\psi(t) = (t - \frac{1}{3}t^3)e^{i\theta}$ and introduce the transformation

$$u = \psi(t) - \psi(1) = (t - \frac{1}{3}t^3 - \frac{2}{3})e^{i\theta}.$$

This maps the integration path into a positively orientated loop surrounding the positive real u-axis, to find

$$e^{2z/3}\mathrm{Ai}(z^{2/3}) = \frac{z^{1/3}}{2\pi i} \int_0^\infty e^{-xu} \left(\frac{dt^+}{du} - \frac{dt^-}{du}\right) du,$$

where $t^{\pm}(u)$ denote the upper and lower halves of the steepest descent path through $t_s = 1$. The inverse of the above transformation can be obtained by a slight modification of that given in (3.3.14) to yield

$$\frac{dt^{\pm}}{du} = \pm i e^{-i\theta} \sum_{k=0}^{\infty} \frac{(\mp i)^k c_{k,0}}{\Gamma(\frac{1}{2}k + \frac{1}{2})} (ue^{-i\theta})^{(k-1)/2}, \qquad c_{k,0} = \frac{\Gamma(\frac{3}{2}k + \frac{1}{2})}{2 \cdot 3^k \, k!}. \tag{3.3.29}$$

The circle of convergence of this expansion is controlled by the presence of the saddle at $t = -1$, where $u(-1) = -\frac{4}{3}e^{i\theta}$, and so is the disc $|u| < \Lambda_0$, where $\Lambda_0 = \frac{4}{3}$.

We again adopt the subdivision Scheme B in §3.2.3. The contribution from the zeroth interval $0 \leq u \leq \omega_0$ containing the saddle point $t_s = 1$, where we set $\omega_0 = \Lambda_0 \vartheta = \frac{4}{3}\vartheta$ with $0 < \vartheta < 1$, is

$$\frac{z^{1/3}}{\pi} S_0(z),$$

where

$$S_0(z) = \sum_{k=0}^{\infty} \frac{(-)^k c_{2k,0}}{z^{k+\frac{1}{2}}} P(k + \tfrac{1}{2}, \tfrac{4}{3}\vartheta x); \tag{3.3.30}$$

compare (3.3.15). We remark that, because we are following a steepest descent path, the argument of the incomplete gamma function is positive real, even though z is complex.

The points $t_n^{\pm}$ on the steepest descent path in the t-plane that correspond to the points $u = \Omega_n^{\pm}$ on the real u-axis are given by appropriate solution of the cubic $u(t_n^{\pm}) = \Omega_n^{\pm}$, that is,

$$t_n^{\pm} - \tfrac{1}{3}(t_n^{\pm})^3 - \tfrac{2}{3} = \Omega_n^{\pm} e^{-i\theta};$$

when $\arg z \neq 0$, we no longer have the simple conjugate result $t_n^- = (t_n^+)^*$ as in Example 3.3. The expansion of dt/du about $t = t_n^{\pm}$ is obtained by putting $t = t_n^{\pm} + \tau$ in (3.3.16) to find

$$\tau(1 - (t_n^{\pm})^2) - \tfrac{1}{3}\tau^3 - t_n^{\pm}\tau^2 = we^{-i\theta},$$

where $w = u - \Omega_n^{\pm}$. The inversion has the form

$$\tau = t - t_n^{\pm} = \sum_{k=1}^{\infty} d_{k,n}^{\pm}(we^{-i\theta})^k,$$

where the coefficents $d_{k,n}^{\pm}$ are specified[9] by the recursion (3.3.18). Thus we have, for $n \geq 1$,

$$\frac{dt^{\pm}}{dw} = ie^{-i\theta} \sum_{k=0}^{\infty} \frac{c_{k,n}^{\pm}}{k!} (we^{-i\theta})^k, \qquad c_{k,n}^{\pm} = -i(k+1)!\, d_{k+1,n}^{\pm} \qquad (3.3.31)$$

valid in $|w| < \Lambda_n^{\pm}$; the determination of the radii of convergence $\Lambda_n^{\pm}$ is discussed below.

Provided the interval lengths $\omega_n^{\pm}$ are chosen to lie in the domains of validity of (3.3.31), the contribution from the nth interval $-\tfrac{1}{2}\omega_n^{\pm} \leq w \leq \tfrac{1}{2}\omega_n^{\pm}$ is then

$$\frac{z^{1/3}}{2\pi i} e^{-\Omega_n^{\pm}x} \int_{-\frac{1}{2}\omega_n^{\pm}}^{\frac{1}{2}\omega_n^{\pm}} e^{-xw} \frac{dt^{\pm}}{dw} \, dw = \frac{z^{1/3}}{2\pi} e^{-\Omega_n^{\pm}x} S_n^{\pm}(z),$$

where

$$S_n^{\pm}(z) = \sum_{k=1}^{\infty} \frac{c_{k,n}^{\pm}}{z^{k+1}} \, \Delta P(k+1, \tfrac{1}{2}\omega_n^{\pm}x) \qquad (3.3.32)$$

and ΔP is defined in (3.2.22). The Hadamard expansion for $\mathrm{Ai}(z^{2/3})$ then becomes

$$e^{2z/3}\,\mathrm{Ai}(z^{2/3}) = \frac{z^{1/3}}{\pi}\left\{ S_0(z) + \frac{1}{2}\sum_{n=1}^{\infty}\{e^{-\Omega_n^+ x} S_n^+(z) - e^{-\Omega_n^- x} S_n^-(z)\}\right\} \qquad (3.3.33)$$

when $0 \leq \arg z < \pi$. In the case $z > 0$, the series $S_n^-(z) = -(S_n^+(z))^*$ (where the asterisk denotes the complex conjugate) and the $\Omega_n^{\pm}$ are equal, so that (3.3.33) reduces to (3.3.19).

---

[9] We remark that the $d_{k,n}^{\pm}$ are explicitly independent of θ; they are implicitly dependent on θ through the presence of $t_n^{\pm}$.

It now remains to consider the radii of convergence $\Lambda_n^{\pm}$ of the expansions (3.3.31). For $0 \leq \theta < \pi$, the expansions about $u = \Omega_n^+$ ($n \geq 1$) on the upper half of the steepest descent path are controlled by the saddle at $t = 1$, so that $\Lambda_n^+ = \Omega_n^+$ can be chosen according to (3.2.25). The expansion of dt/dw at $u = \Omega_n^-$ on the lower half of the steepest descent path, however, can be controlled by either saddle $t = \pm 1$ depending on the value of θ, and we find

$$\Lambda_n^- = \min\{\Omega_n^-, |\Omega_n^- + \tfrac{4}{3}e^{i\theta}|\} \qquad (n \geq 1). \tag{3.3.34}$$

Thus, when $0 \leq \theta \leq \tfrac{1}{2}\pi$, it is seen that $\Lambda_n^{\pm}$ are equal and the expansion points $\Omega_n^{\pm}$ can be taken to be the same on both halves of the steepest descent path and given by (3.2.25). When $\tfrac{1}{2}\pi < \theta < \pi$, the Λ_n^- become θ-dependent. For example, with the zeroth interval length set at $\omega_0 = \tfrac{4}{3}\vartheta$ we have from (3.2.2), together with the requirement of a common geometric decay factor ϑ at each level,

$$\tfrac{1}{2}\omega_1^- = \Omega_1^- - \tfrac{4}{3}\vartheta, \qquad \frac{\omega_1^-}{2\Lambda_1^-} = \vartheta.$$

Using these equations combined with (3.3.34), we find upon some straightforward algebra that

$$\Omega_1^- = \min\left\{\frac{4\vartheta}{3(1-\vartheta)}, \frac{8\vartheta(1+\vartheta\cos\theta)}{3(1-\vartheta^2)}\right\}.$$

This yields

$$\Omega_1^- = \begin{cases} \dfrac{4\vartheta}{3(1-\vartheta)} = \Omega_1^+ & (\tfrac{1}{2}\pi \leq \theta \leq \theta_0) \\[2ex] \dfrac{2(1+\vartheta\cos\theta)}{1+\vartheta}\,\Omega_1^+ & (\theta_0 \leq \theta < \pi), \end{cases}$$

where

$$\theta_0 = \arccos\left(\frac{-1+\vartheta}{2\vartheta}\right).$$

In the particular case $\vartheta = \tfrac{1}{2}$, we therefore have $\theta_0 = \tfrac{2}{3}\pi$ and $\Omega_1^- = \tfrac{4}{3}$ ($\tfrac{1}{2}\pi \leq \theta \leq \tfrac{2}{3}\pi$), $\Omega_1^- = \tfrac{8}{9}(2+\cos\theta)$ ($\tfrac{2}{3}\pi \leq \theta < \pi$). The other expansion points Ω_n^- ($n \geq 2$) can be obtained similarly from $\omega_n^-/(2\Lambda_n^-) = \vartheta$ combined with the geometrical constraint (3.2.2) expressed in the form

$$\tfrac{1}{2}\omega_n^- = \Omega_n^- - (\Omega_{n-1}^- + \tfrac{1}{2}\omega_{n-1}^-).$$

In Table 3.5, we show the first four expansion points Ω_n^- for different values of θ in the range $\tfrac{1}{2}\pi \leq \theta < \pi$ when $\vartheta = \tfrac{1}{2}$ (and with $\omega_0 = \tfrac{4}{3}\vartheta$). When $0 \leq \theta < \tfrac{1}{2}\pi$, the values of Ω_n^- are independent of θ and are equal to the values when $\theta = \tfrac{1}{2}\pi$. It is clear that if the geometric decay factor ϑ is maintained at a fixed value the expansion points Ω_n^- (with $n \geq 1$) on the lower half of the steepest descent path shrink progressively as $\theta \to \pi$. Put another way, we may say that the different

Table 3.5 *The variation of Ω_n^- in the range $\frac{1}{2}\pi \leq \theta < \pi$ when $\vartheta = \frac{1}{2}$*

θ/π	Ω_1^-	Ω_2^-	Ω_3^-	Ω_4^-
0.50	1.33333	4.00000	12.00000	36.00000
0.60	1.33333	3.81647	10.93003	32.06761
0.70	1.25530	3.13938	8.16435	23.03018
0.80	1.05865	2.09059	4.47172	11.37688
0.90	0.93239	1.41721	2.10088	3.89497
0.95	0.89983	1.24355	1.48945	1.96539

Table 3.6 *The absolute error in $e^{2z/3}\mathrm{Ai}(z^{2/3})$ as a function of $\theta = \arg z$ for different $|z|$ when $\vartheta = \frac{1}{2}$; the truncation indices are indicated in the text*

θ/π	$|z| = 15$	$|z| = 20$
0	6.109×10^{-26}	2.080×10^{-28}
0.10	6.105×10^{-26}	2.111×10^{-28}
0.30	6.102×10^{-26}	2.369×10^{-28}
0.50	5.769×10^{-26}	2.984×10^{-28}
0.60	2.491×10^{-26}	3.492×10^{-28}
0.70	2.786×10^{-25}	3.019×10^{-28}
0.80	3.953×10^{-26}	3.143×10^{-28}

exponential levels in the Hadamard expansion corresponding to this half of the path 'sense' the approaching Stokes line at $\theta = \pi$ by 'squeezing up' progressively, and that the angular zone over which this takes place is limited to $\frac{1}{2}\pi$ either side of the Stokes line. This has the consequence that the expansion (3.3.33) will suffer from a progressive loss of accuracy (for a fixed number of levels n) as $\theta \to \pi$.

We present in Table 3.6 the results of computation of $e^{2z/3}\mathrm{Ai}(z^{2/3})$ using (3.3.33) for different $\theta = \arg z$ when $\vartheta = \frac{1}{2}$. We have used $n \leq 2$ and taken the truncation indices $M_0 = M_1 = 60$, $M_2 = 30$ throughout, except when $\theta = 0.8\pi$ where we included the additional level $n = 3$ to maintain accuracy and took $M_2 = 40$, $M_3 = 20$. We have not gone beyond $\theta = 0.8\pi$ in this table since the 'squeezing up' effect of the expansion points becomes too severe and would require a prohibitive increase in the number of levels used to maintain accuracy as $\theta \to \pi$.

Hadamard expansion near the Stokes line

To overcome the problem of the progressive loss of accuracy at a fixed number of levels n as we approach the Stokes line $\theta = \pi$, we can proceed as follows. In Fig. 3.10(a)

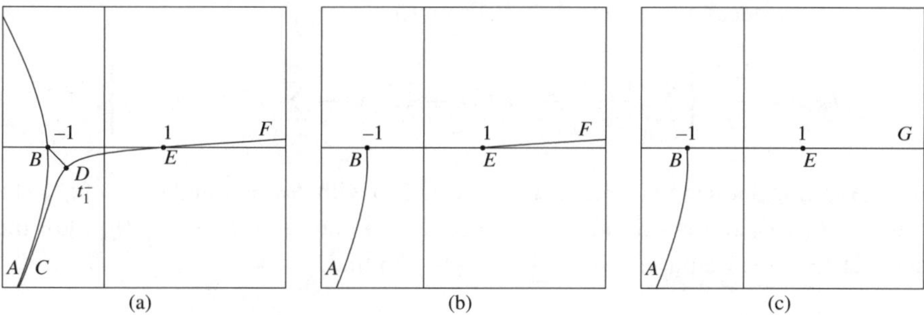

Figure 3.10 Possible integration paths when $\theta \simeq \pi$.

we show the topology of the steepest descent paths when $\theta = \pi - \epsilon$, where ϵ is small and positive. The steepest descent path through $t = 1$ is denoted by $CDEF$; the point labelled D on the lower half of this path corresponding to $u = \frac{4}{3}\vartheta$ in the u-plane is denoted by t_1^-. As $\theta \to \pi$, the portion CD of the path steadily approaches the steepest descent path AB through the saddle at $t = -1$, with the result that, when $\theta = \pi$, the steepest descent path in (3.3.27) becomes the path AB followed by the real t-axis over $[-1, \infty)$. Accordingly, when θ is in the vicinity of the Stokes line, we could take the integration path to be the path $ABDEF$, with the portion BD being a ray from $t = -1$ to the point t_1^-. This approach was employed in Paris (2004b), where it was shown that the contribution from BD is

$$I_{BD} = \frac{z^{1/3}}{2\pi i}\, e^{4z/3} \sum_{k=0}^{\infty} \frac{c_{k,0}}{z^{(k+1)/2}}\, P(\tfrac{1}{2}k + \tfrac{1}{2}, \tfrac{4}{3}(z + \vartheta x)),$$

where $c_{k,0}$ are defined in (3.3.29). To ensure the rapid convergence of this series, it is best to choose $\vartheta = 1$ for the zeroth interval; in this case, the series converges in $|1 + e^{-i\theta}| \le 1$, that is in the sector $|\theta - \pi| \le \pi/3$. The choice $\vartheta = 1$ means that the zeroth interval containing the saddle point $t = 1$ is a maximal integration interval, with the consequent necessity of using the modified form of the expansion given in (3.2.14) for the zeroth series.

Alternative possibilities are to follow the steepest descent paths AB and EF with the inter-saddle path being taken along the real axis, or to follow the path AB and integration along the portion of the real axis BEG. Both these choices, which correspond to the steepest descent path when $\theta = \pi$, are illustrated in Fig. 3.10(b) and (c). Here we have relaxed the condition of strict adherence to following steepest descent paths, as with the segment BD above, which has the consequence that some of the associated Hadamard series will no longer involve incomplete gamma functions of positive real argument.

We present the details only for the choice of integration path shown in Fig. 3.10(b). We write

$$e^{2z/3}\mathrm{Ai}(z^{2/3}) = I_{EF} + I_{BE} + I_{AB} \tag{3.3.35}$$

and choose the decay parameter $\vartheta = \frac{1}{2}$. From (3.3.33)

$$I_{EF} = \frac{z^{1/3}}{2\pi}\left\{\sum_{k=0}^{\infty}\frac{(-i)^k c_{k,0}}{z^{(k+1)/2}}\,P(\tfrac{1}{2}k+\tfrac{1}{2},\tfrac{2}{3}x) + \sum_{n=1}^{\infty}e^{-\Omega_n^+ x}S_n^+(z)\right\},$$

where $S_n^+(z)$ is defined in (3.3.32), Ω_n^+ by (3.2.25) with $\Lambda_0 = \frac{4}{3}$ and $\omega_n^+ = \Omega_n^+$. The treatment of I_{AB} follows an identical procedure to that used to derive I_{EF}. For the saddle at $t = -1$, we put $u = (t - \frac{1}{3}t^3 + \frac{2}{3})e^{i\theta}$ to find

$$\frac{dt^{\pm}}{du} = e^{-i\theta}\sum_{k=0}^{\infty}\frac{(\pm)^{k+1}c_{k,0}}{\Gamma(\tfrac{1}{2}k+\tfrac{1}{2})}\,(ue^{-i\theta})^{(k-1)/2} \qquad (|u| < \tfrac{4}{3}),$$

where the coefficients $c_{k,0}$ are given in (3.3.29) and $t^{\pm}(u)$ refer to the *lower and upper* halves respectively of the steepest descent path through $t = -1$. Then we obtain

$$I_{AB} = -\frac{z^{1/3}}{2\pi i}\,e^{4z/3}\left\{\sum_{k=0}^{\infty}\frac{c_{k,0}}{z^{(k+1)/2}}\,P(\tfrac{1}{2}k+\tfrac{1}{2},\tfrac{2}{3}x) + \sum_{n=1}^{\infty}e^{-\Omega_n^+ x}\hat{S}_n^+(z)\right\},$$

where the Ω_n^+ have the same values as in I_{EF}. The Hadamard series $\hat{S}_n^+(z)$ have the same form as in (3.3.32), except that the coefficients $c_{k,n}^+$ are determined for the path AB and we have introduced the circumflex to distinguish these series from those in I_{EF}. The contribution from the real axis taken between $[-1, +1]$ is obtained as in Example 3.1 to find the 2-stage expansion

$$I_{BE} = \frac{z^{1/3}}{2\pi i}\,e^{2z/3}\sum_{k=0}^{\infty}\frac{c_{k,0}}{z^{(k+1)/2}}\{e^{2z/3}P(\tfrac{1}{2}k+\tfrac{1}{2},\tfrac{2}{3}z) + e^{-2z/3}P^*(\tfrac{1}{2}k+\tfrac{1}{2},-\tfrac{2}{3}z)\},$$

where $P^*(a, -z)$ is defined in (A.3). We have chosen $\vartheta = \frac{1}{2}$ in order that each Hadamard series is associated with the same geometric decay factor. Note that the incomplete gamma functions appearing in I_{BE} have complex argument since BE is not a steepest descent path (when $\theta \neq \pi$).

In Table 3.7 we show the results of computation of $e^{2z/3}\mathrm{Ai}(z^{2/3})$ using (3.3.35) for $\theta = \arg z$ in the neighbourhood of the Stokes line $\theta = \pi$. We have used $n \leq 2$ in I_{EF} and $n \leq 1$ in I_{AB}, since $t = -1$ is a subdominant saddle when $\theta \simeq \pi$. The truncation indices employed [10] are $M_0 = 100$, $M_1 = 60$ and $M_2 = 30$ for I_{EF}, $M_0 = 50$, $M_1 = 30$ for I_{AB} and $M_0 = 100$ for I_{BE}. The other choices of integration path in Fig. 3.10(a) and (c) produce similar results, except that for the choice of path $ABDEF$ it is necessary to use the modified form of the expansion for the zeroth interval about $t = 1$, which requires a greater computational effort.

---

[10] We observe that the zeroth series in I_{EF}, I_{AB} and I_{BE} all involve $\frac{1}{2}k+\frac{1}{2}$, rather than $k+\frac{1}{2}$ as in (3.3.30), and so require the use of a greater number of terms.

Table 3.7 *The absolute error in $e^{2z/3}\mathrm{Ai}(z^{2/3})$ as a function of $\theta = \arg z$ near the Stokes line for different $|z|$ when $\vartheta = \frac{1}{2}$*

θ/π	$|z| = 15$	$|z| = 20$
0.80	8.518×10^{-23}	1.164×10^{-25}
0.90	3.488×10^{-24}	1.102×10^{-25}
0.95	4.924×10^{-24}	1.099×10^{-25}
0.99	2.058×10^{-24}	1.103×10^{-25}
1.00	3.377×10^{-24}	1.097×10^{-25}

3.4 Coalescence problems

We now discuss how the Hadamard expansion process has to be modified in order to deal with various types of coalescence problems. The treatment of such problems has undergone a steady development since the seminal paper of Chester, Friedman and Ursell (1957) on Laplace-type integrals possessing a pair of coalescing saddle points; see §1.5.3. Techniques for dealing with a variety of saddle-point uniformity problems have appeared, including means for handling a saddle point near a pole, a saddle point near an endpoint of an integration contour and three or more saddle points coalescing to a single point; see §1.5 for a discussion of these approximations.

The principal effect of such coalescence on the Hadamard expansion procedure is the loss of exponential separation in the different levels of the expansion, which in turn results from the progressive shrinkage of the integration intervals ω_n. Without any corrective treatment, this would lead to excessive loss of precision when employing a fixed number of levels and so render the procedure unusable in such situations. In the following section, we explain the modification required to deal with this problem.

3.4.1 Expansion scheme for coalescence problems

It has been shown in (3.2.8) and (3.2.24) that the Hadamard expansion of the integral $J(z)$ defined in (3.1.1) taken along a steepest descent path issuing from a saddle point t_s and passing down a valley to infinity is

$$J(z) = e^{-z\psi(t_s)} \sum_{n=0}^{\infty} e^{-\Omega_n x} S_n(z),$$

where the $S_n(z)$ are specified in (3.2.9), or (3.2.20) and (3.2.23) according to the expansion scheme employed. The proximity in the u-plane of either neighbouring saddles or singularities of $f(t)$ determines the choice of the zeroth interval ω_0. In the case of one of these neighbouring saddle points (or singularities) coalescing with the saddle at t_s, then $\omega_0 \to 0$ and there will be a progressive loss of exponential separation between the different levels in the above Hadamard expansion. As a result of this, we would require progressively more levels n to be retained as coalescence takes place in order to maintain a prescribed level of accuracy.

To overcome this problem we make a small modification to the expansion Scheme B described in §3.2.3: instead of forward expansion for the zeroth interval we use *reverse expansion*, with the remaining intervals being dealt with by forward-reverse expansion as described in §3.2.3. For this, we choose arbitrarily [11] the point t_0' on the steepest descent path through t_s, with map $u = \Omega_0'$ in the u-plane determined by

$$\omega_0 = \Omega_0' = e^{i\theta}\{\psi(t_0') - \psi(t_s)\}; \tag{3.4.1}$$

see Fig. 3.1. Then, instead of expanding about t_s for the zeroth interval, we expand about the right-hand endpoint t_0' to obtain

$$\left(f(t)\frac{dt}{dw}\right)_{t_0'} = e^{-i\theta}\sum_{k=0}^{\infty}\frac{c_{k,0}'}{k!}(e^{-i\theta}w)^k, \qquad u = \Omega_0' + w; \tag{3.4.2}$$

compare (3.2.6). We shall assume the simplest case here where the radius of convergence of this expansion about t_0' is controlled by the saddle at t_s (and not by the coalescing saddle or singularity), so that integration using (3.4.2) is valid for $-\Omega_0' \leq w \leq 0$.

The contribution to $J(z)$ from the zeroth interval is then given by

$$e^{-z\psi(t_s)}\int_0^{\Omega_0'} e^{-xu} f(t)\frac{dt}{du}\,du = e^{-z\psi(t_s)-\Omega_0'x}\int_{-\omega_0}^0 e^{-xw} f(t)\frac{dt}{dw}\,dw$$

$$= e^{-z\psi(t_s)-\Omega_0'x}\sum_{k=0}^{\infty}\frac{c_{k,0}'}{k!}e^{-i(k+1)\theta}\int_{-\omega_0}^0 e^{-xw}w^k\,dw$$

$$= e^{-z\psi(t_s)}\,\mathsf{S}_0(z) \tag{3.4.3}$$

upon use of (2.2.23), where

$$\mathsf{S}_0(z) = -e^{-\Omega_0'x}\sum_{k=0}^{\infty}\frac{c_{k,0}'}{z^{k+1}}\,P(k+1,-\omega_0 x). \tag{3.4.4}$$

---

[11] Actually the choice is a compromise between a desired exponential separation between the different levels of the expansion and the rate of decay of the tail of the zeroth Hadamard series.

Since the contributions from the remaining intervals with $n \geq 1$ are the same as in (3.2.21), the Hadamard expansion of the integral $J(z)$ in (3.1.1) is therefore given by

$$J(z) = e^{-z\psi(t_s)} \sum_{n=0}^{\infty} e^{-\Omega_n x} \mathbf{S}_n(z), \qquad (3.4.5)$$

where $\mathbf{S}_n(z) \equiv S_n(z)$ $(n \geq 1)$ as defined in (3.2.23). The expansion points Ω_n $(n \geq 1)$ can be chosen to be associated with the common geometric decay factor ϑ, as in (3.2.25), so that $\omega_n = 2\vartheta\Omega_n$ and

$$\Omega_n = \omega_0 \frac{(1+\vartheta)^{n-1}}{(1-\vartheta)^n} \qquad (n \geq 1). \qquad (3.4.6)$$

The series $\mathbf{S}_0(z)$ has involved a maximal integration interval and accordingly the behaviour of its late terms will be associated with the familiar slow algebraic decay. This requires the series to be used in its modified form which, for a suitably chosen truncation index M_0, becomes

$$\mathbf{S}_0(z) = -e^{-\Omega_0' x} \sum_{k=0}^{M_0-1} \frac{c_{k,0}'}{z^{k+1}} P(k+1, -\omega_0 x) + \mathbf{T}_{M_0,0}(z), \qquad (3.4.7)$$

where, as in an analogous fashion in (3.2.15),

$$\begin{aligned}
\mathbf{T}_{M_0,0}(z) &= -e^{-\Omega_0' x} \sum_{k=M_0}^{\infty} \frac{c_{k,0}'}{z^{k+1}} P(k+1, -\omega_0 x) \\
&= \sum_{k=M_0}^{\infty} (-)^k c_{k,0}' (\omega_0 e^{-i\theta})^{k+1} \sum_{r=0}^{\infty} \frac{(-\omega_0 x)^r}{(k+r+1)!} \\
&= \sum_{r=0}^{\infty} \sigma_{r,0}' (-\xi_0)^r, \qquad \xi_0 = \frac{\omega_0 x}{M_0}
\end{aligned} \qquad (3.4.8)$$

with

$$\sigma_{r,0}' = M_0^r \sum_{k=M_0}^{\infty} \frac{(-)^k c_{k,0}'}{(k+r+1)!} (\omega_0 e^{-i\theta})^{k+1}.$$

We remark that the tail $\mathbf{T}_{M_0,0}(z)$ does not contain the exponentially small factor $\exp(-\Omega_0' x)$, in contrast to the tails $T_{M,n}(z)$ in (3.2.15), since this has cancelled with the factor $\exp(\omega_0 x)$ arising from the incomplete gamma function.

The coefficients $\sigma_{r,0}'$ can be computed in the form

$$\sigma_{r,0}' = M_0^r \left\{ s_0' - \sum_{k=0}^{M_0-1} \frac{(-)^k c_{k,0}'}{(k+r+1)!} (\omega_0 e^{-i\theta})^{k+1} \right\}, \qquad (3.4.9)$$

where

$$s_0' = \sum_{k=0}^{\infty} \frac{(-)^k c_{k,0}'}{(k+r+1)!} (\omega_0 e^{-i\theta})^{k+1}.$$

This sum can be expressed as an integral similar to that in (3.2.18). If we set $u = \omega_0 v$ $(0 \le v \le 1)$, so that from (3.4.2) $w = \omega_0(v - 1)$, then

$$s_0' = \frac{1}{r!} \int_{-\omega_0}^{0} v^r f(t) \frac{dt}{dw} \, dw = \frac{1}{r!} \int_{t_0}^{t_0'} v^r f(t) \, dt, \tag{3.4.10}$$

where, from (3.2.4) with $n = 0$,

$$v \equiv v(t) = \frac{\psi(t) - \psi(t_s)}{\omega_0 e^{-i\theta}}.$$

This representation for the infinite sum s_0' can be established by substituting the expansion (3.4.2) into the first integral appearing in (3.4.10) and evaluating the resulting integral as a beta function in an analogous manner to that used to obtain (3.2.18).

3.5 Examples of coalescence

To conclude this chapter, we present a set of examples that illustrates how the difficulty of coalescence can be overcome. The examples selected involve coalescence between saddles, a saddle and a pole, and a saddle with an endpoint of integration as discussed in §1.5. We demonstrate that high-precision results can still be obtained in the presence of such coalescence, but at the cost of the unavoidable use of the computationally more expensive modified Hadamard expansions discussed in §3.4.1.

Example 3.7 Consider the function $I(z; a)$ defined in $|\arg z| \le \frac{1}{2}\pi$ by the integral

$$I(z; a) = \int_C e^{-z\psi(t)} dt, \qquad \psi(t) = \tfrac{1}{4}t^4 - \tfrac{2}{3}t^3 + \tfrac{1}{2}(1 - a)t^2, \tag{3.5.1}$$

where $z = xe^{i\theta}$, with $x > 0$ and $\theta = \arg z$, and C is a path that starts at $t = 0$ and passes to infinity parallel to the positive imaginary axis. This integral, with the integration path running from $-\infty$ to ∞i, was used by Berry and Howls (1991) as an example of hyperasymptotic expansion. The quantity a is the coalescence parameter (here assumed to be real) and the integrand is characterised by a remote saddle at $t = 0$ and a cluster of two saddles at $1 \pm a^{1/2}$ that coalesce to form a double saddle when $a = 0$. For $a > 0$, the saddles in the cluster are situated on the real axis and are labelled $t_{sj} = 1 \mp a^{1/2}$ $(j = 1, 2)$, respectively; for $a < 0$, they move into the complex t-plane as a complex conjugate pair.

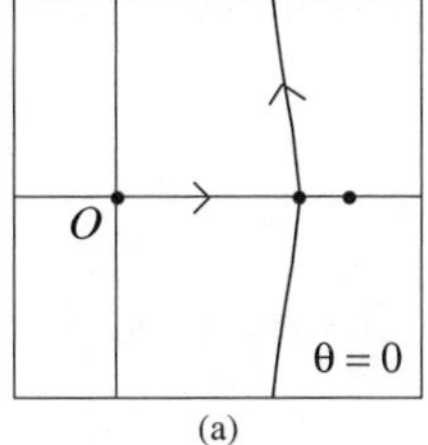

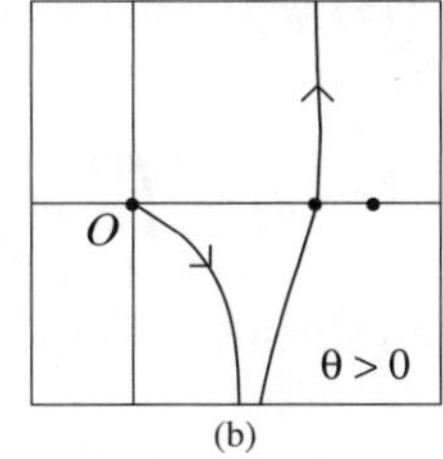

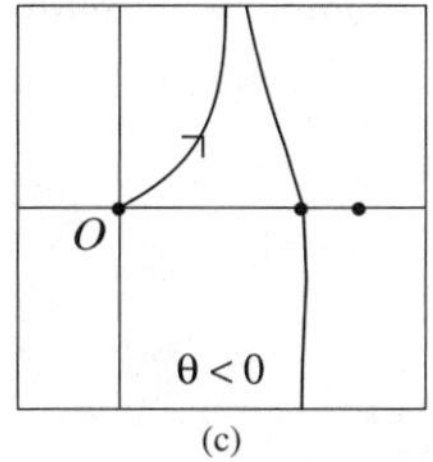

Figure 3.11 The steepest descent paths for (3.5.1) in the t-plane for different values of θ when $0 < a < 1$. The heavy dots are the saddle points and the arrows denote the direction of integration from O. Only the relevant paths are shown.

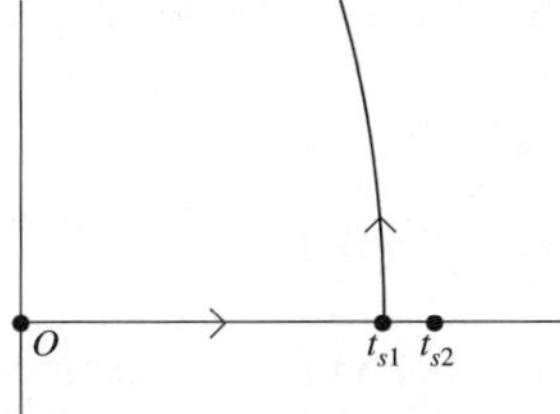

Figure 3.12 The modified integration path when $0 < a < 1$.

The function $I(z; a)$ is associated with a Stokes phenomenon[12] on arg $z = 0$ when $0 < a < 1$; see Fig. 3.11. Accordingly, we shall consider here only the case $a \to 0^+$ and small values of $|\theta|$, thereby concentrating our attention on two very different types of asymptotic phenomena, namely coalescence and the Stokes phenomenon. Typical paths of steepest descent that emanate from the origin and pass to infinity along the ray arg $t = \frac{1}{2}\pi - \frac{1}{4}\theta$ are illustrated for different θ and small $a > 0$.

The normal approach to take when z is complex would be to represent $I(z; a)$ as a sum of steepest descent contours and thence proceed to construct Hadamard expansions associated with each contour integral. This process results in series whose associated incomplete gamma functions have positive arguments, which produces the most rapid rate of convergence for the expansion. If we are willing to accept slightly less rapidly converging series, then much of the computational effort of the above path decomposition can be avoided, when a and $|\theta|$ are small, by taking as our modified integration path that illustrated in Fig. 3.12. This consists of replacing C by the segment of the real axis $[0, t_{s1}]$ (*independently of the value of* θ) followed by the upper half of the steepest descent path through the saddle t_{s1} (which, of course, is θ-dependent). In this manner, we can cover the three situations depicted in Fig. 3.11 when $0 < a < 1$ with a single path decomposition. We shall label the contributions to $I(z; a)$ from these two paths by I_1 and I_2, respectively.

---

[12] A Stokes phenomenon also occurs on portions of the curves $\theta = \theta_c(a)$ when $a < 0$, where $\tan \theta_c(a) = \frac{8}{3}(-a)^{3/2}/(\frac{1}{3} - 2a - a^2)$; see Paris (2007a).

The change of variable $u = \psi(t) - \psi(0)$ yields the contribution from the interval $[0, t_{s1}]$ in the form

$$I_1 = \int_0^{t_{s1}} e^{-z\psi(t)} dt = \int_0^{\omega_0} e^{-zu} \frac{dt}{du} du,$$

where $\omega_0 = \psi(t_{s1}) - \psi(0)$. The inversion of this change of variable leads to the expansion

$$\frac{dt}{du} = \sum_{k=0}^{\infty} \frac{c_{k,0}}{\Gamma(\frac{1}{2}k + \frac{1}{2})} u^{(k-1)/2} \qquad (|u| < \omega_0)$$

with $c_{0,0} = \{\pi/(2(1 - a))\}^{1/2}$. This expansion converges in the disc $|u| < \omega_0$ since the inversion about the saddle $t = 0$ is controlled by the saddle t_{s1}. Then we obtain the Hadamard series

$$I_1 = \sum_{k=0}^{\infty} \frac{c_{k,0}}{z^{(k+1)/2}} P(\tfrac{1}{2}k + \tfrac{1}{2}, \omega_0 z). \tag{3.5.2}$$

We note that, because the path $[0, t_{s1}]$ is not a steepest descent path (except when $\theta = 0$), the argument of the normalised incomplete gamma functions is complex. The modification of the expansion (3.5.2) into a finite main sum and a rapidly convergent tail follows the procedure detailed in §3.2.2 with an obvious modification to take into account the fact that the argument of the incomplete gamma function contains z and not x. Then we find from (3.2.14) and (3.2.15)

$$I_1 = \sum_{k=0}^{M_0-1} \frac{c_{k,0}}{z^{(k+1)/2}} P(\tfrac{1}{2}k + \tfrac{1}{2}, \omega_0 z) + e^{-\omega_0 z} \sum_{r=0}^{\infty} \sigma_r \xi^r, \tag{3.5.3}$$

where $\xi = \omega_0 z / M_0$. From (3.2.17), (3.2.18) and (3.2.19), the coefficients σ_r are specified by

$$\sigma_r = M_0^r \left\{ \frac{1}{r!} \int_0^{t_{s1}} (1 - v)^r dt - \sum_{k=0}^{M_0-1} \frac{c_{k,0}}{\Gamma(\frac{1}{2}k + r + \frac{3}{2})} \right\},$$

where, since $\psi(0) = 0$, we have $v = \psi(t)/\omega_0$.

The normal procedure for dealing with the contribution I_2 from the steepest descent path through the saddle t_{s1} follows a decomposition of the steepest descent path into segments, which leads to an expansion of type (3.2.8) or (3.2.24) depending on the choice of subdivison scheme chosen. As pointed out in §3.4.1, however, in the small a-limit the saddle t_{s2} will control the disc of convergence of the inversion about t_{s1}. This has the consequence that all the convergence intervals ω_n shrink to zero as $a \to 0$, which in turn results in a progressive loss of exponential separation between the different levels in these expansions. To overcome this difficulty, we employ the expansion procedure described in §3.4.1, which consists of reverse expansion for the zeroth interval followed by forward-reverse expansion for the remaining intervals.

We choose a point [13] t_0' on the steepest descent path through t_{s1}, with map $u = \Omega_0'$ in the u-plane determined by

$$\omega_0 = \Omega_0' = e^{i\theta}\{\psi(t_0') - \psi(t_{s1})\}, \tag{3.5.4}$$

and expand about t_0' to obtain the inversion expansion of the form (3.4.2) with the coefficients $c_{k,0}'$. In order not to overburden the notation, we shall not distinguish between the coefficients $c_{k,n}$, nor the quantities ω_0, resulting from the different integration paths. It is understood that their precise values will be different for the integrals I_1 and I_2. Then, from (3.4.4) and (3.4.5), we obtain

$$I_2 = e^{-z\psi(t_{s1})} \sum_{n=0}^{\infty} e^{-\Omega_n x} \mathbf{S}_n(z), \tag{3.5.5}$$

where

$$\mathbf{S}_0(z) = -e^{-\Omega_0' x} \sum_{k=0}^{\infty} \frac{c_{k,0}'}{z^{k+1}} P(k+1, -\omega_0 x)$$

and, for $n \geq 1$, $\mathbf{S}_n(z) \equiv S_n(z)$ as defined in (3.2.23). The quantities Ω_n ($n \geq 1$) are chosen according to (3.2.25) with $\Lambda_0 = \omega_0$. Since the zeroth integration interval is maximal, it is necessary to use the modified form of $\mathbf{S}_0(z)$ given in (3.4.7) and (3.4.8). It must be emphasised that the expansion (3.5.5) for I_2 is valid provided the circles of convergence of the various inversion expansions are controlled by the saddle t_{s1}, and not by the saddle t_{s2}. In Paris (2007a), it is shown that this is the case in the present example when (at least) $|\theta| \leq \frac{1}{2}\pi$.

To illustrate, we consider $a = 10^{-4}$ so that the saddles in the cluster are situated at $t_{s1} = 0.99$ and $t_{s2} = 1.01$. We take $|z| = 15$ and compute the expansions in (3.5.3) and (3.5.5) for different θ in the range $|\theta| \leq \frac{1}{2}\pi$. The Hadamard series in (3.5.3) and (3.5.5) (with $n = 0$) are appropriately truncated (so that the variable $|\xi| < 1$), with their corresponding modified tails computed from (3.5.3), (3.4.7) and (3.4.8). The tails of the modified expansions are truncated after N_n terms commensurate with the level of precision required. The points t_0' and t_n ($n \geq 1$) on the steepest descent path through t_{s1} are obtained from (3.5.4) with the value of $\omega_0 = 1$, and from (3.4.6) with $\vartheta = \frac{1}{2}$, $\Omega_1 = 2$, $\Omega_2 = 6, \dots$. The inversion coefficients are determined by the numerical inversion routine in *Mathematica*.

For the integral I_1, which consists of only a single Hadamard series, we employed the truncation indices $(M_0, N_0) = (25, 25)$. For the integral I_2, we consider only the first three levels $0 \leq n \leq 2$ in (3.5.5), with $(M_0, N_0) = (30, 40)$, $M_1 = 50$ and $M_2 = 10$. An example of the absolute values of the terms in the expansions (3.5.3) and (3.5.5) against ordinal number k is displayed in Fig. 3.13. The results of our computations are summarised in Tables 3.8 and 3.9 in which we present the absolute value of the error in $I(z; a)$ for different values of θ and $a \geq 0$. The exact

---

[13] In computations, it is easier to select a value of ω_0 and obtain t_0' by appropriate solution of (3.5.4).

Table 3.8 *Absolute values of the error in $I(z; a)$ for different values of θ in the range $|\theta| \leq \frac{1}{2}\pi$ when the parameter $a = 10^{-4}$ and $|z| = 15$; the Stokes line is $\theta = 0$, and the truncation indices employed are indicated in the text*

θ/π	|Error|	θ/π	|Error|
0	1.812×10^{-25}		
0.05	1.814×10^{-25}	-0.05	1.868×10^{-25}
0.10	1.873×10^{-25}	-0.10	1.989×10^{-25}
0.15	1.994×10^{-25}	-0.15	2.183×10^{-25}
0.20	2.183×10^{-25}	-0.20	2.467×10^{-25}
0.25	2.452×10^{-25}	-0.25	2.865×10^{-25}
0.50	5.662×10^{-25}	-0.50	9.508×10^{-25}

Table 3.9 *Absolute values of the error in $I(z; a)$ for different values of the parameter a as a function of θ when $|z| = 15$; the truncation indices employed are indicated in the text*

θ/π	|Error|			
	$a = 10^{-2}$	$a = 10^{-4}$	$a = 10^{-6}$	$a = 0$
0	1.986×10^{-25}	1.812×10^{-25}	1.810×10^{-25}	1.810×10^{-25}
0.05	1.200×10^{-25}	1.814×10^{-25}	1.811×10^{-25}	1.811×10^{-25}
0.10	2.077×10^{-25}	1.873×10^{-25}	1.871×10^{-25}	1.871×10^{-25}
0.15	2.216×10^{-25}	1.994×10^{-25}	1.991×10^{-25}	1.991×10^{-25}
0.20	2.431×10^{-25}	2.183×10^{-25}	2.180×10^{-25}	2.180×10^{-25}
0.25	2.736×10^{-25}	2.452×10^{-25}	2.449×10^{-25}	2.449×10^{-25}
0.50	6.198×10^{-25}	5.662×10^{-25}	5.657×10^{-25}	5.657×10^{-25}

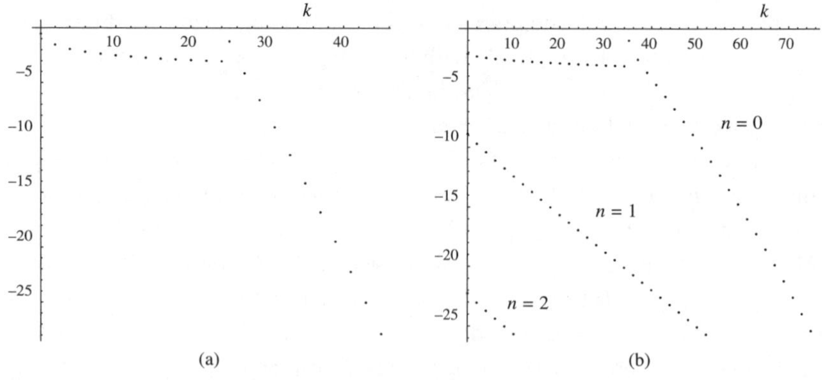

Figure 3.13 The magnitude of the terms (on a $\log_{10}$ scale) as a function of ordinal number k when $a = 10^{-4}$, $|z| = 15$ and $\theta = \pi/10$: (a) the finite main sum and modified tail of I_1 and (b) the finite main sum and modified tail (for $n = 0$) and the levels $n = 1, 2$ of I_2.

value of $I(z; a)$ was obtained by numerical quadrature using *Mathematica* with the integration path being the ray $\arg t = \frac{1}{2}\pi - \frac{1}{4}\theta$.

It is seen that the accuracy level (with fixed truncation indices) remains uniform both in θ across the Stokes line $\theta = 0$ and as the parameter $a \to 0+$. The uniformity in a is a consequence of *neither zeroth-level Hadamard series in I_1 and I_2 involving coefficients that result from expansion about the coalescing saddle t_{s1}*.

Example 3.8 Our second coalescence example involves a cluster of three saddle points. We consider the function $I(x) \equiv I(x; a, \kappa)$ defined by the Fourier integral

$$I(x) = \int_C (1+t)^{-1/2} e^{-ix\psi(t)} dt, \tag{3.5.6}$$

where

$$\psi(t) = \tfrac{1}{4}t^4 - \tfrac{1}{3}\kappa a t^3 + a^2(\tfrac{1}{2}t^2 - \kappa a t)$$

and x will be taken to be a large positive variable. The exponential factor is associated with a cluster of three saddle points situated at $t_{s1} = \kappa a$ and $t_{sj} = \pm ia$ $(j = 2, 3)$, respectively. We take a as the (real) coalescence parameter and $\kappa \geq 0$ is a parameter that alters the shape of the cluster. Here, we are primarily interested in the limit $a \to 0$ when the saddles coalesce to form a single third-order saddle at the origin. The amplitude function $f(t) = (1+t)^{-1/2}$ possesses a remote (as $a \to 0$) branch-point singularity at $t = -1$.

The exponential factor $\exp\{-ix\psi(t)\}$ has valleys at infinity centred on the rays $\arg t = (4j-1)\pi/8$ $(j = 1, 2, 3, 4)$ which we label V_j. The paths of steepest descent and ascent through the saddles are displayed in Fig. 3.14 for different values of κ. For small κ, the saddles in the cluster are not connected by a common steepest descent path. As κ increases the saddles become connected when $\mathrm{Im}\{i\psi(\pm ia) - i\psi(\kappa a)\} = 0$; that is, when $\kappa = \kappa_s$ where

$$\kappa_s = (2\sqrt{3} - 3)^{1/2} \doteq 0.68125.$$

For $\kappa > \kappa_s$, the saddle t_{s1} becomes disconnected while the other two saddles t_{s2} and t_{s3} remain connected. To illustrate our procedure, we take the integration contour C in (3.5.6) to be a path connecting the valleys V_1 and V_3 at infinity. It is then easily seen that as κ increases through the value κ_s the number of contributing saddles changes by one, with the contribution from t_{s1} being switched off. This corresponds to a Stokes phenomenon which can certainly occur for real variables; see also the treatment of the Pearcey integral in §4.3.

As in Example 3.7, we shall avoid decomposing the integration path C entirely into steepest descent paths. The paths from infinity leading to the saddles t_{s2} and t_{s3} will be chosen to coincide with the steepest descent paths through these saddles, but integration through the cluster will be along straight line segments. The actual choice of path through the cluster depends on the circle of convergence of the expansion (3.2.6) about t_{s2} or t_{s3}. For fixed a and sufficiently large κ, the saddle t_{s1} is more

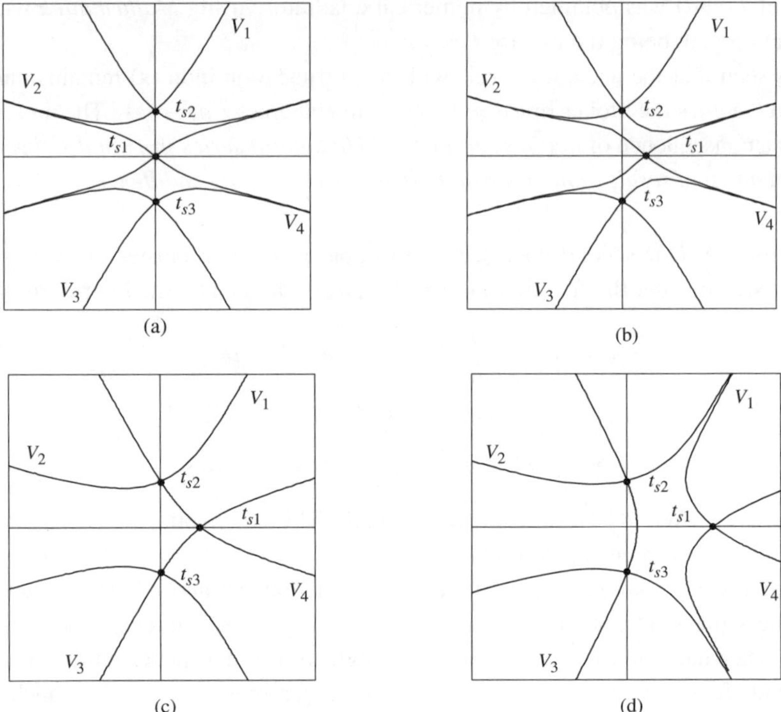

Figure 3.14 The steepest descent and ascent paths for $I(x)$ when (a) $\kappa = 0$, (b) $0 < \kappa < \kappa_s$, (c) $\kappa = \kappa_s$ and (d) $\kappa > \kappa_s$. The saddles are denoted by heavy dots and labelled t_{sj} $(j = 1, 2, 3)$. The valleys (V_j) at infinity are indicated.

distant and consequently the disc of convergence about t_{s3} will be controlled by the neighbouring saddle t_{s2}, and vice versa. As κ decreases, the saddle t_{s1} steadily approaches the other two saddles with the consequence that the disc of convergence about t_{s3} will eventually be controlled by t_{s1}. This happens when $|(\psi(ia) - \psi(-ia)|$ and $|\psi(-ia) - \psi(\kappa a)|$ are equal; that is, when

$$\tfrac{4}{3}\kappa a^4 = \tfrac{1}{12}a^4|(3 + i\kappa)(\kappa + i)|.$$

The desired solution of this equation is given by $\kappa = \kappa_*$, where $\kappa_* = \sqrt{3}$. Thus, when $0 \leq \kappa < \kappa_*$, the disc of convergence about t_{s3} is controlled by t_{s1}, while when $\kappa > \kappa_*$ the disc of convergence is controlled by t_{s2}. Based on these considerations, we take as our paths of integration when $0 \leq \kappa \leq \kappa_*$ and $\kappa > \kappa_*$ those illustrated in Fig. 3.15. Since the value of $\mathrm{Re}\{-i\psi(t)\}$ at the saddles t_{sj} $(j = 1, 2, 3)$ is 0 and $\mp 2\kappa a^4/3$, respectively, it is seen that in both cases the integration path chosen progresses from a more dominant to a less dominant saddle.

We now consider the details of the Hadamard expansion for $I(x)$ as $a \to 0$. The contributions from the steepest descent paths DA and BE are similar to that

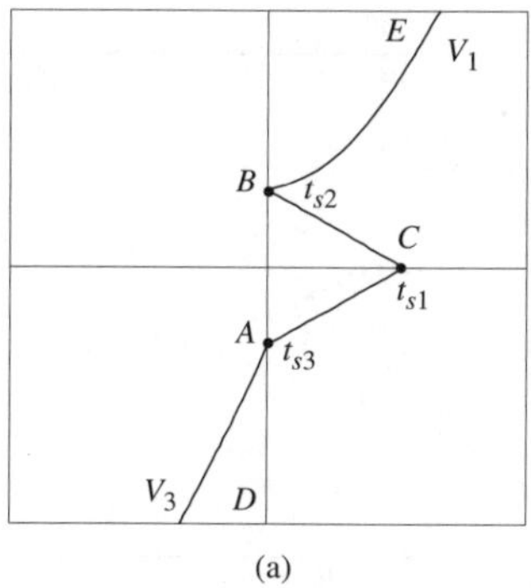

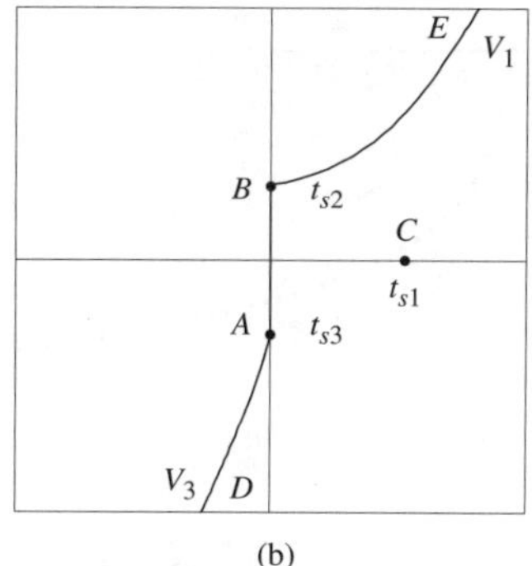

Figure 3.15 The integration paths through the cluster: (a) the path $DACBE$ when $0 \leq \kappa \leq \kappa_*$ and (b) the path $DABE$ when $\kappa > \kappa_*$.

described in Example 3.7 and are given by an expansion of the type (3.4.5), where the point t_0' on each path can be chosen independently of the coalescence parameter a. When $\kappa \geq \kappa_*$, the integration path through the cluster is AB shown in Fig. 3.15(b). With the new variable $u = i\{\psi(t) - \psi(t_{s3})\}$ the inversion expansion about the saddle point t_{s3} yields

$$(1+t)^{-1/2}\frac{dt}{du} = \sum_{k=0}^{\infty} \frac{c_{k,0}}{\Gamma(\frac{1}{2}k + \frac{1}{2})} u^{(k-1)/2} \qquad (|u| < \omega_0), \qquad (3.5.7)$$

where

$$\omega_0 = i\{\psi(t_{s2}) - \psi(t_{s3})\} = \tfrac{4}{3}\kappa a^4.$$

The contribution I_{AB} from the path AB is then given by

$$I_{AB} = \int_{t_{s3}}^{t_{s2}} (1+t)^{-1/2} e^{-ix\psi(t)} dt = e^{-ix\psi(t_{s3})} \int_0^{\omega_0} e^{-xu}(1+t)^{-1/2}\frac{dt}{du}\, du$$

$$= e^{-ix\psi(t_{s3})} \sum_{k=0}^{\infty} \frac{c_{k,0}}{x^{(k+1)/2}} P(\tfrac{1}{2}k + \tfrac{1}{2}, \omega_0 x). \qquad (3.5.8)$$

Since this contribution involves a maximal integration interval, the modified form of this Hadamard series, as given in (3.2.14) and (3.2.15), must be employed.

When $0 \leq \kappa \leq \kappa_*$, the integration path through the cluster is the path ACB shown in Fig. 3.15(a). The inter-cluster contribution I_{AB} is now replaced by $I_{AC} + I_{CB}$. The contribution I_{AC} is given by (3.5.8), with the interval ω_0 replaced by $\omega_0 = i\{\psi(t_{s1}) - \psi(t_{s3})\}$, and

$$I_{CB} = e^{-ix\psi(t_{s1})} \sum_{k=0}^{\infty} \frac{c_{k,0}'}{x^{(k+1)/2}} P(\tfrac{1}{2}k + \tfrac{1}{2}, \omega_0' x),$$

where $\omega_0' = i\{\psi(t_{s2}) - \psi(t_{s1})\}$ and the coefficients $c_{k,0}'$ result from expansion of $(1+t)^{-1/2}dt/du$ about the saddle t_{s1}. As maximal integration intervals are being

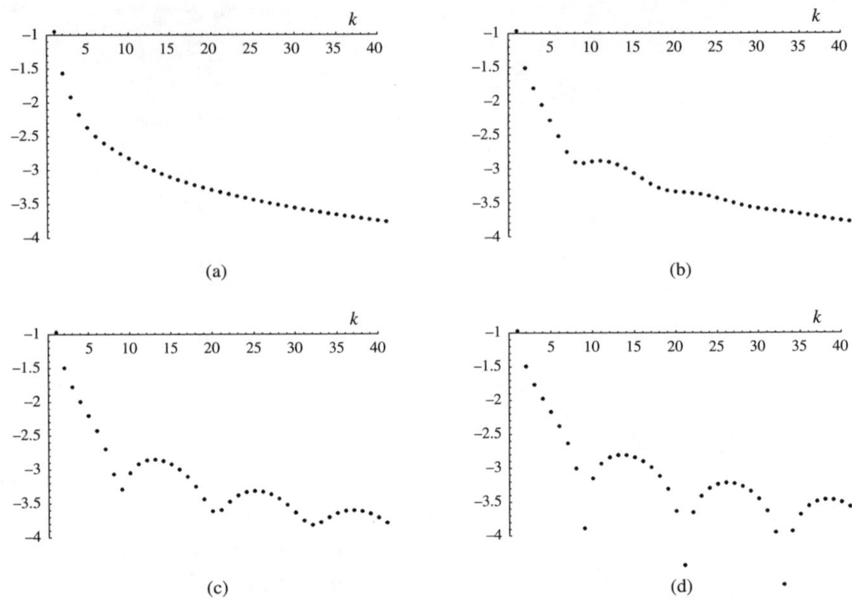

Figure 3.16 The magnitude of the terms (on a $\log_{10}$ scale) of the inter-saddle series I_{AB} in (3.5.8) as a function of ordinal number k and varying κ when $a = 0.1$ and $x = 15$: (a) $\kappa = 3.0$, (b) $\kappa = 2.0$, (c) $\kappa = 1.8$ and (d) $\kappa = \sqrt{3}$.

used in both cases, it is necessary to employ the Hadamard expansions for I_{AC} and I_{CB} in their modified forms.

In Fig. 3.16 we show the behaviour of the terms in the Hadamard expansion for I_{AB} in (3.5.8) for a fixed value of a and different values of the parameter $\kappa \geq \kappa_*$. The familiar slow algebraic decay of the terms is apparent. As κ decreases, the influence of the approaching saddle t_{s1} on the decay of these terms results in a series of oscillations which become progressively more pronounced as $\kappa \to \kappa_*$. When $\kappa = \kappa_*$, the convergence about t_{s3} is controlled by both t_{s1} and t_{s2}; for $\kappa < \kappa_*$, the disc of convergence about t_{s3} is controlled by t_{s1} and the terms in (3.5.8) then reach a global minimum before ultimately diverging.

The results of numerical calculations of $I(x)$ are presented in Table 3.10, where the absolute value of the error is displayed for $x = 15$ and different values of $a \to 0$. The exact value of $I(x)$ was obtained by numerical quadrature based on steepest descent paths. The truncation indices employed in the inter-cluster contributions (which consist of single Hadamard expansions) were $(M_0, N_0) = (20, 5)$, while those in the steepest descent contributions involved only the first two levels $n \leq 2$ in (3.4.5) with $(M_0, N_0) = (30, 40)$, $M_1 = 50$ and $M_2 = 10$. As in Example 3.7, the points t_n on the steepest descent paths through t_{s2} and t_{s3} are obtained with the value $\omega_0 = 1$ and Ω_n $(n \geq 1)$ determined by (3.4.6) with the geometric decay factor $\vartheta = \frac{1}{2}$. The level of accuracy of the computations (at fixed truncation indices) is

Table 3.10 *Absolute values of the error in*
$I(x)$ *for different values of* a *and* κ *when*
$x = 15;$ *the value of* $\kappa_S \doteq 0.68125$

| | |Error| | |
κ	$a = 0.1$	$a = 0.01$
0	1.051×10^{-21}	1.056×10^{-21}
0.5	1.034×10^{-21}	1.054×10^{-21}
κ_S	1.020×10^{-21}	1.053×10^{-21}
1.0	1.166×10^{-21}	1.051×10^{-21}
1.5	1.916×10^{-20}	1.049×10^{-21}
$\sqrt{3}$	9.983×10^{-22}	1.048×10^{-21}
3.0	9.750×10^{-22}	1.043×10^{-21}

uniform as both $a \to 0$ and as κ passes through the critical value κ_s (corresponding to a Stokes phenomenon).

When $a = 0$, we have a single third-order saddle at the origin and the Hadamard expansion of $I(x)$ in this case could, of course, be evaluated by expansion about the origin as described in §3.2.1. However, the expansion procedure described above will result in a uniform level of accuracy as $a \to 0$ since the zeroth-level Hadamard series employed do not involve coefficients related to the coalescing saddles.

Example 3.9 In this example we examine the effect of a simple pole approaching a stationary saddle point. We consider the integral (Lauwerier, 1966, p. 60)

$$J(x) = \int_{-\infty}^{\infty} e^{-x(t^2 - 2it)} \frac{dt}{t^2 + a^2} = e^{-x} \int_{-\infty}^{\infty} e^{-x(t-i)^2} \frac{dt}{t^2 + a^2}, \qquad (3.5.9)$$

where x is assumed to be a large positive variable and a is the coalescence parameter satisfying $0 < a \leq 1$. The phase function $\psi(t) = (t - i)^2$ has a saddle point at $t_s = i$ and the steepest descent path through t_s is the line $\mathrm{Im}(t) = 1$ parallel to the real t-axis. The amplitude function $f(t) = (t^2 + a^2)^{-1}$ has simple poles at $t = \pm ia$ on the imaginary axis. We are primarily concerned here with the case when the pole at ia is close to the saddle; that is, in the limit $a \to 1$.

When $a < 1$, the integration path can be displaced over the pole at $t = ia$ and made to coincide with the steepest descent path through t_s to yield

$$J(x) = \frac{\pi}{a} e^{-a(2-a)x} + e^{-x} \int_{-\infty+i}^{\infty+i} e^{-x(t-i)^2} \frac{dt}{t^2 + a^2} \qquad (a < 1). \qquad (3.5.10)$$

In the limit $a \to 1$ we must use the reverse-expansion procedure described in §3.4.1, since the radius of convergence of the expansion of the integrand about t_s will be controlled by the nearby pole at $t = ia$ with the consequence that the expansion intervals $\omega_n \to 0$. With the new variable $u = \psi(t) - \psi(t_s) = \psi(t)$, we choose the zeroth interval ω_0 according to (3.4.1) and the remaining intervals according to (3.4.6). In this example the inversion is particularly simple and we have $t^{\pm}(u) = i \pm \sqrt{u}$, so that $dt^{\pm}/du = \pm\frac{1}{2}u^{-1/2}$. The radius of convergence of the expansion [14] (3.2.6) for $f(t)dt/du$ about the points $t_n = 1 + \sqrt{\Omega_n}$ $(n \geq 1)$ is

$$\min\{\psi(t_n) - \psi(t_s), \ |\psi(t_n) - \psi(ia)|\} = \min\{\Omega_n, \ \Omega_n + 1 - a^2\}$$

and so is controlled by the saddle point; a similar consideration applies to the expansion (3.4.2) about $t_0' = 1 + \sqrt{\Omega_0'}$ (where $\Omega_0' = \omega_0$).

From (3.4.5), the Hadamard expansion of $J(x)$ is then given by

$$J(x) = \frac{\pi}{a}e^{-a(2-a)x} + 2e^{-x}\operatorname{Re}\sum_{n=0}^{\infty}e^{-\Omega_n x}\,\mathbf{S}_n(x) \qquad (a < 1), \qquad (3.5.11)$$

where $\mathbf{S}_0(x)$ is the series defined in (3.4.4) and $\mathbf{S}_n(x) = S_n(x)$ $(n \geq 1)$ are defined in (3.2.23). Since $x > 0$, the contribution to $J(x)$ from the half of the path situated in $\operatorname{Re}(t) \leq 0$ is the conjugate of that situated in $\operatorname{Re}(t) \geq 0$, and it is sufficient therefore to take twice the real part of the contribution from the positive half of the integration path. In computations the modified form of $\mathbf{S}_0(x)$ must be employed where, from (3.4.7) and (3.4.8),

$$\mathbf{S}_0(x) = -e^{-\Omega_0' x}\sum_{k=0}^{M_0-1}\frac{c_{k,0}'}{x^{k+1}}\,P(k+1, -\omega_0 x) + \sum_{r=0}^{\infty}\sigma_{r,0}'\left(\frac{-\omega_0 x}{M_0}\right)^r$$

for a suitably chosen truncation index M_0, with the coefficients $\sigma_{r,0}'$ specified by (3.4.9) and (3.4.10) as

$$\sigma_{r,0}' = M_0^r\left\{\frac{1}{r!}\int_{t_s}^{t_0'}v^r f(t)\,dt - \sum_{k=0}^{M_0-1}\frac{(-)^k c_{k,0}'\omega_0^{k+1}}{(k+r+1)!}\right\}, \qquad v = \frac{\psi(t)}{\omega_0}.$$

In numerical computations, we took $x = 10$ and used three levels with the zeroth interval $\omega_0 = 1$ and $\Omega_1 = 2$, $\Omega_2 = 6$ (corresponding to the geometric decay factor $\vartheta = \frac{1}{2}$). The decay of the terms in the first three series in (3.5.11) against ordinal number k is shown in Fig. 3.17(a) for the particular case $a = 0.9$. It is evident that, as a result of the nearby pole at $t = ia$, the terms in the finite main sum of $\mathbf{S}_0(x)$ (truncated after $M_0 = 30$ terms) decay very slowly, but that those in the modified tail decay rapidly; if M_0 is reduced these latter terms decay more slowly. The decay of the terms at levels $n = 1$ and $n = 2$ is geometric controlled by the factor ϑ.

---

[14] In the expansions (3.2.6) and (3.4.2) we set $\theta = 0$, since $x > 0$.

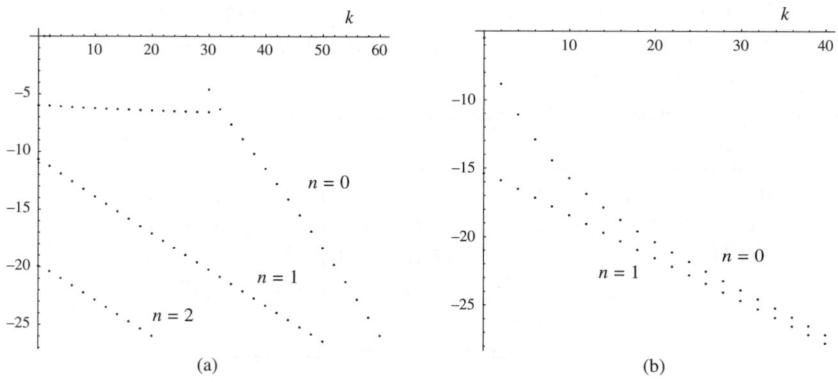

Figure 3.17 Magnitude of the terms (on a $\log_{10}$ scale) against ordinal number k when $x = 10$ in the Hadamard expansions (a) in (3.5.11) with $a = 0.9$ and (b) in (3.5.13) with $a = 1$.

When $a = 1$, the pole in the upper half-plane becomes coincident with the saddle point. In this case, we take the integration path in (3.5.10) to be the same horizontal line, $\mathrm{Im}(t) = 1$, but indented *above* the point $t = i$, and adopt the approach discussed in Van der Waerden (1951). Here we apply the expansion Scheme B in §3.2.3 with suitably chosen intervals ω_n and expansion points Ω_n. The inversion expansion about $t = t_s$ has the form

$$\left(f(t) \frac{dt}{du} \right)^{\pm} = \frac{1}{4iu} \left(1 \pm \frac{u^{\frac{1}{2}}}{2i} \right)^{-1} = \frac{1}{4iu} + \sum_{k=0}^{\infty} \frac{(\pm)^{k-1} c_{k,0}}{\Gamma(\frac{1}{2}k + \frac{1}{2})} u^{(k-1)/2}, \qquad (3.5.12)$$

where the coefficients $c_{k,0}$ are given explicitly by

$$c_{k,0} = 2^{-k-3} i^k \Gamma(\tfrac{1}{2}k + \tfrac{1}{2}).$$

The radius of convergence of this expansion is controlled by the pole at $t = -i$ and so is $|u| < 4$; this means that we must choose $\omega_0 \leq 4$. The indented integration path corresponds to a negatively oriented loop [15] in the u-plane surrounding the positive real u-axis. As a consequence, the term $(4iu)^{-1}$ contributes $-\frac{1}{2}\pi e^{-x}$ to the integral in (3.5.10) which, with the residue term of πe^{-x}, yields the contribution $\frac{1}{2}\pi e^{-x}$. From (3.5.11) and (3.2.24), we then obtain the Hadamard expansion when $a = 1$

$$J(x) = \tfrac{1}{2}\pi e^{-x} + e^{-x} \left\{ S_0(x) + 2\mathrm{Re} \sum_{n=1}^{\infty} e^{-\Omega_n x} S_n(x) \right\}, \qquad (3.5.13)$$

---

[15] If the indentation had been below the point $t = i$, the contour in the u-plane would be a positively oriented loop.

Table 3.11 *Absolute values of the error in $J(x)$ in (3.5.9) for different values of a and x; the truncation indices employed are indicated in the text*

$1-a$	$x=10$	$x=15$	$x=20$
10^{-1}	3.970×10^{-27}	3.079×10^{-24}	1.456×10^{-22}
10^{-2}	2.596×10^{-27}	7.600×10^{-26}	3.595×10^{-24}
10^{-3}	3.130×10^{-27}	2.665×10^{-25}	1.261×10^{-23}
10^{-4}	3.183×10^{-27}	3.005×10^{-25}	1.421×10^{-23}
0	7.764×10^{-25}	6.231×10^{-31}	1.093×10^{-31}

where, from (3.2.12),

$$S_0(x) = 2 \sum_{k=0}^{\infty} \frac{c_{2k,0}}{x^{k+\frac{1}{2}}} P(k + \tfrac{1}{2}, \omega_0 x)$$

and the series $S_n(x)$ $(n \geq 1)$ are defined in (3.2.23).

In the calculations with $a = 1$ we used the levels $n \leq 1$ where, from (3.2.25) with $\vartheta = \frac{1}{2}$ and $\Lambda_0 = 4$, we have $\omega_0 = 2$ and $\Omega_1 = \omega_1 = 4$. The decay of the terms in $S_0(x)$ and $S_1(x)$ in (3.5.13) against ordinal number k is displayed in Fig. 3.17(b). The values of the absolute error in the computation of $J(x)$ for different values of x and a in the limit $a \to 1$ are presented in Table 3.11. For convenience, we used fixed truncation indices when $a < 1$; better accuracy will result, of course, if these indices are tailored to the particular value of x under consideration. The values employed are $(M_0, N_0) = (30, 30)$, $M_1 = 50$, $M_2 = 20$ for $a < 1$ and $M_0 = M_1 = 30$ when $a = 1$ (for $x = 20$ we used $M_0 = M_1 = 20$ for this latter calculation). The error was obtained by comparison with the exact evaluation of $J(x)$ in terms of the complementary error function given by

$$J(x) = \frac{\pi}{a} e^{-a(2-a)x} + \frac{\pi}{2a} e^{-x}(E_+ - E_-),$$

where $E_\pm = \exp(\alpha_\pm^2 x)\,\mathrm{erfc}(\alpha_\pm x^{\frac{1}{2}})$ and $\alpha_\pm = 1 \pm a$; see (1.5.11).

Example 3.10 Our final example involves a saddle point coalescing with an endpoint of the integration path. We consider the integral

$$I(x) = \int_0^1 e^{-x\psi(t)}\,dt, \tag{3.5.14}$$

where

$$\psi(t) = \tfrac{1}{3}t^3 + \tfrac{1}{2}(b-a)t^2 - abt + c, \tag{3.5.15}$$

$$b \geq 0, \quad c = \tfrac{1}{6}a^3 + \tfrac{1}{2}ab$$

and $0 \leq a < 1$ is the coalescence parameter; examples of this integral have been considered in Paris and Kaminski (2005). The parameter b controls the position of the 'remote' saddle and the constant c is chosen to make $\psi(a) = 0$. The phase function $\psi(t)$ has saddle points at $t = a$ and $t = -b$, so that in the small-a limit, the saddle at $t = a$ coalesces with the endpoint $t = 0$. The variable x will initially be taken to be positive, but subsequently will be extended to a complex variable.

When $x > 0$, the path $[0, 1]$ is the steepest descent path through $t = a$. The path of integration is decomposed into the intervals $[0, a]$ and $[a, 1]$ so that, with the usual change of variable $u = \psi(t) - \psi(a) = \psi(t)$, we obtain

$$I(x) = \left\{ \int_0^a + \int_a^1 \right\} e^{-x\psi(t)} dt = \left\{ -\int_0^{\omega_0} + \int_0^{\omega_1} \right\} e^{-xu} \frac{dt}{du} \, du, \qquad (3.5.16)$$

where

$$\omega_0 = \psi(0) = c, \qquad \omega_1 = \psi(1) = \tfrac{1}{3} + \tfrac{1}{2}(b - a) - ab + c.$$

The inversion of the change of variable $t \mapsto u$ is controlled by the presence of the remote saddle at $t = -b$, so that the series in u obtained by inversion will converge in a disc centred at $u = 0$ of radius

$$\Lambda_0 = |\psi(-b) - \psi(a)| = \tfrac{1}{6}b^3 + \tfrac{1}{2}ab^2 + c.$$

This leads to the expansion

$$\frac{dt^{\pm}}{du} = \sum_{k=0}^{\infty} \frac{(\pm)^{k-1} c_k}{\Gamma(\tfrac{1}{2}k + \tfrac{1}{2})} u^{(k-1)/2} \qquad (|u| < \Lambda_0), \qquad (3.5.17)$$

where $c_0 = \{\pi/(2\psi''(a))\}^{1/2}$ and the $\pm$ superscripts refer to the upper ($t > a$) and lower ($t < a$) halves of the steepest descent path through $t = a$. The coefficients c_k in specific cases can be determined by numerical inversion using *Mathematica*.

Substitution of (3.5.17) into (3.5.16), followed by reversal of the order of summation and integration, then yields the Hadamard expansion

$$I(x) = \sum_{k=0}^{\infty} \frac{(-)^k c_k}{x^{(k+1)/2}} P(\tfrac{1}{2}k + \tfrac{1}{2}, \omega_0 x) + \sum_{k=0}^{\infty} \frac{c_k}{x^{(k+1)/2}} P(\tfrac{1}{2}k + \tfrac{1}{2}, \omega_1 x) \equiv S_0(x) + S_1(x)$$

$$(3.5.18)$$

when $x > 0$. Notice that as $a \to 0$, the factor $\omega_0 \to 0$ in the argument of the first incomplete gamma function, whereas in the second incomplete gamma function $\omega_1 \to \tfrac{1}{3} + \tfrac{1}{2}b$.

From (3.2.10), the decay of the late terms of the series $S_0(x)$ and $S_1(x)$ is controlled by

$$e^{-\omega_j x} k^{-3/2} (\omega_j / \Lambda_0)^{k/2} \qquad (k \to \infty; \ j = 0, 1). \qquad (3.5.19)$$

Then, as $a \to 0$, the decay of $S_0(x)$ will be rapid provided b is bounded away from zero. For $S_1(x)$, the ratio $\omega_1 / \Lambda_0 \to (2 + 3b)/b^3$ as $a \to 0$. Thus, when $b \gg 2$, the decay of the terms of $S_1(x)$ will also be rapid, but when $b = 2$ the ratio $\omega_1/\Lambda_0 \to 1$ in the limit and we will then have the familiar slow algebraic decay controlled by $k^{-3/2}$. In this case we must use the modified form of the expansion for $S_1(x)$ given in (3.2.14)–(3.2.19) by

$$S_1(x) = \sum_{k=0}^{M_0-1} \frac{c_k}{x^{(k+1)/2}} P(\tfrac{1}{2}k + \tfrac{1}{2}, \omega_1 x) + e^{-\omega_1 x} \sum_{r=0}^{\infty} \sigma_{r,1} \left(\frac{\omega_1 x}{M_0}\right)^r, \qquad (3.5.20)$$

with

$$\sigma_{r,1} = M_0^r \left\{ \frac{1}{r!} \int_a^1 (1-v)^r \, dt - \sum_{k=0}^{M_0-1} \frac{c_k \omega_1^{(k+1)/2}}{\Gamma(\tfrac{1}{2}k + r + \tfrac{3}{2})} \right\}, \qquad v = \frac{\psi(t)}{\omega_1}$$

and the truncation index M_0 chosen such that $\omega_1 x/M_0 < 1$.

We illustrate in Fig. 3.18 the decay of the terms in the Hadamard series $S_0(x)$ and $S_1(x)$ for a small value of the parameter a when $x = 5$ and two values of b, one corresponding to rapid decay of both series ($b = 5$) and the second ($b = 2$) necessitating the use of the modified expansion for $S_1(x)$. The results of numerical computations when $b = 2$ and $x > 0$ are displayed in the left half of Table 3.12, where the absolute error in the calculation of $I(x)$ by means of (3.5.18) is given for different values of a and fixed x. The exact value was obtained by high-precision numerical quadrature. The modified tail in (3.5.20) is truncated at $r = N_0$. We take $x = 10$ and employ the *fixed* truncation indices $M_0 = 20$, $N_0 = 20$ or 30 for $S_1(x)$ and up to a maximum of 20 terms (according to the value of a) for $S_0(x)$.

When x becomes complex the approach to take would be to represent $I(x)$ as a sum of steepest descent contours and then proceed to construct Hadamard expansions associated with each contour integral. An example of these paths is shown in

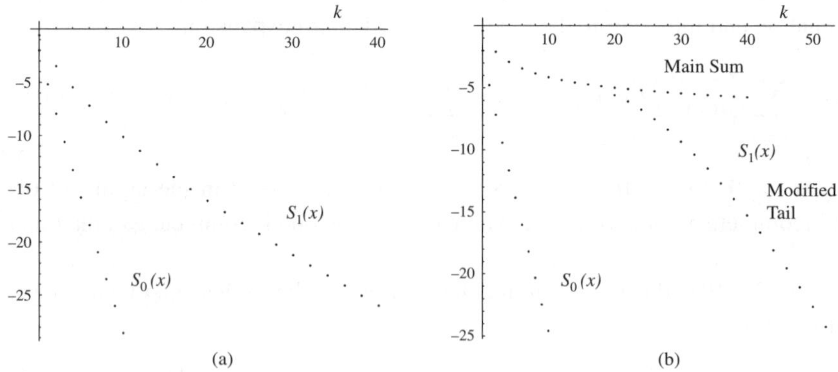

Figure 3.18 Magnitude of the terms (on a $\log_{10}$ scale) against ordinal number k in the Hadamard series $S_0(x)$ and $S_1(x)$ when $x = 5$ and $a = 10^{-2}$: (a) $b = 5$ and (b) $b = 2$ with the modified form of $S_1(x)$.

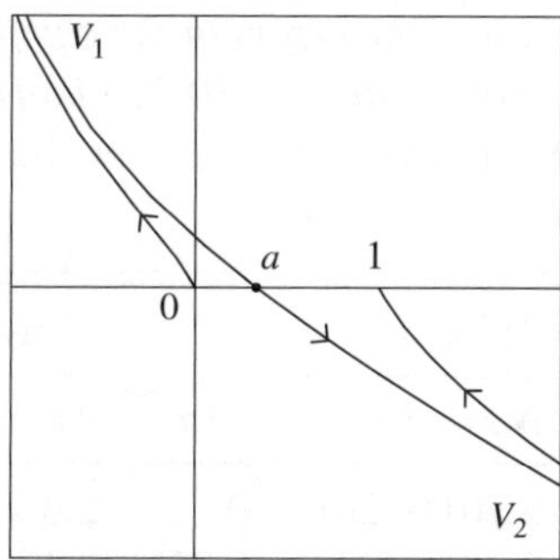

Figure 3.19 The paths of steepest descent for $I(x)$ when arg $x = \frac{1}{4}\pi$. The arrows denote the direction of integration.

Fig. 3.19 for the case $\theta = $ arg $x = \pi/4$. If $0 < \theta \le \pi$, the path issuing from the origin passes to infinity down the valley V_1 centred on the ray arg $t = (2\pi - \theta)/3$, whereas that through the saddle $t = a$ commences in V_1 and passes down the valley V_2 centred on the ray arg $t = -\theta/3$. The third path in this decomposition commences in V_2 and ends at the point $t = 1$. If instead we have $-\pi \le \theta < 0$, then a similar structure exists with the asymptote from the origin now being given by arg $t = -(2\pi + \theta)/3$.

This process produces series where the associated incomplete gamma functions have arguments that are positive, guaranteeing that the functions P have the desirable 'cut-off' property of dropping quickly and monotonically from 1 to 0 as k increases, which in turn produces the most rapid possible convergence for the expansion. If we are willing to obtain slightly less rapidly converging series, then much of the computational cost of the above path decomposition can be avoided. Furthermore, the approach taken in developing (3.5.18) works without modification if x is allowed to be complex. The expansion (3.5.18) will, however, now involve incomplete gamma functions whose arguments $\omega_j x$ are complex. In the right half of Table 3.12 we show results for complex x using (3.5.18) when $a = 10^{-2}$ and with the same truncation indices.

As b decreases further in the interval $0 \le b < 2$, the rate of decay of the terms in $S_0(x)$ progressively decreases. However, expansion about the saddle point $t = a$ will not cover the interval $[a, 1]$ in a single Hadamard series, since now $\omega_1/\Lambda_0 > 1$ as $a \to 0$. We illustrate the situation corresponding to the case $b = 0$, where the integral $I(x)$ takes the form

$$I(x) = \int_0^1 e^{-x\psi(t)}dt, \qquad \psi(t) = \tfrac{1}{3}t^3 - \tfrac{1}{2}at^2 + \tfrac{1}{6}a^3; \qquad (3.5.21)$$

a similar example is described in Paris and Kaminski (2005). As $a \to 0$, the saddle at $t = a$ coalesces with the saddle at the endpoint $t = 0$ to form a double saddle.

Table 3.12 *Absolute values of the errors in the computation of $I(x)$ when $b = 2$ for different values of a with $x = 10$ (left half) and for different values of θ with $a = 10^{-2}$ (right half) when $x = 10e^{i\theta}$ for truncation indices $M_0 = 20$ and $N_0 = 20$ or 30*

	|Error| ($\theta = 0$ fixed)			|Error|^a ($a = 10^{-2}$ fixed)	
a	$N_0 = 20$	$N_0 = 30$	θ/π	$N_0 = 20$	$N_0 = 30$
10^{-1}	$8.119(-21)$	$6.701(-26)$	0	$2.320(-17)$	$6.361(-23)$
10^{-2}	$2.320(-17)$	$6.361(-23)$	0.25	$9.175(-16)$	$2.623(-21)$
10^{-4}	$5.312(-17)$	$1.752(-22)$	0.50	$7.415(-12)$	$2.234(-17)$
10^{-6}	$5.356(-17)$	$1.770(-22)$	0.75	$6.254(-12)$	$1.933(-17)$
0	$5.356(-17)$	$1.770(-22)$	1.00	$5.910(-12)$	$1.839(-17)$

^{a} Entries in the right half of the table corresponding to $\mathrm{Re}(x) < 0$ have been scaled by a factor of $e^{x\psi(1)}$. The notation $x(-y)$ denotes $x \times 10^{-y}$.

The contribution from $[0, a]$ is given by $S_0(x)$ in (3.5.18), but with the coefficients c_k corresponding to $\psi(t)$ in (3.5.21) and $\omega_0 = \frac{1}{6}a^3$.

For the interval $[a, 1]$, it is no longer practical in the small-a limit to expand about the saddle $t = a$, since the associated radius of convergence $\Lambda_0 = \omega_0$ shrinks to zero as $a \to 0$. With forward expansion, this would result in a sequence of many Hadamard series to cover the interval $[a, 1]$. Accordingly, we use reverse expansion about the endpoint $t = 1$ and make the change of variable $u = \psi(t) - \psi(1)$. This yields the inversion

$$\frac{dt}{du} = \sum_{k=0}^{\infty} \frac{b_k u^k}{k!},$$

where $b_0 = 2(1 - a)$, valid in the disc $|u| < \omega_1 = \psi(1)$ whose radius is controlled by the saddle at $t = a$. Then we find

$$\int_a^1 e^{-x\psi(t)}\,dt = -e^{-x\psi(1)} \int_0^{-\omega_1} e^{-xu} \frac{dt}{du}\,du$$

$$= -e^{-\omega_1 x} \sum_{k=0}^{\infty} \frac{b_k}{x^{k+1}} P(k+1, -\omega_1 x) \equiv \mathbf{S}_1(x). \qquad (3.5.22)$$

This last Hadamard series is of the type defined in §3.4.1 and, of course, must be employed in its modified form, since its derivation has involved the maximal integration interval. From (3.4.7)–(3.4.10), we have

$$\mathbf{S}_1(x) = -e^{-\omega_1 x} \sum_{k=0}^{M_0-1} \frac{b_k}{x^{k+1}} P(k+1, -\omega_1 x) + \sum_{r=0}^{\infty} \sigma'_{r,1}(-\omega_1 x/M_0)^r,$$

where

$$\sigma'_{r,1} = M_0^r \left\{ \frac{1}{r!} \int_a^1 v^r \, dt - \sum_{k=0}^{M_0-1} \frac{(-)^k c_k \omega_1^{k+1}}{(k+r+1)!} \right\}, \qquad v = \frac{\psi(t)}{\omega_1}$$

with M_0 chosen so that $\omega_1 x/M_0 < 1$.

Since $\omega_0/\Lambda_0 = 1$ when $b = 0$, it follows from (3.5.19) that the decay of the terms in $S_0(x)$ will be now controlled by $k^{-3/2}$ so that the modified form of this expansion is required.[16] The terms in the finite sum (truncated after M_0 terms), however, are found to decay slowly as $a \to 0$; consequently, in this example, we set $M_0 = 0$ and represent $S_0(x)$ by means of its rearranged tail which, from (3.2.14)–(3.2.19), takes the form

$$S_0(x) = e^{-\omega_0 x} \sum_{r=0}^{\infty} \sigma_{r,0}(\omega_0 x)^r,$$

where

$$\sigma_{r,0} = \sum_{k=0}^{\infty} \frac{(-)^k c_k \omega_0^{(k+1)/2}}{\Gamma(\frac{1}{2}k + r + \frac{3}{2})} = -\frac{1}{r!} \int_0^{\omega_0} (1-v)^r \frac{dt^-}{du} \, du$$

$$= \frac{1}{r!} \int_0^a (1-v)^r \, dt, \qquad v = \frac{\psi(t)}{\omega_0}.$$

An example of the decay of the terms in $S_0(x)$ and $\mathbf{S}_1(x)$ when $b = 0$ is displayed in Fig. 3.20. It can be seen that in the small-a limit the terms in the (modified) inter-saddle contribution $S_0(x)$ decay very rapidly so that computation of the coalescence to high accuracy presents no especial difficulty.

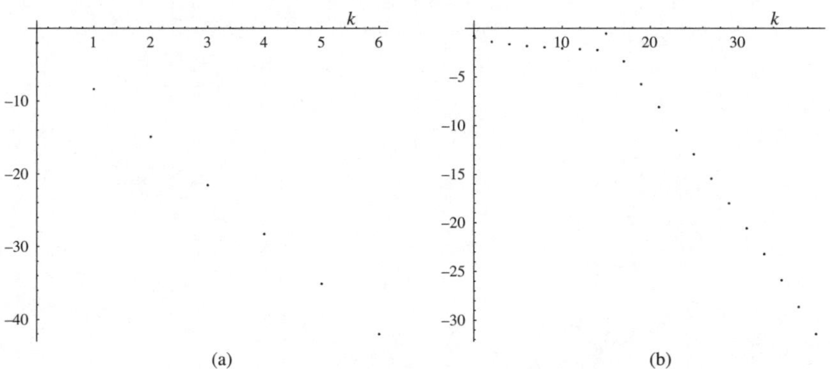

Figure 3.20 Magnitude of the terms (on a $\log_{10}$ scale) against ordinal number k in the Hadamard series for $I(x)$ when $x = 5$, $a = 10^{-2}$ and $b = 0$: (a) the modified form of $S_0(x)$ with $M_0 = 0$, $N_0 = 6$ and (b) $\mathbf{S}_1(x)$ with $M_0 = 15$, $N_0 = 24$.

---

[16] An alternative procedure is to obtain a 2-stage expansion for the inter-saddle contribution $S_0(x)$, as described in Example 3.1.

To conclude, we mention that an example of a cluster of two saddle points and two poles coalescing at an endpoint is given in Paris (2007b). This consists of taking $b = a$ in (3.5.15), so that the saddles are situated at $t = \pm a$, and including the amplitude factor $f(t) = t^2/\{t^2 + (\kappa a)^2\}$, where κ is a positive parameter. This function has two simple poles at $t = \pm i\kappa a$; as $a \to 0$, the saddles and poles approach the endpoint of integration $t = 0$ as a cluster to produce a double saddle at the origin (with $f(t) \equiv 1$) when $a = 0$.

4

Applications

4.1 Introduction

In Chapter 3 we introduced two basic modes of expansion using Hadamard series, namely the forward expansion Scheme A and the forward-reverse expansion Scheme B. The first scheme uses the Hadamard series $S_n(z)$, defined in (3.2.9), and is suitable for isolated saddle points when adjacent saddles or other singularities are sufficiently remote to result in a sequence of well-separated exponential levels. If maximal exponential separation is employed, the resulting convergence of the Hadamard series has to be accelerated through use of the modified form of the series. This involves the computation of coefficients expressed in terms of one-dimensional integrals in (3.2.18) of a common form at each level of the expansion. If one is prepared to accept a reduced exponential separation, however, it is possible, through judicious choice of the expansion points Ω_n, to produce Hadamard series that converge rapidly at a geometric rate without the need for the computationally more expensive modified form.

In Scheme B, the zeroth interval is dealt with by forward expansion as in Scheme A, but with forward-reverse expansion about the points Ω_n for the intervals with $n \geq 1$. This has the advantage of covering a given interval on the integration path with fewer evaluations of the inversion expansions. By careful choice of the Ω_n it is similarly possible to arrange for the Hadamard series at all levels to converge at a geometric rate.

A small, but significant, variation of Scheme B is indispensable for coalescence problems and involves the Hadamard series of the second kind $\mathbf{S}_0(z)$ defined in (3.4.4). This mode of expansion deals with the contribution from the zeroth interval by reverse expansion about the right-hand endpoint Ω_0' of the interval (see Fig. 3.1), instead of by forward expansion about the left-hand endpoint $\Omega_0 = 0$. The other intervals corresponding to $n \geq 1$ are treated as in Scheme B. This avoids the loss of exponential separation that results in forward expansion when either another saddle or singularity approaches the saddle point under consideration. The zeroth interval contribution in this variant of Scheme B requires use of the modified form of $\mathbf{S}_0(z)$

to accelerate convergence, with the series associated with the remaining intervals converging sufficiently rapidly to make use of their modified form unnecessary. Thus, the coalescence translates into an unavoidable use of the modified form for the zeroth Hadamard series, thereby requiring greater computational effort.

In this chapter we apply these different Hadamard expansion procedures to the hyperasymptotic evaluation of some well-known special functions in domains ordinarily regarded as problematic. The functions we consider are the Bessel function $J_\nu(\nu x)$ of large order and argument, the Pearcey integral $\mathrm{Pe}(x, y)$ in the neighbourhood of the two cusps in the x, y-plane (across which there is either a coalescence of saddle points or a Stokes phenomenon), the parabolic cylinder function $U(a, z)$ of large order and argument and, finally, $\log \Gamma(z)$.

4.2 The Bessel function $J_\nu(\nu z)$

We consider the Hadamard expansion of the Bessel function $J_\nu(\nu z)$ where, for simplicity in presentation, we shall assume throughout this section that $\nu > 0$ and z is a complex variable. Since $J_\nu(z^*) = J_\nu^*(z)$, where the asterisk denotes the complex conjugate, and

$$J_\nu(z) = e^{\pi i \nu} J_\nu(z e^{-\pi i}),$$

it is sufficient to take $0 \leq \arg z \leq \frac{1}{2}\pi$. The well-known representation of the Bessel function $J_\nu(\nu z)$ as a Laplace-type integral has the form (Watson, 1952, p. 238)

$$J_\nu(\nu z) = \frac{1}{2\pi i} \int_{\infty-\pi i}^{\infty+\pi i} e^{-\nu z \psi(t)} dt \qquad (|\arg z| < \tfrac{1}{2}\pi), \qquad (4.2.1)$$

where the phase function

$$\psi(t) = t \cosh \gamma - \sinh t, \quad z = \operatorname{sech} \gamma$$

and $\gamma = \alpha + i\beta$, with α real and $0 \leq \beta < \frac{1}{2}\pi$. Saddle points of the integrand occur at $\psi'(t) = 0$, that is, at the points

$$t = \pm\gamma \pm 2\pi k i \qquad (k = 0, 1, 2, \ldots).$$

4.2.1 The case z positive

We take $z = x > 0$ to be a real variable. When $0 < x \leq 1$, the parameter $\gamma = \alpha \geq 0$ whereas when $x > 1$ we have $\gamma = i\beta$, with $0 < \beta < \frac{1}{2}\pi$. For large ν and bounded x, the above integral involves only two *active* saddles at $t = \pm\gamma$. As x varies, these saddles move in the complex plane: when $x < 1$, the active saddles lie on the real t-axis symmetrically either side of the origin, and when $x > 1$ they are situated symmetrically on the imaginary axis. When $x = 1$, the active saddles coalesce to

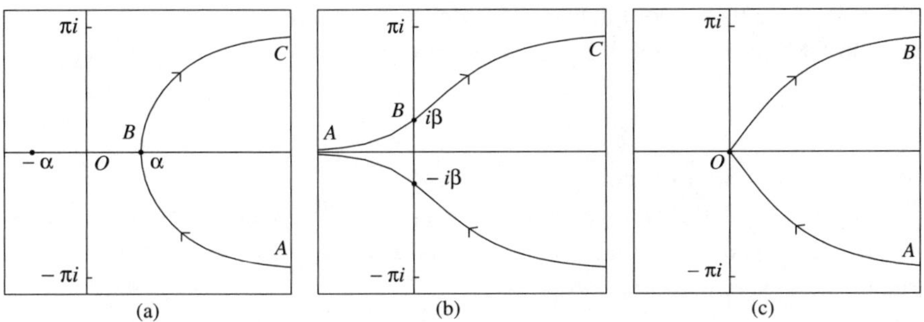

Figure 4.1 Paths of steepest descent through the saddles when (a) $x < 1$, (b) $x > 1$ and (c) $x = 1$. The active saddles are at (a) $t = \pm\alpha$, (b) $t = \pm i\beta$ and (c) $t = 0$, where they have coalesced to form a double saddle. The arrows indicate the direction of integration.

form a double saddle point at the origin. The steepest descent paths in each case are discussed in detail in Watson (1952, §8.31) and are shown in Fig. 4.1.

We first consider the case $0 < x < 1$ and take the integration path in (4.2.1) to be the steepest descent path ABC passing through the saddle $t_s = \alpha$ shown in Fig. 4.1(a). In terms of the new variable u defined by

$$u = \psi(t) - \psi(t_s),\qquad(4.2.2)$$

we find

$$J_\nu(\nu x) = \frac{e^{\nu(\tanh\alpha - \alpha)}}{2\pi i}\int_0^\infty e^{-\nu x u}\left(\frac{dt^+}{du} - \frac{dt^-}{du}\right)du,\qquad(4.2.3)$$

where $t^\pm$ refer to the upper and lower halves BC and BA of the steepest descent path. The positive u-axis is subdivided into intervals of length ω_n ($n \geq 0$) with the expansion points Ω_n (with $\Omega_0 = 0$) chosen according to Scheme B in §3.2.3. From (3.2.25), this corresponds to the choice

$$\omega_n = 2\vartheta\,\Omega_n,\qquad \Omega_n = \omega_0\frac{(1+\vartheta)^{n-1}}{(1-\vartheta)^n}\qquad (n \geq 1),\qquad(4.2.4)$$

where $0 < \vartheta < 1$ is the geometric decay factor and $\omega_0 = \Lambda_0\vartheta$. The quantity Λ_0 is the shortest distance in the u-plane from the saddle t_s to the nearest saddles at $t = -\alpha$ and $t = \alpha \pm 2\pi i$, where

$$u(-\alpha) = 2(\sinh\alpha - \alpha\cosh\alpha),\qquad u(\alpha \pm 2\pi i) = \pm 2\pi i\cosh\alpha.$$

Provided $\alpha - \tanh\alpha < \pi$, that is, provided $x > x^*$, where $x^* \doteq 0.03180$, the closest saddle is that at $t = -\alpha$ and

$$\Lambda_0 = 2|(\alpha\cosh\alpha - \sinh\alpha)|.\qquad(4.2.5)$$

We shall assume throughout that (4.2.5) applies. The corresponding points t_n on the upper half BC of the steepest descent path are obtained by solution of

$$\psi(t_n) - \psi(t_s) = \Omega_n; \tag{4.2.6}$$

a conjugate set of points subdivides the lower half BA.

The inversion of the change of variable $t \mapsto u(t)$ about the points t_n (where $t_0 = \alpha$) is obtained by writing $t = t_n + \tau$ and $u = \Omega_n + w$, where τ is a complex variable and w is real, to find

$$
\begin{aligned}
u - \Omega_n &= \psi(t_n + \tau) - \psi(t_n) \\
&= \tau \cosh\alpha - \sinh(t_n + \tau) + \sinh t_n \\
&= \tau(\cosh\alpha - \cosh t_n) - \tfrac{1}{2}\tau^2 \sinh t_n + \cdots .
\end{aligned}
$$

This yields the inversion [1] for the upper half of the steepest descent path

$$
t - t_n = \sum_{k=1}^{\infty} a_{k,n}^{+} w^{\mu_n k} \qquad |w| < \begin{cases} \Lambda_0 & (n = 0) \\ \Omega_n & (n \geq 1), \end{cases}
$$

where $\mu_0 = \tfrac{1}{2}$ and $\mu_n = 1$ $(n \geq 1)$. The explicit representation of the first few coefficients with $n = 0$ can be written in the form $a_{k,0}^{+} = i^k \hat{a}_k$, $\hat{a}_k = (2\operatorname{cosech}\alpha)^{k/2} b_k$, where the b_k are real, with $b_1 = 1$ and

$$
b_2 = -\tfrac{1}{6}\coth\alpha, \quad b_3 = \tfrac{1}{72}(-3 + 5\coth^2\alpha), \quad b_4 = -\tfrac{1}{270}(-9 + 10\coth^2\alpha)\coth\alpha,
$$
$$
b_5 = \tfrac{1}{17280}(81 - 462\coth^2\alpha + 385\coth^4\alpha), \dots . \tag{4.2.7}
$$

Conjugate expansions hold for the points t_n situated on the lower half of the steepest descent path.

We then find that

$$
\frac{dt^{+}}{du} - \frac{dt^{-}}{du} = 2i \sum_{k=0}^{\infty} \frac{c_{k,n}(\alpha)}{\Gamma((k+1)\mu_n)} w^{(k+1)\mu_n - 1}
$$

valid in $|w| < \Lambda_0$ $(n = 0)$ and $|w| < \Omega_n$ $(n \geq 1)$, where the coefficients $c_{k,n}(\alpha)$ are given by

$$
\frac{c_{k,n}(\alpha)}{\Gamma((k+1)\mu_n + 1)} = \frac{1}{2i}(a_{k+1,n}^{+} - a_{k+1,n}^{-}) = \operatorname{Im}(a_{k+1,n}^{+}). \tag{4.2.8}
$$

In the case $n = 0$, we have

$$
c_{2k,0}(\alpha) = (-)^k \hat{a}_{2k+1} \Gamma(k + \tfrac{3}{2}),
$$

---

[1] When $n \geq 1$, the convergence disc has radius Ω_n since the nearest singularity in the mapping when expanding about Ω_n is at $u = 0$ (corresponding to the saddle at $t = \alpha$).

with the coefficients corresponding to odd index being zero. Then, from (3.2.12) and (3.2.21), the contribution to the integral in (4.2.3) from the zeroth interval $0 \leq u \leq \omega_0$ and the nth interval $-\frac{1}{2}\omega_n \leq u - \Omega_n \leq \frac{1}{2}\omega_n$ is $2ie^{-\Omega_n \nu x}S_n(\nu x)$, where $\Omega_0 = 0$,

$$S_0(\nu x) = \sum_{k=0}^{\infty} \frac{c_{2k,0}(\alpha)}{(\nu x)^{k+\frac{1}{2}}} P(k + \tfrac{1}{2}, \omega_0 \nu x) \tag{4.2.9}$$

and

$$S_n(\nu x) = \sum_{k=0}^{\infty} \frac{c_{k,n}(\alpha)}{(\nu x)^{k+1}} \Delta P(k + 1, \tfrac{1}{2}\omega_n \nu x) \tag{4.2.10}$$

for $n \geq 1$. The Hadamard expansion for $J_\nu(\nu x)$ when $x < 1$ therefore takes the form

$$J_\nu(\nu x) = \frac{e^{\nu(\tanh \alpha - \alpha)}}{\pi} \sum_{n=0}^{\infty} e^{-\Omega_n \nu x} S_n(\nu x). \tag{4.2.11}$$

When $x > 1$, the quantity $\gamma = i\beta$ is purely imaginary and the active saddles are now situated on the imaginary axis at $\pm i\beta$ with the paths of steepest descent illustrated in Fig. 4.1(b). Then, since the variables are positive, we can write[2]

$$J_\nu(\nu x) = \mathrm{Re}\left\{ \frac{1}{\pi i} \int_{-\infty}^{\infty + \pi i} e^{-\nu x \psi(t)} dt \right\} \tag{4.2.12}$$

which involves only the single saddle at $t = i\beta$. With the change of variable $t \mapsto u$ in (4.2.2) the positive u-axis is again subdivided at the points Ω_n according to (4.2.4). The corresponding points $t_n^{\pm}$ on the steepest descent path in the t-plane (where the $\pm$ signs now refer to the parts of this path situated in $\mathrm{Re}(t) > 0$ and $\mathrm{Re}(t) < 0$, respectively) are obtained by solution of (4.2.6). The same procedure as described above then leads to the Hadamard expansion when $x > 1$ in the form

$$J_\nu(\nu x) = \frac{2}{\pi} \mathrm{Re}\left\{ e^{\nu(\tanh \alpha - \alpha)} \sum_{n=0}^{\infty} e^{-\Omega_n \nu x} S_n(\nu x) \right\}, \tag{4.2.13}$$

where $S_n(\nu x)$ have the forms specified in (4.2.9) and (4.2.10).

Finally, when $x = 1$, the active saddles at $t = \pm \alpha$ coalesce to form a double saddle at the origin. In the expansion process the radius of convergence about $t = 0$ is now controlled by the adjacent double saddles situated at $t = \pm 2\pi i$, so that $\Lambda_0 = 2\pi$. The inversion of the change of variable $u = t - \sinh t$ about $t = 0$ yields

$$t^{\pm} = \sum_{k=1}^{\infty} a_{k,0}(6ue^{\pm\pi i})^{(2k-1)/3} \qquad (|u| < 2\pi)$$

---

[2] The imaginary part of this expression yields the Bessel function $Y_\nu(\nu x)$.

on the two steepest descent paths situated in $\mathrm{Re}(t) \geq 0$. The coefficients $a_{k+1,0} \equiv b_k/(2k-1)$, where

$$b_1 = 1, \quad b_2 = -\tfrac{1}{20}, \quad b_3 = \tfrac{1}{280}, \quad b_4 = -\tfrac{1}{3600}, \quad b_5 = \tfrac{387}{17\,248\,000}, \cdots .$$

Then we have

$$\frac{dt^+}{du} - \frac{dt^-}{du} = 2i \sum_{k=0}^{\infty} \frac{c_{k,0}(0)}{\Gamma(\tfrac{2}{3}k + \tfrac{1}{3})} u^{(2k-2)/3}$$

valid in $|u| < 2\pi$, where

$$c_{k,0}(0) = 6^{(2k+1)/3} a_{k+1,0}\, \Gamma(\tfrac{2}{3}k + \tfrac{4}{3}) \sin(\tfrac{2}{3}k + \tfrac{1}{3})\pi.$$

This leads to the zeroth series when $x = 1$ given by

$$S_0(\nu) = \sum_{k=0}^{\infty} \frac{c_{k,0}(0)}{\nu^{(2k+1)/3}}\, P(\tfrac{2}{3}k + \tfrac{1}{3}, \omega_0 \nu).$$

The Hadamard expansion of $J_\nu(\nu)$ is then

$$J_\nu(\nu) = \frac{1}{\pi} \sum_{n=0}^{\infty} e^{-\Omega_n \nu} S_n(\nu), \qquad (4.2.14)$$

where the $S_n(\nu)$ with $n \geq 1$ have the form specified in (4.2.10) with $x = 1$.

4.2.2 Numerical results

The standard asymptotic procedure for the determination of the expansion of $J_\nu(\nu x)$ for large ν is to apply Watson's lemma in §1.2.4 to the integral (4.2.4); see, for example, (Copson, 1965, Chapter 6) and Olver (1997, p. 112). When $0 < x < 1$, this is equivalent to extending the upper limit of integration ω_0 in the integral for the zeroth contribution to $+\infty$; see (3.2.7) and (3.2.13). This corresponds to the Hadamard series $S_0(\nu x)$ with the normalised incomplete gamma functions replaced by unity, namely

$$J_\nu(\nu \operatorname{sech} \alpha) \sim \frac{e^{\nu(\tanh\alpha - \alpha)}}{\pi} \sum_{k=0}^{\infty} \frac{c_{2k,0}(\alpha)}{(\nu x)^{k+\frac{1}{2}}}$$

$$= \frac{e^{\nu(\tanh\alpha - \alpha)}}{\sqrt{2\pi\nu\tanh\alpha}} \sum_{k=0}^{\infty} \frac{(\tfrac{1}{2})_k A_k}{(\tfrac{1}{2}\nu\tanh\alpha)^k}, \qquad (4.2.15)$$

where the coefficients A_k are defined by

$$A_k = \frac{c_{2k,0}(\alpha)}{\Gamma(k + \tfrac{1}{2})} (\tfrac{1}{2}\sinh\alpha)^{k+\frac{1}{2}}.$$

From (4.2.7), we find

$$A_0 = 1, \quad A_1 = \tfrac{1}{8} - \tfrac{5}{24}\coth^2\alpha,$$

$$A_2 = \tfrac{3}{128} - \tfrac{77}{576}\coth^2\alpha + \tfrac{385}{3456}\coth^4\alpha, \ldots,$$

so that the zeroth series of the expansion in (4.2.11) contains the standard Poincaré asymptotic series given in Watson (1952, p. 243)). The zeroth level series of the expansions in (4.2.13) and (4.2.14) can similarly be shown to contain the standard asymptotic series in these cases; see Paris (2004c).

As a numerical illustration we take $x = 0.7$ and $\nu = 30$, for which from (4.2.5) $\Lambda_0 \doteq 0.518415$. The geometrical decay factor is taken to be $\vartheta = \tfrac{1}{2}$ so that $\omega_0 = \tfrac{1}{2}\Lambda_0$ and, from (4.2.4), $\Omega_1 = 2\omega_0$, $\Omega_2 = 6\omega_0$, $\Omega_3 = 18\omega_0, \ldots$, with $\omega_n = \Omega_n$ ($n \geq 1$). The terms in each level $n \leq 3$ are displayed against ordinal number k in Fig. 4.2(a). Using the levels $n \leq 2$ and the truncation indices $M_0 = M_1 = 50$ and $M_2 = 40$, we obtain the value of $J_{30}(21)$ to an accuracy of order 10^{-24}; this is to be compared to the error in the optimally truncated asymptotic expansion (4.2.15) of order 10^{-10}.

In Table 4.1 we show the absolute error in the computation of $J_\nu(\nu x)$ using (4.2.11) and (4.2.13) for different ν and x and the *fixed* truncation indices mentioned above. It is apparent that the accuracy at fixed truncation indices deteriorates as $x \to 1$ caused by the coalescence of the active saddles. This results in the quantity $\Lambda_0 \to 0$ in this limit, which engenders a loss of exponential separation between the different levels of the Hadamard expansions. In Paris (2004c) the Hadamard expansions of both $J_\nu(\nu x)$ and $Y_\nu(\nu x)$ were given using the subdivision Scheme A described in §3.2.2 with maximal integration intervals. This required the use of

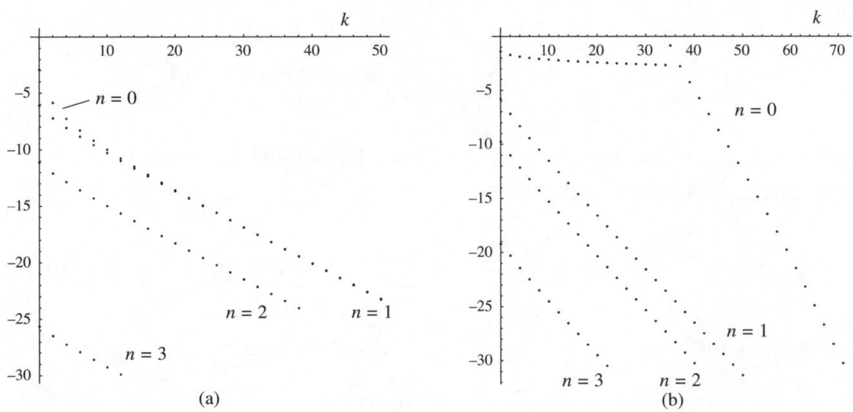

Figure 4.2 The behaviour of the terms (on a $\log_{10}$ scale) against ordinal number k of the levels $n \leq 3$ in the Hadamard expansion for $J_\nu(\nu x)$; (a) in (4.2.11) when $x = 0.7$, $\nu = 30$ with $\vartheta = \tfrac{1}{2}$ and (b) in (4.2.16) when $x = 0.99$, $\nu = 20$ with $\vartheta = \tfrac{1}{3}$.

Table 4.1 *The absolute error in $J_\nu(\nu x)$ for different ν and x with $\vartheta = \frac{1}{2}$ using the levels $n \leq 2$ and the truncation indices given in the text*

x^a	$\nu = 20$	$\nu = 30$	$\nu = 50$
0.5	5.256×10^{-27}	6.240×10^{-31}	9.246×10^{-39}
0.6	2.031×10^{-24}	5.022×10^{-27}	3.185×10^{-32}
0.7	1.036×10^{-18}	4.948×10^{-24}	3.297×10^{-27}
0.8	2.263×10^{-19}	5.665×10^{-21}	2.075×10^{-23}
0.9	2.301×10^{-10}	2.465×10^{-14}	5.925×10^{-19}
1.3	1.452×10^{-14}	7.342×10^{-19}	2.038×10^{-22}
1.4	1.108×10^{-18}	3.989×10^{-22}	6.534×10^{-24}
1.5	3.152×10^{-22}	2.158×10^{-26}	1.007×10^{-25}

[a] For $x = 0.8$ and $x = 0.9$ the additional level $n = 3$ has been employed with truncation index $M_3 = 30$.

the computationally more expensive modified forms of the associated Hadamard series.

4.2.3 The case $x \simeq 1$

In the neighbourhood of $x = 1$, we employ the expansion procedure described in §3.4.1. For $x \lesssim 1$, we select the length of the zeroth interval ω_0 and expand about its right-hand endpoint Ω_0' (equal to ω_0); the corresponding endpoint t_0' on the upper half of the steepest descent path in Fig. 4.1(a) is obtained by solution of (4.2.6). This yields

$$\frac{dt^+}{du} - \frac{dt^-}{du} = 2i \sum_{k=0}^{\infty} \frac{c_{k,0}'(\alpha)}{k!} w^k \qquad (|w| < \omega_0),$$

where $u = \Omega_0 + w$ and the coefficients $c_{k,0}'(\alpha)$ are defined as in (4.2.8).

Then, from (3.4.5), we find

$$J_\nu(\nu x) = \frac{e^{\nu(\tanh \alpha - \alpha)}}{\pi} \sum_{n=0}^{\infty} e^{-\Omega_n \nu x} \mathbf{S}_n(\nu x), \qquad (4.2.16)$$

where $\mathbf{S}_n(\nu x) \equiv S_n(\nu x)$ $(n \geq 1)$ are defined in (4.2.10) and

$$\mathbf{S}_0(\nu x) = -e^{-\Omega_0' \nu x} \sum_{k=0}^{\infty} \frac{c_{k,0}'(\alpha)}{(\nu x)^{k+1}} P(k + 1, -\omega_0 \nu x).$$

The modified form of $\mathbf{S}_0(\nu x)$ must be used, where the above sum is truncated after M_0 terms and the tail $T_{M_0,0}(\nu x)$ is evaluated according to (3.4.8), (3.4.9) and

Table 4.2 *The absolute error in $J_\nu(\nu x)$ for different x in the neighbourhood of $x = 1$ when $\nu = 20$, with $\vartheta = \frac{1}{3}$ using the levels $n \leq 3$ and the truncation indices given in the text*

x	|Error|	x	|Error|
0.900	3.213×10^{-29}	0.990	1.542×10^{-30}
0.950	6.523×10^{-30}	0.999	1.071×10^{-30}
0.980	2.257×10^{-30}	1.000	1.026×10^{-30}

(3.4.10). This takes the form

$$T_{M_0,0}(\nu x) = \sum_{r=0}^{\infty} \sigma'_{r,0}(-\omega_0 \nu x/M_0)^r,$$

$$\sigma'_{r,0} = M_0^r \left\{ \frac{1}{r!} \mathrm{Im} \int_{t_s}^{t'_0} v^r \, dt - \sum_{k=0}^{M_0-1} \frac{(-)^k c'_{k,0} \omega_0^{k+1}}{(k+r+1)!} \right\}, \qquad v = \frac{\psi(t) - \psi(t_s)}{\omega_0};$$

in computations the tail is truncated after N_0 terms.

An example of the typical behaviour of the terms in (4.2.16) is shown in Fig. 4.2(b) for $\nu = 20$ and $x = 0.99$, with the levels $1 \leq n \leq 3$ computed using $\vartheta = \frac{1}{3}$. The value chosen for the zeroth interval is $\omega_0 = 0.5$ and, from (4.2.4), the expansion points are $\Omega_1 = 1.5\omega_0$, $\Omega_2 = 3\omega_0$, $\Omega_3 = 6\omega_0, \ldots$. It should be noted that the leading terms in the tail $T_{M_0,0}(\nu x)$ are considerably larger in magnitude than the final terms of the associated finite main sum. We show in Table 4.2 the absolute error in the computation of $J_\nu(\nu x)$ for different values of x in the neighbourhood of $x = 1$ when $\nu = 20$ using the fixed truncation indices $(M_0, N_0) = (35, 38)$, $M_1 = 50$, $M_2 = 40$ and $M_3 = 20$. When $x \gtrsim 1$, the expansion of the integral in (4.2.12) is dealt with in a similar manner; we omit these details.

4.2.4 The case z complex

We briefly consider the case when z is complex satisfying $|\arg z| < \frac{1}{2}\pi$ and $\nu > 0$. At the active saddles $t = \pm\gamma$, the integrand in (4.2.1) has the values $\exp\{\mp\nu(\gamma - \tanh\gamma)\}$; these saddles therefore connect when $\mathrm{Im}(\gamma - \tanh\gamma) = 0$, that is, when

$$\cosh 2\alpha = \frac{\sin 2\beta}{\beta} - \cos 2\beta. \tag{4.2.17}$$

The curve in the γ-plane on which (4.2.17) is satisfied is shown in Fig. 4.3(a); a similar curve exists in the strip $0 \leq \beta < \frac{1}{2}\pi$, $\alpha < 0$ for conjugate values

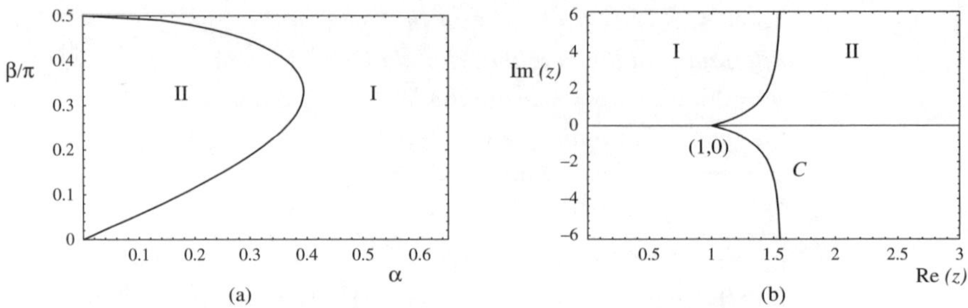

Figure 4.3 The solution of (4.2.17) (a) in the γ-plane and (b) in the z-plane.

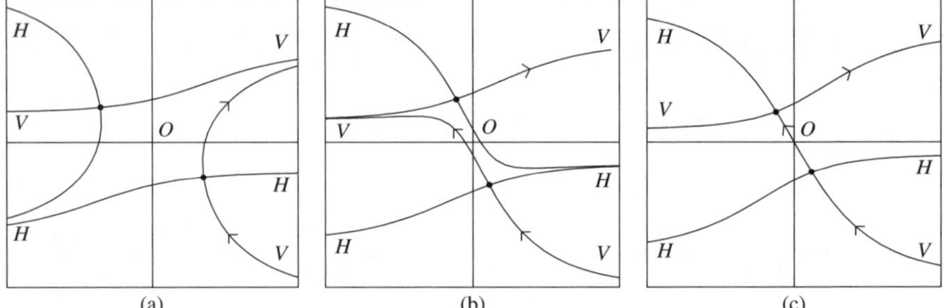

Figure 4.4 Paths of steepest descent and ascent through the saddles at $t = \pm\gamma$ when (a) $z = 0.5 + 0.5i$ in zone I, (b) $z = 1.5 + i$ in zone II and (c) $z \doteq 1.18417 + 0.41055i$ on the cusped curve C. The valleys (V) and hills (H) at infinity are indicated and the arrows denote the direction of integration.

of z. The corresponding cusped curve C in the z-plane is shown in Fig. 4.3(b), which has the asymptote[3] $\mathrm{Re}(z) = \frac{1}{2}\pi$. Note that this division of the γ-plane is not the same as that discussed in Watson (1952, p. 265). To the left of C (zone I) only the saddle at $t = \gamma$ contributes to the evaluation of $J_\nu(\nu x)$, whereas to the right (zone II) both saddles contribute. For values of z on C, we have a Stokes phenomenon, with the second saddle switching either on or off depending in which sense we traverse the curve. Typical paths of steepest descent and ascent for z situated in the two zones and on C are illustrated in Fig. 4.4.

In zone I, the expansion in (4.2.11) applies provided we make the obvious modification to take account of the fact that the upper and lower halves of the steepest descent path are no longer conjugates. In zone II, the expansion (4.2.11) must be used about each saddle, while on or near C the relevant half paths from each saddle are used together with the inter-saddle contribution.

---

[3] When $\beta = \frac{1}{2}\pi - \epsilon$, $\epsilon \to 0^+$, it is easily established from (4.2.17) that $\alpha \simeq (2\epsilon/\pi)^{1/2}$. It then follows that $z = \mathrm{sech}\,\gamma \simeq \frac{1}{2}\pi \pm i(2\epsilon/\pi)^{-1/2}$ in this limit.

4.3 The Pearcey integral

The Pearcey integral is defined[4] by

$$\text{Pe}(x, y) = \int_{-\infty}^{\infty} \exp\{i(t^4 + xt^2 + yt)\}dt \tag{4.3.1}$$

for real x and y and belongs to the hierarchy of canonical diffraction integrals used in the treatment of many short-wavelength problems, such as wave propagation and optical diffraction. From the symmetry of the integral it follows that $\text{Pe}(x, y) = \text{Pe}(x, -y)$ so that we need only consider $y \geq 0$. The asymptotic expansion of $\text{Pe}(x, y)$ for large x and y, including the asymptotics of its zeros, has been given in Kaminski (1989), Paris (1991) and Kaminski and Paris (1999); various numerical schemes for the evaluation of $\text{Pe}(x, y)$ and its derivatives have been proposed in Connor and Curtis (1982).

By application of Jordan's lemma and a simple change of variable (when $x \neq 0$) it follows that we can write

$$\text{Pe}(\pm x, y) = x^{1/2} \int_{\Gamma} e^{ix^2 \psi(t)} dt \qquad (x > 0), \tag{4.3.2}$$

where

$$\psi(t) = t^4 \pm t^2 + \alpha t, \qquad \alpha = y/x^{3/2}$$

and Γ is a path linking $-\infty e^{\pi i/8}$ to $\infty e^{\pi i/8}$. In the right-hand (resp. left-hand) half of the x, y-plane we take the upper (resp. lower) sign in $\psi(t)$. The saddle points t_{sj} ($j = 1, 2, 3$) are given by the roots of the equation $\psi'(t) = 0$ and can be evaluated either numerically or by application of the elementary theory of cubics.

The curve on which $\psi'(t) = \psi''(t) = 0$ is denoted by the cusp (or caustic) C_1, given by $y^2 + (\frac{2}{3}x)^3 = 0$, and corresponds to the coalescence of two saddles; see Fig. 4.5. For values of x and y in the left-hand plane situated inside C_1 all three saddles are real and ordered such that $t_{s1} < t_{s2} \leq t_{s3}$; outside C_1, t_{s1} remains on the

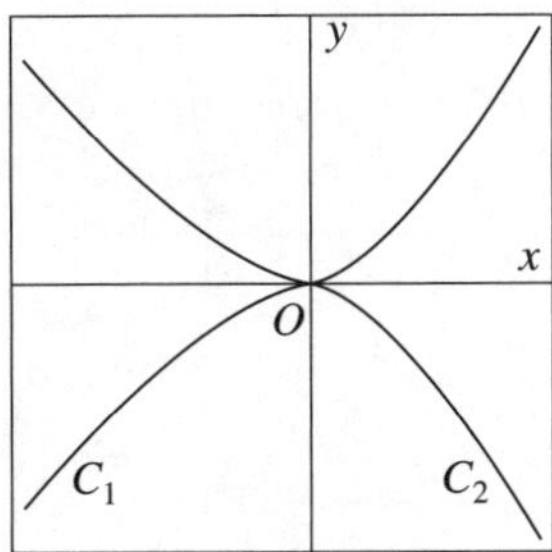

Figure 4.5 The x, y-plane showing the left-hand and right-hand cusps C_1 and C_2.

[4] The notation $\text{Pe}(x, y)$ for the Pearcey integral is adopted here to avoid a notational clash with the normalised incomplete gamma function $P(a, x)$.

negative real axis with the other two saddles forming a complex conjugate pair in the right-hand plane. It has been shown in Wright (1980) that inside the second cusp (or Stokes set) C_2 in the right-hand plane, given by

$$y^2 - (\tfrac{2}{3}x)^3(\tfrac{5}{4} + \tfrac{3}{4}\sqrt{3}) = 0,$$

only the real saddle t_{s1} is active in the asymptotics of $\mathrm{Pe}(x, y)$; in the domains containing the y-axis between these two cusps, the real saddle t_{s1} and only one of the complex saddles are contributory. The division of the x, y-plane by the cusps C_1 and C_2 is shown in Fig. 4.5. The topology of the paths of steepest descent Γ_j ($j = 1, 2, 3$) and ascent through the saddles t_{sj} in the different regions of the x, y-plane is illustrated in Fig. 4.6. Thus, when (x, y) is situated in the right-hand cusp the integration path Γ can be taken to be $\Gamma = \Gamma_1$, in between the cusps $\Gamma = \Gamma_1 \cup \Gamma_2$ and in the left-hand cusp $\Gamma = \Gamma_1 \cup \Gamma_2 \cup \Gamma_3$.

A hyperasymptotic expansion for complex variables – applied to the analytic continuation of (4.3.1) – has been given by Berry and Howls (1991) as an example of their treatment of very accurate asymptotics of Laplace-type integrals with several saddles; see §1.8.6. The example chosen by these authors corresponds to a situation when only one saddle contributes to the first stage of the expansion process, although the other saddles (*adjacent* saddles) are found to make a contribution in the higher levels of the expansion. Here we obtain the Hadamard expansion of $\mathrm{Pe}(x, y)$ and present numerical results in the three different regions of the x, y-plane, including the neighbourhoods of the cusps C_1 and C_2. Our presentation is based on Paris and Kaminski (2006).

4.3.1 The Hadamard expansion of Pe(x, y)

In the derivation of the Hadamard expansion for $\mathrm{Pe}(x, y)$ away from the neighbourhood of the cusps C_1 and C_2 it is sufficient to present only the details of the

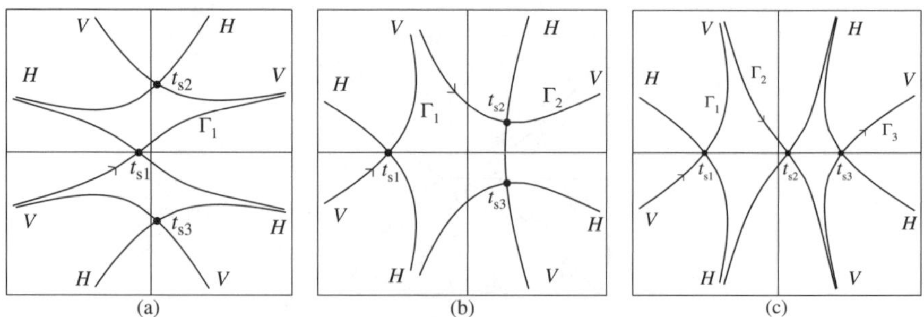

Figure 4.6 Paths of steepest descent Γ_j and ascent when the point (x, y), with $y \geq 0$, is situated (a) inside the right-hand cusp C_2, (b) in between the cusps and (c) inside the left-hand cusp C_1. The saddles t_{sj} are represented by heavy dots and the asymptotic valleys (V) and hills (H) are indicated. The arrows denote the direction of integration.

contribution from one of the saddles, as the details of the other contributions are similar. The modification of the procedure required to deal with the neighbourhood of the cusps will be presented in §4.3.2.

The contribution I_j to (4.3.2) that results from the steepest descent path Γ_j passing through the saddle t_{sj} can be written as

$$I_j = x^{1/2} \int_{\Gamma_j} e^{ix^2 \psi(t)} \, dt = x^{1/2} e^{ix^2 \psi(t_{sj})} \int_0^\infty e^{-x^2 u} \left(\frac{dt^+}{du} - \frac{dt^-}{du} \right) du,$$

where we have defined the new variable u by

$$iu = \psi(t) - \psi(t_{sj}), \tag{4.3.3}$$

which is real and non-negative on Γ_j. For each non-zero value of u, there are two values of t: $t^+(u)$ on the 'upper' half of the steepest descent path emanating from t_{sj}, and $t^-(u)$ on the 'lower' half leading into t_{sj}. In the u-plane (the so-called Borel plane), the images of the saddles t_{sj} are denoted by u_j and these points are situated on different Riemann sheets. On the sheet corresponding to u_j, the path Γ_j maps into the negatively oriented loop surrounding the positive real u-axis.

In this section, we follow Paris and Kaminski (2006) and adopt the subdivision Scheme A discussed in §3.2.1. The positive real u-axis is subdivided into intervals of length $\omega_0, \omega_1, \omega_2, \ldots$ with left-hand endpoints at $\Omega_0 = 0, \Omega_1, \Omega_2, \ldots,$ where

$$\Omega_n = \sum_{r=0}^{n-1} \omega_r \qquad (n \geq 1);$$

see Fig. 3.1(a) and (3.2.1). The corresponding endpoints on Γ_j are denoted by $t_0, t_1^\pm, t_2^\pm, \ldots$ which are obtained by appropriate solution of

$$\psi(t_n^\pm) - \psi(t_{sj}) = i\Omega_n \qquad (n \geq 1), \tag{4.3.4}$$

where $t_0 \equiv t_{sj}$. Then, for arbitrary complex τ and $n \geq 0$, we have from (4.3.3) and (4.3.4) (the superscripts $\pm$ are omitted)

$$\psi(t_n + \tau) - \psi(t_n) = i(u - \Omega_n) = h_n(\tau), \tag{4.3.5}$$

where

$$h_n(\tau) = \tau \{ \tau^3 + 4t_n \tau^2 + \tfrac{1}{2} \psi''(t_n) \tau + \psi'(t_n) \}.$$

Upon inversion we find on the upper and lower halves of the steepest descent path Γ_j

$$\tau^\pm(u) = \sum_{k=1}^\infty a_{k,n} \, w^{\mu_n k}, \qquad w = \begin{cases} \pm (iu)^{\frac{1}{2}} & (n = 0) \\ u - \Omega_n & (n \geq 1), \end{cases} \tag{4.3.6}$$

where $\mu_0 = \tfrac{1}{2}$ and $\mu_n = 1$ ($n \geq 1$). The expansion (4.3.6) is valid in a disc centred at $w = 0$ whose radius is controlled by the nearest singularity. For the zeroth interval, we now choose ω_0 to be the radius of convergence, that is

$$\omega_0 = \min\{|\psi(t_{sj}) - \psi(t_{sr})|\} \qquad (j = 1, 2, 3; \ r \neq j), \tag{4.3.7}$$

where only those values of r that *correspond to saddles adjacent*[5] *to the saddle* t_{sj} are admissible. The coefficients $a_{k,0}$ are determined in specific cases by a numerical inversion procedure using *Mathematica*. Then, we obtain for the zeroth interval

$$\frac{dt^+}{du} - \frac{dt^-}{du} = \sum_{k=0}^{\infty} \frac{c_{k,0}^{(j)}}{\Gamma(k+\frac{1}{2})} u^{k-\frac{1}{2}}, \qquad c_{k,0}^{(j)} = 2i^{k+\frac{1}{2}} a_{2k+1,0}$$

valid in $0 \leq u < \omega_0$. For the intervals with $n \geq 1$, we choose ω_n to be the radius of convergence of (4.3.6) by setting

$$\omega_n = \min\{|\psi(t_n) - \psi(t_{sr})|\} \qquad (r = 1, 2, 3; \ n \geq 1),$$

where, when $r \neq j$, only the adjacent saddles are considered. This leads to the result

$$\frac{dt^+}{du} - \frac{dt^-}{du} = \sum_{k=0}^{\infty} \frac{c_{k,n}^{(j)}}{k!} w^k, \qquad c_{k,n}^{(j)} = a_{k+1,n}^+ - a_{k+1,n}^-$$

valid in $0 \leq u - \Omega_n < \omega_n$.

From (3.2.8), (3.2.9) and (3.2.12), the Hadamard expansion for I_j takes the form

$$I_j = x^{1/2} e^{ix^2 \psi(t_{sj})} \sum_{n=0}^{\infty} e^{-\Omega_n x^2} S_n^{(j)}(x) \qquad (j = 1, 2, 3), \tag{4.3.8}$$

where

$$S_n^{(j)}(x) = \sum_{k=0}^{\infty} \frac{c_{k,n}^{(j)}}{x^{2(k+\mu_n)}} P(k + \mu_n, \omega_n x^2) \qquad (n \geq 0). \tag{4.3.9}$$

Since the interval lengths have been chosen to be maximal integration intervals, it is necessary to use the Hadamard series $S_n^{(j)}(x)$ in their modified form given in (3.2.14) and (3.2.15). Thus

$$S_n^{(j)}(x) = \sum_{k=0}^{M_n-1} \frac{c_{k,n}^{(j)}}{x^{2(k+\mu_n)}} P(k + \mu_n, \omega_n x^2) + e^{-\omega_n x^2} \sum_{r=0}^{\infty} \sigma_{r,n}^{(j)} \xi_n^r,$$

for $n = 0, 1, 2, \ldots$, where $\xi_n = \omega_n x^2 / M_n$ and the truncation indices M_n are free to be suitably chosen. From (3.2.17), (3.2.18) and (3.2.19), the coefficients $\sigma_{r,n}^{(j)}$ are given by

$$\sigma_{r,n}^{(j)} = M_n^r \left\{ \frac{1}{r!} \left(\int_{t_n^+}^{t_{n+1}^+} + \int_{t_{n+1}^-}^{t_n^-} \right) (1-v)^r \, dt - \sum_{k=0}^{M_n-1} \frac{c_{k,n}^{(j)}}{(k+r+\mu_n)!} \right\},$$

---

[5] For example, in the case of Fig. 4.6(c) where the saddles t_{sj} all lie on the real t-axis, the saddle t_{s1} is adjacent to t_{s2} and the saddle t_{s3} is adjacent to t_{s2}, while the saddle t_{s2} is adjacent to *both* t_{s1} and t_{s3}.

where $v = \{\psi(t) - \psi(t_n)\}/\omega_n$. For the zeroth interval the two integrals reduce to a single integral over $[t_1^-, t_1^+]$. In computations the tails of $S_n^{(j)}(x)$ are truncated after N_n terms.

We note that the standard Poincaré asymptotic expansion for I_j is obained by replacing the normalised incomplete gamma functions in the zeroth Hadamard series $S_0^{(j)}(x)$ by unity, to obtain

$$I_j \sim e^{ix^2\psi(t_{sj})} \sum_{k=0}^{\infty} \frac{c_{k,0}^{(j)}}{x^{2k+\frac{1}{2}}} \qquad (x \to \infty).$$

To illustrate we consider the evaluation of $Pe(x, y)$ in different regions of the x, y-plane corresponding to one, two or three saddle-point contributions. We first take (x, y) situated in the right-hand cusp C_2 in Fig. 4.5 and choose $x = 7$, $y = 1$, so that the steepest descent path Γ consists of the single path Γ_1 passing through the saddle t_{s1} on the real axis; see Fig. 4.6(a). In Fig. 4.7 we display the typical behaviour of the terms against ordinal number k in the different levels of the expansion I_1 in (4.3.8) for the levels $n \leq 2$ and a given choice of truncation indices.

In the numerical computations we employ the truncation indices $(M_0, N_0) = (30, 35)$ and $(M_1, N_1) = (30, 25)$, $(M_2, N_2) = (30, 0)$ (that is, no terms from the tail at level 2 are used) to evaluate $Pe(7, 1)$ to an accuracy of order 10^{-24}. The results are presented in the first entry of Table 4.3. The exact value of $Pe(x, y)$ was computed from the series expansion in ascending powers of y given by [6]

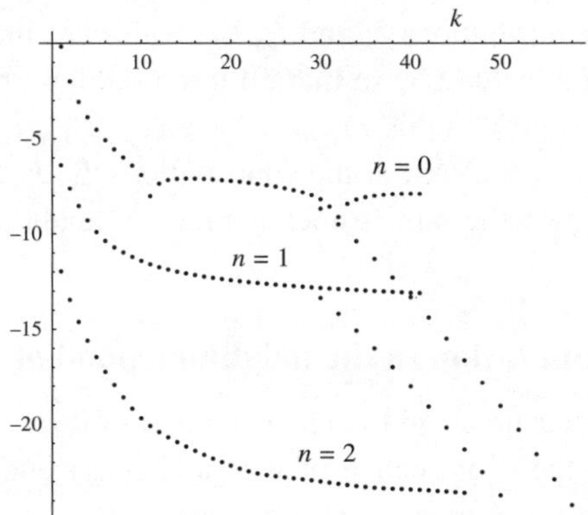

Figure 4.7 The behaviour of the absolute value of the terms in I_1 (on a $\log_{10}$ scale) as a function of ordinal number k for the levels $n \leq 2$ of the Hadamard expansion (4.3.8) when $x = 7$, $y = 1$. The terms in the modified tails at levels $n = 0, 1$ are indicated with truncation index $M_n = 30$.

---

[6] When computing $Pe(-x, y)$ with $x > 0$, it is necessary to take into account the branch structure of $U(a, b, z)$, since we have $\arg z = \frac{3}{2}\pi$, and employ the analytic continuation relation in Abramowitz and Stegun (1965, Eq. (13.1.10)).

Table 4.3 *Absolute values of the error in* $\mathrm{Pe}(x, y)$ *for different values of* x *and* y *at different levels* n; *the truncation indices employed are indicated in the text*

n	$\mathrm{Pe}(7, 1)$ I_1	$\mathrm{Pe}(-4, 12)$ $I_1 + I_2$	$\mathrm{Pe}(-7, 1)$ $I_1 + I_2 + I_3$
0	3.898×10^{-7}	1.910×10^{-7}	2.772×10^{-6}
1	1.036×10^{-12}	2.123×10^{-11}	5.678×10^{-11}
2	8.198×10^{-24}	3.310×10^{-19}	2.775×10^{-20}

$$\mathrm{Pe}(\pm x, y) = 2e^{\pi i/8} \int_0^\infty \exp\{-t^4 \pm xe^{-\pi i/4}t^2\} \cos(e^{\pi i/8}yt)\, dt$$

$$= e^{\pi i/8} \sum_{k=0}^\infty \frac{(-)^k}{(2k)!} \frac{(ye^{\pi i/8})^{2k}}{2^{k+\frac{1}{2}}} \Gamma(k+\tfrac{1}{2})\, U(\tfrac{1}{2}k + \tfrac{1}{4}, \tfrac{1}{2}, \tfrac{1}{4}i\zeta^2), \quad \zeta = e^{\mp\frac{1}{2}\pi i}x$$

upon expansion of the cosine term, where $U(a, b, z)$ denotes the confluent hypergeometric function (Abramowitz and Stegun, 1965, p. 504).

The results of similar calculations for $\mathrm{Pe}(x, y)$ on both sides of the caustic C_1 are also presented in Table 4.3. In the second example, we take $x = -4$, $y = 12$ which is situated outside the caustic C_1, so that we have one real and one complex saddle (labelled by t_{s1} and t_{s2}, respectively) contributing to the asymptotics as shown in Fig. 4.6(b). Then the integration path is $\Gamma = \Gamma_1 \cup \Gamma_2$ and $\mathrm{Pe}(-4, 12)$ is given by the sum of the associated contributions I_1 and I_2. In the third example, we take $x = -7$, $y = 1$ which is situated inside C_1, so that all three saddles are situated on the real t-axis as shown in Fig. 4.6(c). In this case, we have $\Gamma = \Gamma_1 \cup \Gamma_2 \cup \Gamma_3$, with the result that $\mathrm{Pe}(-7, 1)$ is given by the sum of the contributions I_1, I_2 and I_3. For simplicity in presentation, we employ the same truncation indices about each saddle point.

4.3.2 Computation in the neighbourhood of the cusps

We first deal with the caustic C_1 in the left-hand half of the x, y-plane. The caustic C_1 is associated with the coalescence of the saddles t_{s2} and t_{s3}: as we cross the upper half of C_1 in the sense of increasing y, the saddles t_{s2} and t_{s3} situated on the positive real t-axis steadily approach one another, coalesce to yield a double saddle point and then move off the real axis to form a complex conjugate pair; see Fig. 4.8. In the expansion of I_2 in (4.3.8), the zeroth convergence interval ω_0 is determined by the proximity of the saddle t_{s3} according to (4.3.7), and so steadily decreases to zero as C_1 is approached. A similar remark applies to the contribution I_3 for points just inside C_1. This will result in a corresponding progressive loss of exponential separation in the different levels of these expansions, with the consequence that an

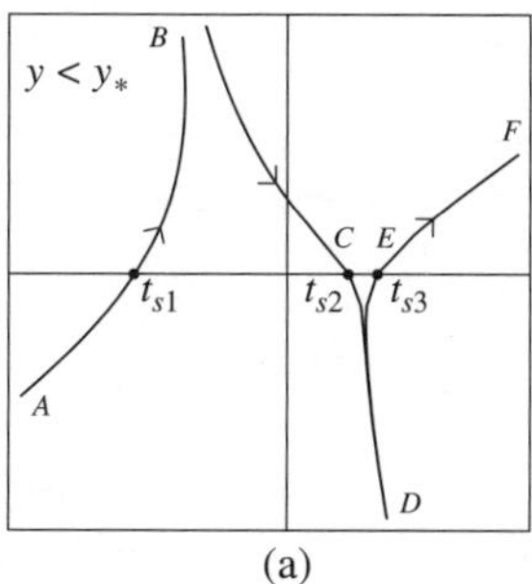

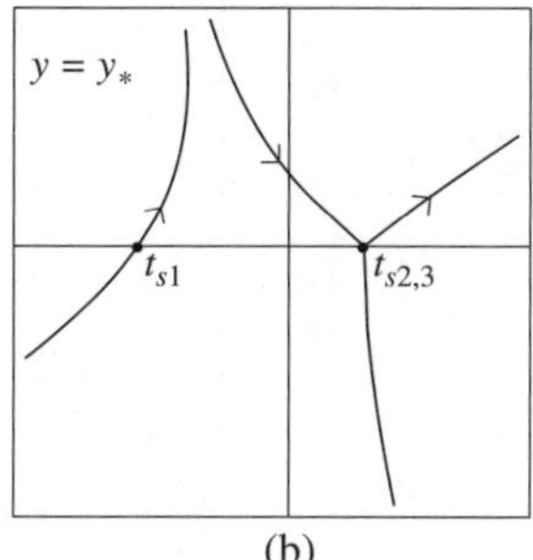

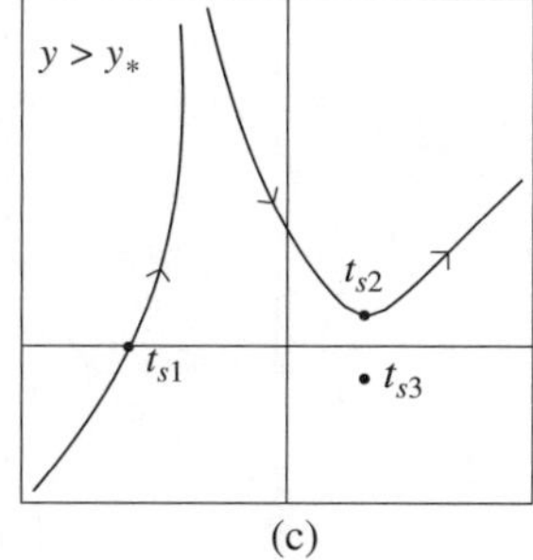

Figure 4.8 The steepest descent paths in the neighbourhood of the caustic C_1: (a) inside C_1 ($y < y_*$); (b) on C_1 ($y = y_*$) and (c) outside C_1 ($y > y_*$). The definition of y_* is given in (4.3.11). The heavy dots are the saddle points and the arrows denote the direction of integration.

increasing number of levels would need to be considered to maintain a given level of accuracy.

Let us consider the situation arising when we are just inside C_1, where the saddles t_{s2} and t_{s3} are situated on the real axis; see Fig. 4.8(a). The procedure for the contribution I_1 remains unchanged and is as described in §4.3.1, since the saddle t_{s1} is not affected by the coalescence. For the contributions I_2 and I_3, we choose a suitable zeroth interval ω_0, with right-hand endpoint Ω'_0, and employ the mode of expansion described in §3.4.1 to obtain, from (3.4.4) and (3.4.5),

$$I_j = x^{1/2} e^{ix^2 \psi(t_{sj})} \sum_{n=0}^{\infty} e^{-\Omega_n x^2} \mathbf{S}_n^{(j)}(x) \qquad (j = 2, 3), \tag{4.3.10}$$

where $\mathbf{S}_n^{(j)}(x) = S_n^{(j)}(x)$ $(n \geq 1)$ defined in (4.3.9),

$$\mathbf{S}_0^{(j)}(x) = -e^{-\Omega'_0 x^2} \sum_{k=0}^{\infty} \frac{c'_{k,0}}{x^{2k+2}} P(k+1, -\omega_0 x^2),$$

and $c'_{k,0}$ are the coefficients that arise in the inversion expansion about the image points of Ω'_0 $(= \omega_0)$. The modified form of $\mathbf{S}_0^{(j)}(x)$ is obtained from (3.4.7).

We remark that, near coalescence on the inside of C_1, the contributions from those parts of the steepest descent paths Γ_2 and Γ_3 situated in $\mathrm{Im}(t) < 0$, the path CDE in Fig. 4.8(a), almost cancel. In this case, an alternative integration path which involves less labour, is to take only the half-contributions to I_2 and I_3 that result from the parts of these paths situated in $\mathrm{Im}(t) \geq 0$, together with the line segment on the real axis joining t_{s2} and t_{s3}. This latter contribution is obtained by expanding about the saddle t_{s2} in the manner described in Example 3.1. The resulting integration path Γ is then given by the path $ABCEF$.

We now consider the cusp C_2 in the right-hand half of the x, y-plane. In the neighbourhood of this cusp we are not confronted with a coalescence problem but with a

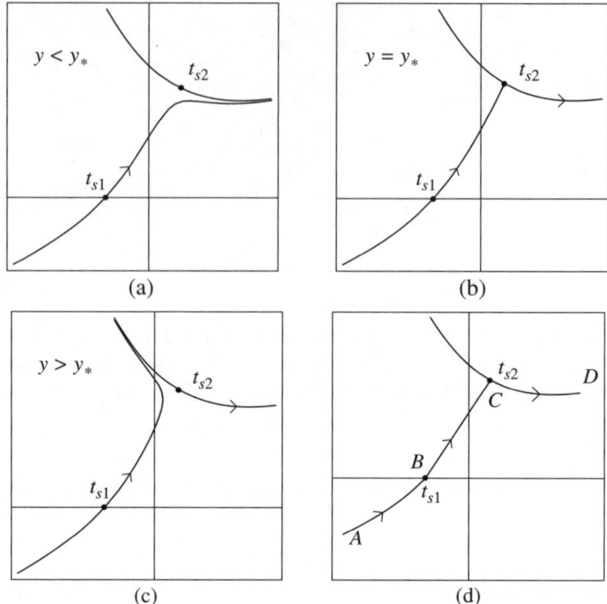

Figure 4.9 The steepest descent paths in the neighbourhood of the cusp C_2: (a) inside C_2 $(y < y_*)$; (b) on C_2 $(y = y_*)$; (c) outside C_2 $(y > y_*)$; and (d) the path of integration $ABCD$ containing the line segment joining t_{s1} and t_{s2}. The definition of y_* is given in (4.3.11). The saddle point t_{s3} situated in $\mathrm{Im}(t) < 0$ is not shown.

Stokes phenomenon. Inside C_2, only the real saddle t_{s1} is active and the integration path $\Gamma = \Gamma_1$. On C_2, the steepest descent path through t_{s1} connects with the steepest descent path through the saddle t_{s2}; for points situated above C_2, the integration path is $\Gamma = \Gamma_1 \cup \Gamma_2$. The topology of the steepest descent paths in the neighbourhood of C_2 is shown in Fig. 4.9.

The modification of the basic procedure to deal with a Stokes phenomenon has been discussed in detail in Example 3.6. This consists of taking the integration path Γ to be the half of the steepest descent path Γ_1 leading up to t_{s1}, followed by the line segment joining t_{s2} and t_{s3} and thence along the half of the steepest descent path Γ_2 emanating from t_{s2}; see Fig. 4.9(d). The half-contributions from Γ_1 and Γ_2 are obtained as described in §3.2.2, whereas the inter-saddle contribution is as above.

In Table 4.4, we show the results of numerical computations of $\mathrm{Pe}(\pm x, y)$ in the neighbourhood of the cusps C_1 and C_2. For a given value of x, we compute $\mathrm{Pe}(\pm x, y)$ for a series of y values measured in units of y_*, where y_* is given by

$$y_* = \lambda(\tfrac{2}{3}x)^{3/2}, \qquad \lambda = \begin{cases} 1 & \text{for } C_1 \\ (\tfrac{5}{4} + \tfrac{3}{4}\sqrt{3})^{1/2} & \text{for } C_2. \end{cases} \tag{4.3.11}$$

Table 4.4 *Absolute values of the error in* $\mathrm{Pe}(\pm x, y)$ *in the neighbourhood of the cusps* C_1 *and* C_2 *when* $x = 5$; *the value of* y *is measured in units of* y_* *defined in (4.3.11); the value of the zeroth interval for the coalescing saddles was chosen to be* $\omega_0 = 0.5$; *levels with* $n \leq 2$ *have been employed with the truncation indices given in the text*

Cusp C_1							
y/y_*	0.90	0.95	0.99	1.00	1.01	1.05	1.10
$\lvert\text{Error}\rvert \times 10^{-24}$	4.156	6.537	7.881	1.823	8.221	8.113	7.278
Cusp C_2							
y/y_*	0.90	0.95	0.99	1.00	1.01	1.05	1.10
$\lvert\text{Error}\rvert \times 10^{-23}$	5.844	3.970	3.483	3.420	3.352	4.752	0.190

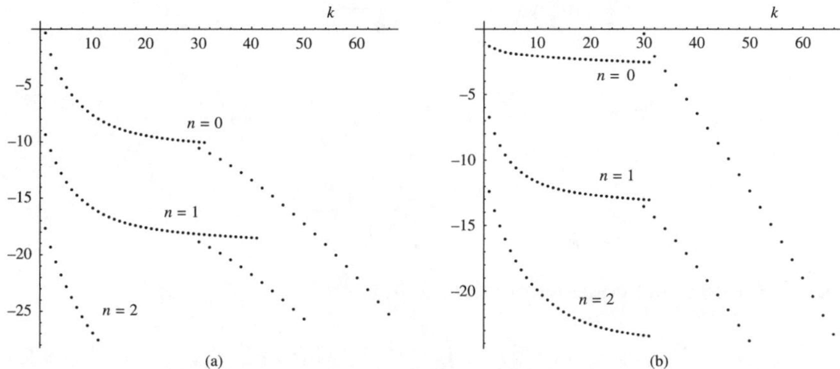

Figure 4.10 The behaviour of the absolute value of the terms (on a $\log_{10}$ scale) as a function of ordinal number k for the levels $n \leq 2$ of the Hadamard expansions in the neighbourhood of the caustic C_1 for (a) I_1 and (b) I_2 with $\omega_0 = 0.5$. The figures correspond to $x = 5$, $y = 0.99y_*$, where y_* is defined in (4.3.11).

We again employ the levels with $n \leq 2$. The truncation indices chosen for C_1 are: $(M_0, N_0) = (30, 35)$, $(M_1, N_1) = (30, 30)$, $(M_2, N_2) = (30, 0)$ for both I_1 and I_2, with inter-saddle indices $(M_0, N_0) = (30, 20)$ when $y < y_*$. The truncation indices for C_2 are: $(M_0, N_0) = (30, 40)$, $(M_1, N_1) = (30, 20)$, $(M_2, N_2) = (10, 0)$ for I_1 and $(M_0, N_0) = (30, 25)$, $(M_1, N_1) = (30, 5)$, $(M_2, N_2) = (5, 0)$ for I_2, with inter-saddle indices $(M_0, N_0) = (30, 40)$. The results show that computation of $\mathrm{Pe}(\pm x, y)$ across the cusps C_1 and C_2 can be achieved with *a uniform level of accuracy using a fixed number of levels*. We present in Fig. 4.10 the typical behaviour of the terms in the different levels of the expansions I_1 and I_2 for a point close to the caustic C_1. It is apparent that the decay of the terms at the zeroth level in the contribution I_2 in (4.3.10) is not initially as rapid as that for the corresponding terms in the contribution I_1.

4.4 The parabolic cylinder function

We consider the parabolic cylinder function $U(a, x)$ of large positive order and argument. From Olver *et al.* (2010, §12.5(i)), we have[7]

$$U(a + \tfrac{1}{2}, X) = \frac{e^{-X^2/4}}{\Gamma(1+a)} \int_0^\infty \tau^a e^{-\frac{1}{2}\tau^2 - X\tau} d\tau$$

and

$$U(-a - \tfrac{1}{2}, X) = \sqrt{\frac{2}{\pi}} e^{X^2/4} \int_0^\infty \tau^a e^{-\frac{1}{2}\tau^2} \cos(X\tau - \tfrac{1}{2}\pi a) d\tau$$

for $a > -1$ and $X > 0$.

For $a > 0$, if we make the change of variables $\tau = a^{1/2}t$ and $X = 2a^{1/2}x$, we find

$$U(a + \tfrac{1}{2}, 2a^{\frac{1}{2}}x) = \frac{a^{\frac{1}{2}a + \frac{1}{2}} e^{-ax^2}}{\Gamma(1+a)} \int_0^\infty e^{-a\psi(t;x)} dt$$

and

$$U(-a - \tfrac{1}{2}, 2a^{\frac{1}{2}}x) = \sqrt{\frac{2}{\pi}} a^{\frac{1}{2}a + \frac{1}{2}} e^{ax^2} \operatorname{Re} \int_0^\infty e^{-a\psi(t;ix) - \frac{1}{2}\pi ia} dt,$$

where

$$\psi(t; z) = \tfrac{1}{2}t^2 + 2zt - \log t.$$

We are therefore led to consideration of the integral

$$I(a; z) = \int_0^\infty e^{-a\psi(t;z)} dt, \qquad (4.4.1)$$

for $a \to +\infty$ and finite values of $|z|$ bounded away from zero. In connection with the above parabolic cylinder functions we shall be concerned primarily with $z = x$ or $z = ix$, but in general z may be taken as a complex variable.

The phase function $\psi(t; z)$ has saddle points at

$$t_{sj} = -z \pm \sqrt{1 + z^2} \qquad (j = 1, 2).$$

When z is real and positive the saddles lie on the real axis, one on either side of the origin. When z assumes complex values the saddles move in the complex plane such that, when $\arg z = \pm \tfrac{1}{2}\pi$, they either straddle the imaginary axis ($|z| < 1$), coalesce to form a double saddle at $t = \mp i$ ($|z| = 1$) or both lie on the imaginary axis ($|z| > 1$). Typical paths of steepest descent and ascent through the saddles are illustrated in Fig. 4.11 for different values of z; it is readily shown that the steepest ascent paths are asymptotic to the line $\operatorname{Re}(t) = -2\operatorname{Re}(z)$ as $\operatorname{Im}(t) \to \pm\infty$. The integration path $[0, \infty)$ can be deformed to pass over the saddle at $t = t_{s1}$ when $|\arg z| < \tfrac{1}{2}\pi$

---

[7] In terms of the older notation we have $D_a(x) = U(-a - \tfrac{1}{2}, x)$.

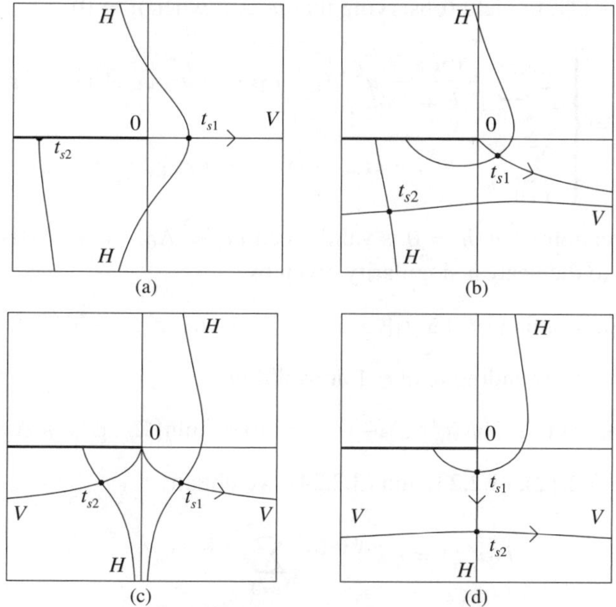

Figure 4.11 The paths of steepest descent and ascent through the saddles t_{sj} for the integral $I(a; z)$ when (a) arg $z = 0$, (b) $0 < $ arg $z < \frac{1}{2}\pi$, (c) arg $z = \frac{1}{2}\pi$, $|z| < 1$ and (d) arg $z = \frac{1}{2}\pi$, $|z| > 1$. The heavy line on the negative real axis is the branch cut and the arrows denote the direction of integration from the origin. The asymptotic valleys (V) and hills (H) at infinity are indicated.

or arg $z = \pm\frac{1}{2}\pi$, $|z| < 1$, and to pass over both saddles when arg $z = \pm\frac{1}{2}\pi$, $|z| \geq 1$.

We introduce the new variable

$$u = \psi(t; z) - \psi(t_{s1}; z)$$

and employ the subdivision Scheme B in §3.2.3. The singular points of this mapping occur at $t = t_{s1}$ and $t = t_{s2}$ where dt/du is singular. The point at $t = t_{s1}$ corresponds to $u = \pm 2\pi ki$, $k = 0, 1, 2, \ldots$ in the u-plane; compare Examples 1.3 and 3.2. We consider the inversion of the mapping $t \mapsto u(t)$ about $u = 0$ and the points $u = \Omega_n$, which correspond in the t-plane to the points $t_n^{\pm}$ on the steepest descent path through t_{s1} given by $\psi(t_n^{\pm}; z) - \psi(t_{s1}; z) = \Omega_n$ $(n \geq 1)$. We have

$$t^{\pm}(u) - \left\{ \begin{matrix} t_{s1} \\ t_n^{\pm} \end{matrix} \right\} = \begin{cases} \displaystyle\sum_{k=1}^{\infty} (\pm)^k a_{k,0} u^{k/2} & (n = 0) \\ \displaystyle\sum_{k=1}^{\infty} a_{k,n}^{\pm} w^k & (n \geq 1), \end{cases}$$

where $w = u - \Omega_n$, to yield (observing that $w \equiv u$ when $n = 0$)

$$\frac{dt^+}{dw} - \frac{dt^-}{dw} = \begin{cases} 2\sum_{k=0}^{\infty} \dfrac{c_{2k,0}}{\Gamma(k+\frac{1}{2})}\, u^{k-\frac{1}{2}}, & c_{2k,0} = a_{2k+1,0}\, \Gamma(k+\tfrac{3}{2}) \quad (n=0) \\[2ex] \sum_{k=0}^{\infty} \dfrac{c_{k,n}}{k!}\, w^k, & c_{k,n} = (k+1)!\{a^+_{k+1,n} - a^-_{k+1,n}\} \quad (n \geq 1). \end{cases}$$

The above inversion when $n = 0$ is valid when $|u| < \Lambda_0$, where Λ_0 is the distance in the u-plane to the nearest singularity given by

$$\Lambda_0 = \min\{2\pi, |\Delta\psi|\}, \qquad \Delta\psi = \psi(t_{s1}; z) - \psi(t_{s2}; z),$$

whereas those corresponding to $n \geq 1$ are valid in

$$|w| < \min\{\Omega_n, |\psi(t_n^{\pm}; z) - \psi(t_{s2}; z)|\} = \min\{\Omega_n, |\Omega_n + \Delta\psi|\}.$$

Then, from (3.2.12), (3.2.23) and (3.2.24), we obtain

$$I(a; z) = e^{-a\psi(t_{s1}; z)} \sum_{n=0}^{\infty} e^{-\Omega_n a}\, S_n(a), \tag{4.4.2}$$

where

$$S_0(a) = 2\sum_{k=0}^{\infty} \frac{c_{2k,0}}{a^{k+\frac{1}{2}}}\, P(k+\tfrac{1}{2}, \omega_0 a)$$

and

$$S_n(a) = \sum_{k=0}^{\infty} \frac{c_{k,n}}{a^{k+1}}\, \Delta P(k+1, \tfrac{1}{2}\omega_n a).$$

The zeroth interval $\omega_0 = \Lambda_0 \vartheta$, where ϑ is the geometric decay factor satisfying $0 < \vartheta < 1$, and $\omega_n = 2\Omega_n \vartheta$ with Ω_n specified in (3.2.25).

In Fig. 4.12 we display the behaviour of the terms in the first four levels $n \leq 3$ of (4.4.2) when $a = 6$, $z = 1.2$ and $\vartheta = \frac{1}{3}$. The first figure shows for comparison the terms in $S_0(a)$ and the asymptotic expansion

$$e^{a\psi(t_{s1})} I(a; z) \sim 2\sum_{k=0}^{\infty} \frac{c_{2k,0}}{a^{k+\frac{1}{2}}},$$

which is obtained by replacing the normalised incomplete gamma functions in the series $S_0(a)$ by unity. The Hadamard expansion (4.4.2) will also hold for complex z provided that the integration path can be deformed to pass over only the saddle at $t = t_{s1}$. Examples of the absolute error in the computation of $\exp\{a\psi(t_{s1})\}I(a; z)$ for different complex z when $a = 6$ and $\vartheta = \frac{1}{2}$ are given in Table 4.5 using the levels with $n \leq 2$ and the truncation indices $M_0 = M_1 = 50$, $M_2 = 25$. The expansion (4.4.2) ceases to be useful, however, when $|\arg z| \to \frac{1}{2}\pi$ since then we are either confronted with a coalescence (or near coalescence) problem when $|z| \leq 1$ or a

Table 4.5 *The absolute errors in* $\exp\{a\psi(t_{s1})\}I(a;z)$
for different z when $a = 6$ *and* $\vartheta = \frac{1}{2}$; *levels with*
$n \le 2$ *have been employed with the truncation indices*
given in the text

θ/π	$z = 0.5e^{i\theta}$	$z = 2e^{i\theta}$
0	1.370×10^{-23}	6.743×10^{-28}
0.1	6.180×10^{-23}	6.850×10^{-28}
0.2	1.623×10^{-20}	7.184×10^{-28}
0.3	4.331×10^{-18}	2.896×10^{-27}
0.4	2.757×10^{-16}	6.768×10^{-24}

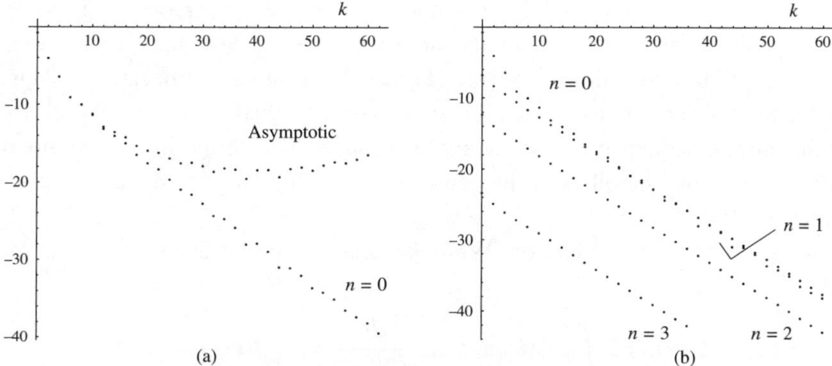

Figure 4.12 The behaviour of the terms (on a $\log_{10}$ scale) in the expansion of $\exp\{a\psi(t_{s1})\}I(a;z)$ in (4.4.2) as a function of ordinal number k when $a = 6$, $z = 1.2$ and $\vartheta = \frac{1}{3}$: (a) the zeroth level and the asymptotic series and (b) the levels $n \le 3$.

Stokes phenomenon when $|z| > 1$. Evaluation of $I(a;z)$ in such situations would necessitate the use of either the expansion scheme described in §3.4.1 for coalescence or the approach adopted in Example 3.6 and §4.3.2 for the Stokes phenomenon. We do not discuss this any further here.

4.5 The expansion for log Γ(z)

Our final application concerns the slowly varying part[8] $\Omega(z)$ of the logarithm of $\Gamma(z)$ defined by

---

[8] No confusion should arise between $\Omega(z)$ and the expansion points Ω_n.

$$\log \Gamma(z) = (z - \tfrac{1}{2}) \log z - z + \tfrac{1}{2} \log 2\pi + \Omega(z), \qquad (4.5.1)$$

where $z = xe^{i\theta}$ with x and θ real. It is well known that $\Omega(z)$ has the asymptotic expansion given by the Stirling series (Whittaker and Watson, 1952, p. 251)

$$\Omega(z) \sim \sum_{r=1}^{\infty} \frac{B_{2r}}{2r(2r-1)z^{2r-1}} \qquad (4.5.2)$$

as $|z| \to \infty$ in $|\arg z| \le \pi - \delta, \delta > 0$, where B_{2r} denote the Bernoulli numbers. From the reflection formula for $\Gamma(z)$ in the form $-z\Gamma(z)\Gamma(-z) = \pi / \sin \pi z$, we have

$$\Omega(z) = -\Omega(ze^{\mp\pi i}) - \log(1 - e^{\pm 2\pi i z}), \qquad (4.5.3)$$

so that it is necessary to consider the calculation of $\Omega(z)$ only in the sector $|\arg z| \le \tfrac{1}{2}\pi$. We observe that since successive even-order Bernoulli numbers have opposite signs, all terms in the above expansion have the same phase on $\arg z = \pm \tfrac{1}{2}\pi$. It then follows (see §1.7.1) that the positive and negative imaginary axes are Stokes lines, across which it is found (Berry, 1991b; Paris and Wood, 1992) that an infinite sequence of exponentially subdominant terms in the asymptotics of $\Gamma(z)$ switch on smoothly in the sense of increasing $|\arg z|$; see also Paris and Kaminski (2001, §6.4).

If we use Binet's representation (Whittaker and Watson 1952, §12.32; Olver 1997, p. 295)

$$\Omega(z) = 2 \int_0^{\infty} \arctan(t/z) \frac{dt}{e^{2\pi t} - 1} \qquad (\mathrm{Re}(z) > 0),$$

expand the inverse tangent about $t = 0$ (valid in $0 \le t < x$) and expand the factor $(e^{2\pi t} - 1)^{-1}$ as a geometric series in powers of $e^{-2\pi t}$, we obtain for the above integral taken over the interval $0 \le t \le x$ (the zeroth contribution)

$$\frac{1}{\pi} \sum_{m=1}^{\infty} \frac{1}{m} \sum_{k=0}^{\infty} \frac{(-)^k (2k)!}{(2\pi m z)^{2k+1}} P(2k+2, 2\pi mx). \qquad (4.5.4)$$

For $z > 0$, this is essentially the result obtained by Hadamard (1912) in his attempt to render the Stirling series (4.5.2) convergent, although he did not identify the integrals as incomplete gamma functions. However, although innovative, his analysis was incomplete for two reasons. First, because he discarded the contribution from the integral over the interval $x \le t < \infty$ as being exponentially small. And second, because each individual Hadamard series in (4.5.4) – given by the inner sum over k – as it stands has late terms behaving like $(-)^k / k^2$ (see (A.6)) and so is of little computational use without further modification to overcome the slow algebraic convergence. The Hadamard expansion of $\Omega(z)$ in this form has been discussed for complex z in the sector $|\arg z| \le \tfrac{1}{2}\pi$ in Paris (2000).

Here we shall derive a Hadamard expansion for $\Omega(z)$ using the alternative Binet representation in the form of a Laplace integral (Whittaker and Watson, 1952, §12.31), namely

$$\Omega(z) = \int_0^\infty f(t)e^{-zt}\,dt \qquad (\mathrm{Re}(z) > 0), \qquad (4.5.5)$$

where the amplitude function $f(t)$ is given by

$$f(t) = \left(\frac{1}{2} - \frac{1}{t} + \frac{1}{e^t - 1}\right)t^{-1}.$$

The function $f(t)$ has singularities on the imaginary axis at $t = \pm 2\pi k i$, $k = 0, 1, 2, \ldots$. Its expansion about the point $t = 0$ is consequently controlled by the closest singularities at $t = \pm 2\pi i$, and we have (Temme, 1996, p. 55)

$$f(t) = \sum_{k=0}^\infty \frac{c_{k,0}}{(2k)!}\, t^{2k} \qquad (|t| < 2\pi),$$

where the coefficients $c_{k,0}$ can be expressed in terms of the Bernoulli numbers by $c_{k,0} = B_{2k+2}/\{(2k+1)(2k+2)\}$.

The procedure follows closely that described for the confluent hypergeometric function in §2.4. We analytically continue the integral in (4.5.5) by rotating the integration path through the acute angle ψ to obtain

$$\Omega(z) = e^{-i\psi} \int_0^\infty f(\tau e^{-i\psi})e^{-z'\tau}\,d\tau, \qquad z' = xe^{i(\theta - \psi)}$$

valid in $|\theta - \psi| \le \frac{1}{2}\pi - \delta$, $\delta > 0$. By suitable choice of ψ, this would enable us to extend the sector of validity of (4.5.5) to $|\arg z| \le \pi - \delta$; however, on account of (4.5.3), it is sufficient to consider the sector $|\arg z| \le \frac{1}{2}\pi$. We shall adopt the subdivision Scheme B described in §3.2.3, which consists of forward expansion for the zeroth interval (of length $\omega_0 = 2\pi\vartheta$, $0 < \vartheta < 1$) and forward-reverse expansion about the midpoints $\tau = \Omega_n$ of the intervals corresponding to $n \ge 1$.

The expansion of $f(\tau e^{-i\psi})$ about the points $\tau = \Omega_n$ ($n \ge 1$) is obtained more easily by recurrence than use of the series expansion facility in *Mathematica*. First, note that if $g(t) := (e^t - 1)^{-1}$ then it is easily seen that

$$\frac{dg}{dt} + g + g^2 = 0.$$

Substitution of the expansion $g(t_n) = \sum_{k=0}^\infty b_{k,n}(ue^{-i\psi})^k$ into this equation leads to the recurrence

$$b_{k+1,n} = -\frac{b_{k,n}}{k+1} - \frac{1}{k+1}\sum_{r=0}^k b_{r,n}b_{k-r,n} \qquad (k \ge 0)$$

subject to the initial value $b_{0,n} = (\exp(\Omega_n e^{-i\psi}) - 1)^{-1}$. If we now set $U_n \equiv (\Omega_n + u)e^{-i\psi}$, some straightforward algebra then produces

$$f(U_n) = \frac{1}{U_n}\left\{\frac{1}{2} - \frac{1}{U_n} + \sum_{k=0}^{\infty} b_{k,n}(ue^{-i\psi})^k\right\}$$

$$= \frac{1}{U_n}\sum_{k=0}^{\infty} a_{k,n}(ue^{-i\psi})^k = \sum_{k=0}^{\infty} \frac{c_{k,n}(\psi)}{k!}(ue^{-i\psi})^k, \qquad (4.5.6)$$

where

$$a_{0,n} = b_{0,n} + \frac{1}{2} - \frac{e^{i\psi}}{\Omega_n}, \qquad a_{k,n} = b_{k,n} + \frac{(-)^{k+1}e^{i\psi}}{\Omega_n^{k+1}} \quad (k \geq 1)$$

and

$$c_{k,n}(\psi) = k!\sum_{r=0}^{k} \frac{(-)^r a_{k-r,n}}{(e^{-i\psi}\Omega_n)^{r+1}}.$$

From the above-mentioned singularity structure of $f(t)$, the expansion (4.5.6) is valid in

$$|u| < \min_{k\geq 0}\{|\Omega_n e^{-i\psi} \pm 2\pi k i|\}. \qquad (4.5.7)$$

Then, applying the arguments used in §§2.4.2 and 2.4.5, we obtain the Hadamard expansion for $\Omega(z)$ in the form

$$\Omega(z) = \sum_{k=0}^{\infty} \frac{c_{k,0}}{z^{2k+1}} P(2k+1, \omega_0 z') + \sum_{n=1}^{\infty} e^{-\Omega_n z'} S_n(z; \psi), \qquad (4.5.8)$$

where

$$S_n(z; \psi) = \sum_{k=0}^{\infty} \frac{c_{k,n}(\psi)}{z^{k+1}} \Delta P(k+1, \tfrac{1}{2}\omega_n z')$$

and ΔP is defined in (2.2.26); when $\psi = \theta$, we note that $z' = x$. To illustrate, we present numerical results in Table 4.6 showing the absolute error in the computation of $\Omega(z)$ by (4.5.8) for varying phase $\theta \geq 0$ (with $x = 8$ fixed). The exact value of $\Omega(z)$ was determined from (4.5.1) using *Mathematica*. For the first set of values, we have used the levels with $n \leq 1$ and the truncation indices $M_0 = M_1 = 25$; we chose $\vartheta = \frac{1}{2}$, so that $\omega_0 = \pi$, $\Omega_1 = 2\pi$. For the second set of values, we have used the levels $n \leq 2$ with the truncation indices $M_0 = M_1 = 50$, $M_2 = 40$. The geometric decay factor was chosen to be $\vartheta = \frac{1}{2}$ for the zeroth interval (so that again $\omega_0 = \pi$), but $\vartheta = \frac{1}{3}$ for the remaining intervals so that $\Omega_1 = \frac{3}{2}\pi$ and $\Omega_2 = 3\pi$. The rotation angle was set at $\psi = -\theta$ for small θ and changed to $\psi = -\frac{1}{6}\pi$ (indicated by the asterisk) and $\psi = -\frac{1}{4}\pi$ (indicated by the dagger) for larger θ-values.

Although we have fixed the values of Ω_1 and Ω_2 to correspond to the chosen values of the geometric decay factor ϑ, it is important to realise that the effective value

Table 4.6 *The absolute error in $\Omega(z)$ for different $\theta = \arg z$ when $|z| = 8$ using different numbers of levels n; the values of the geometric decay factor ϑ, the rotation angle ψ and the truncation indices are given in the text*

θ/π	Levels $n \leq 1$	θ/π	Levels $n \leq 2$
0	2.758×10^{-26}	0	3.836×10^{-45}
0.1	4.283×10^{-24}	0.1	3.321×10^{-45}
0.2*	3.706×10^{-22}	0.2	7.048×10^{-42}
0.3*	2.902×10^{-21}	0.3*	6.549×10^{-41}
0.4*	2.208×10^{-19}	0.4†	2.791×10^{-36}
0.5*	1.076×10^{-16}	0.5†	2.390×10^{-32}

of ϑ will decrease with path rotation as we approach the Stokes lines $\arg z = \pm\frac{1}{2}\pi$. To see this, consider the case of the first set of values in Table 4.6 with $\omega_0 = \pi$ and $\Omega_1 = 2\pi$, which corresponds to the nominal value $\vartheta = \frac{1}{2}$. From (4.5.7) with $k = 0$ and $k = 1$, the expansion of $f(\tau e^{-i\psi})$ about $\tau = \Omega_1$ has radius of convergence

$$2\pi \min\{1, \sqrt{2(1 - |\sin\psi|)}\}.$$

Then, for $|\psi| \leq \frac{1}{6}\pi$ the radius of convergence is 2π and so $\vartheta = (\Omega_1 - \omega_0)/(2\pi) = \frac{1}{2}$, whereas for $|\psi| > \frac{1}{6}\pi$ the effective value of ϑ is

$$\vartheta' = \frac{\Omega_1 - \omega_0}{2\pi\sqrt{2(1 - |\sin\psi|)}} = \frac{2^{-3/2}}{\sqrt{1 - |\sin\psi|}} > \frac{1}{2}.$$

For example, if $\psi = \pm\frac{1}{4}\pi$ then $\vartheta' \doteq 0.653$. This has the effect of reducing the accuracy when computing with a fixed number of levels n. Of course, it is also possible to compute with a prescribed value of ϑ when using path rotation by suitably adjusting the choice of the expansion points Ω_n.

Appendix A
Properties of $P(a, z)$

In this appendix we collect together the main properties of the normalised incomplete gamma function $P(a, z)$ that are required in various sections.

A.1 Definitions

The normalised incomplete gamma function $P(a, z)$ is defined by

$$P(a, z) = \frac{\gamma(a, z)}{\Gamma(a)} = \frac{1}{\Gamma(a)} \int_0^z e^{-t} t^{a-1} dt \qquad (\mathrm{Re}(a) > 0) \qquad (A.1)$$

provided the integration path does not cross the negative real axis. In terms of the confluent hypergeometric function defined by

$$_1F_1(a; b; z) = \sum_{n=0}^{\infty} \frac{(a)_n\, z^n}{(b)_n\, n!} \qquad (|z| < \infty)$$

when $b \neq 0, -1, -2, \ldots$ and $(a)_n = \Gamma(a + n)/\Gamma(a)$, we have

$$P(a, z) = \frac{z^a e^{-z}}{\Gamma(1 + a)}\, _1F_1(1; 1 + a; z)$$

$$= z^a e^{-z} \sum_{n=0}^{\infty} \frac{z^n}{\Gamma(n + a + 1)} \qquad (A.2)$$

valid for all (complex) values of z and a.

The function $P(a, z)$ has a branch cut along the negative real z-axis for general values of a, due to the presence of the factor z^a in (A.2); for integer values of a the function is single valued. Consequently, we define

$$P^*(a, -z) = e^{\pm \pi i a} P(a, -z), \qquad (A.3)$$

where the upper or lower sign is chosen according as $\arg z > 0$ or $\arg z \leq 0$, respectively. When $x > 0$, we therefore have

$$P^*(a, -x) = e^{-\pi i a} P(a, -x). \qquad (A.4)$$

The complementary incomplete gamma function $\Gamma(a, z)$ is defined by

$$\Gamma(a, z) = \Gamma(a) - \gamma(a, z)$$

and has the integral representation

$$\Gamma(a, z) = \int_z^\infty e^{-t} t^{a-1} dt. \tag{A.5}$$

A.2 Asymptotic behaviour

The behaviour of $P(a, z)$ for large z can be determined from

$$P(a, z) = 1 - \frac{\Gamma(a, z)}{\Gamma(a)}$$

combined with the asymptotic expansion of $\Gamma(a, z)$ given by Abramowitz and Stegun (1965, p. 263)

$$\Gamma(a, z) \sim z^{a-1} e^{-z} \left\{ 1 + \frac{a-1}{z} + \frac{(a-1)(a-2)}{z^2} + \cdots \right\}$$

as $z \to \infty$ in the sector $|\arg z| \leq \frac{3}{2}\pi - \delta$, where δ denotes an arbitrarily small positive quantity. The behaviour for large a follows directly from (A.2) since the $_1F_1$ function tends to the value unity as $|a| \to \infty$ in $|\arg a| \leq \pi - \delta$. Hence we have the leading behaviour

$$P(a, z) \sim \begin{cases} 1 + O(e^{-z}) & (|z| \to \infty \text{ in } |\arg z| \leq \frac{1}{2}\pi - \delta) \\ \dfrac{z^a e^{-z}}{\Gamma(1+a)} & (|a| \to \infty \text{ in } |\arg a| \leq \pi - \delta). \end{cases} \tag{A.6}$$

A.3 Descriptive properties

In Fig. A.1 we present plots of $P(a, x)$ as a function of a for both positive and negative values of x. For positive a and x, $P(a, x)$ is a monotonically decreasing function of a and is approximately equal to 1 for $a \lesssim x$ and decays to zero for $a \gtrsim x$. This is the simple cut-off structure shown in Fig. A.1(a) that is fundamental to the Hadamard expansion procedure. When $x < 0$, $P^*(a, x) = e^{-\pi i a} P(a, x)$ has exponential growth proportional to e^x before decaying for sufficiently large a according to (A.6); see Fig. A.1(b).

A proof of the monotonicity property satisfied by $P(a, x)$ for $a > 0$ and $x > 0$ can be established by a slight variation of that given in Tricomi (1950). We have

$$P(a, x) = \frac{\gamma(a, x)}{\Gamma(a)} = \frac{\gamma(a, x)}{\gamma(a, x) + \Gamma(a, x)},$$

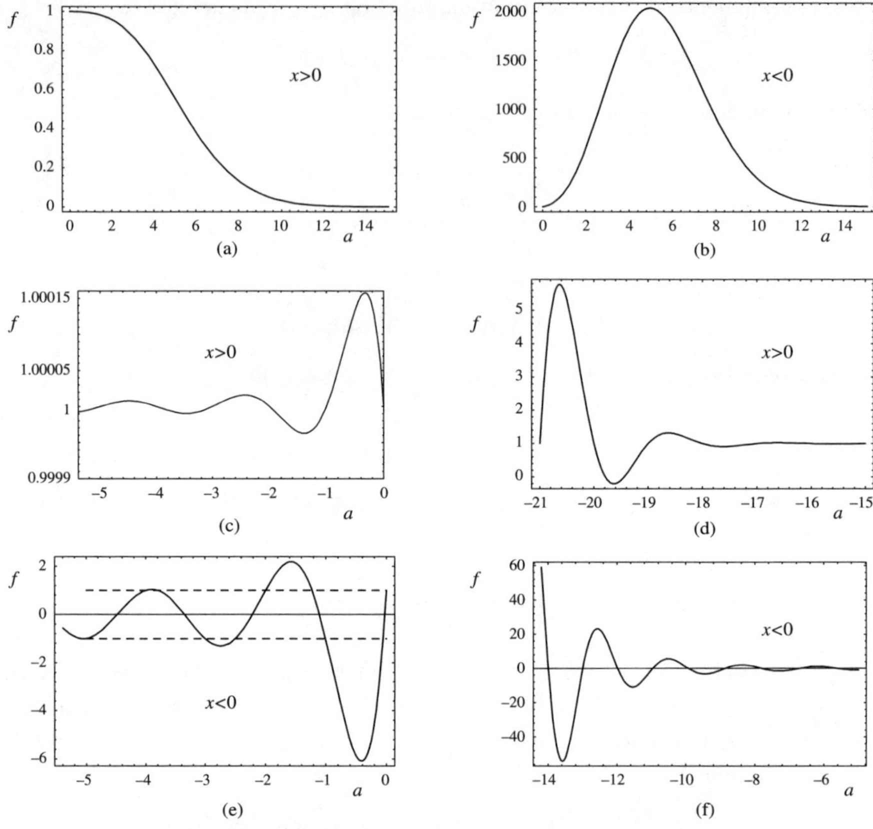

Figure A.1 Plots of $P(a, x)$ as a function of a for fixed positive and negative values of x: (a) $f = P(a, 5)$; (b) $f = P^*(a, -5)$ when $a \geq 0$; (c), (d) $f = P(a, 5)$ when $a \leq 0$; (e), (f) $f = P^*(a, -5)$ when $a \leq 0$. In (e) the dashed lines have ordinate values ± 1.

so that

$$\frac{\partial P(a, x)}{\partial a} = -\frac{1}{\Gamma^2(a)} \left\{ \gamma(a, x) \frac{\partial \Gamma(a, x)}{\partial a} - \Gamma(a, x) \frac{\partial \gamma(a, x)}{\partial a} \right\}.$$

From (A.1) and (A.5), the quantity in curly braces can be written as

$$\int_x^\infty e^{-t} t^{a-1} \log t \, dt \times \int_0^x e^{-u} u^{a-1} du - \int_x^\infty e^{-t} t^{a-1} dt \times \int_0^x e^{-u} u^{a-1} \log u \, du$$

$$= \int_x^\infty dt \int_0^x e^{-(u+t)} (ut)^{a-1} \log (t/u) \, du.$$

This last integral is positive, since $t/u \geq 1$ in the integration domain. Hence $\partial P(a, x)/\partial a < 0$ for $a > 0$ and $x > 0$. $\qquad\square$

The remaining figures in Fig. A.1 show the behaviour of $P(a, x)$ for negative values of a. We first note that, since

$$\mathbf{F}(z) \equiv \frac{{}_1F_1(1; 1 + a; z)}{\Gamma(1 + a)} = \frac{1}{\Gamma(1 + a)} + \frac{z}{\Gamma(2 + a)} + \frac{z^2}{\Gamma(3 + a)} + \cdots,$$

it is easily seen when $a = -n$, where n is a non-negative integer, that $\mathbf{F}(z) = z^n e^z$; hence from (A.2)

$$P(-n, z) = 1 \qquad (n = 0, 1, 2, \ldots) \tag{A.7}$$

for arbitrary z. The behaviour of $P(a, x)$ for $x > 0$ and negative values of a initially exhibits oscillations about the value 1 of small amplitude; these oscillations eventually grow as $-a$ increases, as shown in Fig. A.1(c) and (d). The graph of $P^*(a, x)$ for $x < 0$ and negative values of a exhibits a similar behaviour; see Fig. A.1 (e) and (f).

A.4 Recurrence relation and computation

In the treatment of Laplace-type integrals with saddle points the computation of $P(k+\mu, z)$ with $\mu = \frac{1}{2}$ or $\mu = 1$ is of frequent occurrence. The values of $P(k+\frac{1}{2}, x)$ for positive integer k can be related by the recurrence

$$P(a + 1, z) = P(a, z) - \frac{z^a e^{-z}}{\Gamma(1 + a)} \tag{A.8}$$

to the error function, since $P(\frac{1}{2}, x) = \mathrm{erf}\sqrt{x}$. The values of $P(k + 1, x)$ are particularly simple since they are expressible in terms of the truncated exponential sum (Abramowitz and Stegun, 1965, p. 262)

$$P(k + 1, z) = 1 - e^{-z} e_k(z), \qquad e_k(z) = \sum_{n=0}^{k} \frac{z^n}{n!}. \tag{A.9}$$

We make two remarks at this point on the different nature of the normalised incomplete gamma functions $P(k+1, \pm\frac{1}{2}\omega_n x)$ that appear in the Hadamard series resulting from forward-reverse expansion, as typified by the expansion for the confluent hypergeometric function $U(a, a + b, z)$ discussed in §2.4.2. From (2.4.9), the Hadamard series with index $n \geq 1$ contain the difference

$$\Delta P(k + 1, \tfrac{1}{2}\omega_n x) = P(k + 1, \tfrac{1}{2}\omega_n x) - P(k + 1, -\tfrac{1}{2}\omega_n x),$$

where k is a non-negative integer and $\omega_n > 0$ are the expansion intervals. The first point to be made is that the function $P(k + 1, -\frac{1}{2}\omega_n x)$ has no branch-point structure for non-negative integer k and consequently assumes real values for $x > 0$. The second point is that the ratio of the function with negative argument to that with positive argument is roughly of order $\exp(\omega_n x)$ for large k, since from (A.6) we have

$$\frac{P(k+1, -\frac{1}{2}\omega_n x)}{P(k+1, \frac{1}{2}\omega_n x)} \sim (-)^{k+1} e^{\omega_n x} \qquad (k \to \infty). \tag{A.10}$$

This difference in the magnitude of these two incomplete gamma functions can be displayed alternatively by noting that, from (A.1),

$$\frac{P(k+1, X)}{X^{k+1}} = \frac{1}{k!} \int_0^1 t^k e^{-Xt} dt,$$

$$\frac{e^{-X} P(k+1, -X)}{(-X)^{k+1}} = \frac{1}{k!} \int_0^1 (1-t)^k e^{-Xt} dt.$$

The maximum value of the first integrand occurs at $t = 1$ (when $k \geq X$) whereas that of the second integrand occurs at $t = 0$ (for all $k \geq 0$). It is then evident that the ratio $P(k+1, -X)/P(k+1, X)$ is of order $\exp(2X)$ when $X > 0$.

A.5 An estimate for $P(a, -x)$ for large $x > 0$

Finally, we give an estimate for $P(a, -x)$ for large positive x and finite $a \geq 1$. From (A.2), we have on the upper side of the branch cut, with $-x = xe^{\pi i}$,

$$P(a, -x) = \frac{(e^{\pi i} x)^a e^x}{\Gamma(1+a)} F, \tag{A.11}$$

where, provided $a > 0$ (Abramowitz and Stegun, 1965, p. 505),

$$F \equiv {}_1F_1(1; 1+a; -x) = a \int_0^1 e^{-xt}(1-t)^{a-1} dt = 1 - x \int_0^1 e^{-xt}(1-t)^a dt$$

after an integration by parts. Straightforward estimates resulting from the above integrals show that for $x > 0$

$$e^{-x} \leq F < 1 \quad (a \geq 0), \qquad F < a/x \quad (a \geq 1). \tag{A.12}$$

Moreover, it is also easily seen that F is a monotonically increasing function of a on $[0, \infty)$. This follows from the derivative

$$\frac{\partial F}{\partial a} = -x \int_0^1 e^{-xt}(1-t)^a \log(1-t)\, dt$$

whence, since $\log(1 - t) \leq 0$ on $[0, 1]$, we have $\partial F/\partial a > 0$.

From (A.11) and (A.12), we then obtain the estimate

$$|P(a, -x)| = O(x^{a-1}e^x) \tag{A.13}$$

for large positive x and finite values of $a \geq 1$. As $a \to \infty$, with x held fixed, the behaviour of $P(a, -x)$ is, of course, given by the second result in (A.6).

Appendix B
Convergence of Hadamard series

The type of Hadamard series encountered in applications has the general form

$$S(x) = \sum_{n=0}^{\infty} \frac{A_n}{x^{\alpha n + \mu}} \, P(\alpha n + \mu, \omega x) \qquad (\omega > 0), \tag{B.1}$$

where the coefficients A_n in certain cases can be expressed as a quotient of products of gamma functions

$$A_n = \frac{\prod_{r=1}^{p} \Gamma(a_r n + b_r)}{\prod_{r=1}^{q} \Gamma(c_r n + d_r)} \tag{B.2}$$

with p, q denoting non-negative integers. The parameters α, a_r and c_r are assumed positive, with μ, b_r and d_r being, in general, arbitrary complex constants.

From (A.6), the late terms ($n \gg 1$) in $S(x)$ have the behaviour

$$\frac{\omega^{\alpha n + \mu} e^{-\omega x} A_n}{\Gamma(\alpha n + \mu + 1)} \qquad (n \to +\infty).$$

With the definition of the parameters

$$\kappa = \sum_{r=0}^{q} c_r - \sum_{r=1}^{p} a_r, \quad h = \prod_{r=1}^{p} a_r^{a_r} \prod_{r=0}^{q} c_r^{-c_r} \quad (c_0 \equiv \alpha),$$

$$\lambda = \sum_{r=1}^{p} b_r - \sum_{r=1}^{q} d_r - \mu + \tfrac{1}{2}(q - p),$$

application of Stirling's formula, as described in detail in Paris and Kaminski (2001, §2.2) (see Lemma 2.1 and Eq. (2.2.4) therein), shows that

$$\frac{A_n}{\Gamma(\alpha n + \mu + 1)} \sim \begin{cases} \dfrac{(h \kappa^{\kappa})^n}{\Gamma(\kappa n + 1 - \lambda)} & (\kappa > 0) \\[2mm] h^n n^{\lambda - 1/2} & (\kappa = 0) \end{cases} \tag{B.3}$$

as $n \to +\infty$.

There are two cases to consider: namely when $\kappa > 0$ and when $\kappa = 0$. When $\kappa > 0$, the late terms in $S(x)$ consequently possess the behaviour

$$\frac{e^{-\omega x}(\omega^\alpha h \kappa^\kappa)^n}{\Gamma(\kappa n + 1 - \lambda)} \qquad (n \to +\infty), \tag{B.4}$$

so that convergence of the series in this case is factorial, with no further restriction on the various parameters. On the other hand, when $\kappa = 0$ the late terms possess the behaviour

$$e^{-\omega x}(\omega^\alpha h)^n n^{\lambda - 1/2}. \tag{B.5}$$

If $\omega^\alpha h < 1$, the series $S(x)$ converges geometrically, whereas if $\omega^\alpha h = 1$ we require the additional condition

$$\mathrm{Re}(\lambda) < -\tfrac{1}{2} \tag{B.6}$$

for absolute convergence.

A case of common occurrence in the treatment of Laplace integrals corresponds to $\alpha = a_r = c_r = 1$, so that $\kappa = q - p + 1$ and $h = 1$. Convergence of $S(x)$ is then factorial when $q \geq p$ ($\kappa \geq 1$) and, when $q = p - 1$ ($\kappa = 0$), is either geometric if $\omega < 1$ or requires the condition (B.6) if $\omega = 1$. In the hypergeometric case with $p = 2, q = 1$ ($\kappa = 0$) we have

$$S(x) = \sum_{n=0}^{\infty} \frac{\Gamma(n + b_1)\Gamma(n + b_2)}{\Gamma(n + d_1)\, x^{n+\mu}}\, P(n + \mu, \omega x).$$

This series converges geometrically like ω^n when $\omega < 1$ but requires the convergence condition

$$\mathrm{Re}(b_1 + b_2 - d_1 - \mu) < 0 \tag{B.7}$$

when $\omega = 1$.

An example of a series of the above type is the Hadamard series for the modified Bessel function $I_\nu(x)$ in (2.2.5), namely

$$I_\nu(x) = \frac{e^x}{\sqrt{2\pi x}} \sum_{k=0}^{\infty} \frac{a_k(\nu)}{(2x)^k}\, P(k + \nu + \tfrac{1}{2}, 2x),$$

where the coefficients $a_k(\nu)$ are defined in (2.1.2). Here we have $b_1 = \tfrac{1}{2} + \nu$, $b_2 = \tfrac{1}{2} - \nu$, $d_1 = 1$, $\mu = \nu + \tfrac{1}{2}$ and (on putting $2x \mapsto x$) $\omega = 1$. The condition (B.7) therefore yields $\mathrm{Re}(\nu) > -\tfrac{1}{2}$, as obtained in (2.2.8).

Appendix C
Connection with the exp-arc integrals

We briefly describe the theory developed by Borwein, Borwein and Crandall (2008) and Borwein, Borwein and Chan (2008) for the evaluation of the Bessel functions employing what they call 'exp-arc integrals'. We show, in the case of the $I_\nu(z)$ Bessel function, how their theory is related to the 2-stage Hadamard expansion discussed in §2.2.4.

These authors adopted as their integral representation the definition (Watson, 1952, p. 181)

$$I_\nu(z) = \frac{1}{\pi} \int_0^\pi e^{z \cos t} \cos \nu t \, dt - \frac{\sin \pi \nu}{\pi} \int_0^\infty e^{-z \cosh t - \nu t} dt \qquad (\text{C.1})$$

valid when $\mathrm{Re}(z) > 0$. For the case of integer ν (equal to m), they divided the integration path $[0, \pi]$ into $[0, \frac{1}{2}\pi]$ and $[\frac{1}{2}\pi, \pi]$ and made the substitution $t \to \pi - t$ in the second interval. Then

$$I_m(z) = \frac{1}{\pi} \left\{ \int_0^{\pi/2} e^{z \cos t} \cos mt \, dt + (-)^m \int_0^{\pi/2} e^{-z \cos t} \cos mt \, dt \right\}$$

$$= \frac{1}{2\pi} \left\{ \mathcal{I}(z, m) + (-)^m \mathcal{I}(-z, m) \right\}, \qquad (\text{C.2})$$

where

$$\mathcal{I}(z, m) = 2 \int_0^{\pi/2} e^{z \cos t} \cos mt \, dt = \int_{-\pi/2}^{\pi/2} e^{z \cos t - imt} dt$$

$$= e^z \int_{-\pi/2}^{\pi/2} e^{-2z \sin^2 \frac{1}{2} t} e^{-imt} dt.$$

With the change of variable $u = \sin \frac{1}{2} t$, they then used the series expansion

$$\frac{e^{\tau \arcsin u}}{\sqrt{1 - u^2}} = \sum_{k=0}^\infty r_{k+1}(\tau) \frac{u^k}{k!}$$

valid for $-1 < u < 1$, where

$$r_{2k+1}(\tau) = \prod_{j=1}^{k}\{\tau^2 + (2j-1)^2\}, \quad r_{2k}(\tau) = \tau^{-1}\prod_{j=1}^{k}\{\tau^2 + (2j-2)^2\}.$$

It is through expansion of the above exponential factor involving the arcsin function that Borwein, Borwein and Chan (2008) coined the term 'exp-arc method'. Then

$$\begin{aligned}
\mathcal{I}(z, m) &= 2e^z \int_{-2^{-1/2}}^{2^{-1/2}} \frac{e^{-2im\,\arcsin u}}{\sqrt{1-u^2}}\, e^{-2zu^2}\, du \\
&= 4e^z \sum_{k=0}^{\infty} \frac{r_{2k+1}(-2im)}{(2k)!} \int_{0}^{2^{-1/2}} e^{-2zu^2} u^{2k}\, du \\
&= 2e^z \sum_{k=0}^{\infty} \frac{r_{2k+1}(-2im)}{(2k)!}\, B_k(z),
\end{aligned}$$

where

$$r_{2k+1}(-2im) = \prod_{j=1}^{k}\{(2j-1)^2 - 4m^2\} = 2^{2k}k!\, a_k(m),$$

$$B_k(z) = 2^{-k-\frac{1}{2}} \int_{0}^{1} e^{-zu} u^{k-\frac{1}{2}}\, du = \frac{\Gamma(k+\frac{1}{2})}{(2z)^{k+\frac{1}{2}}}\, P(k+\tfrac{1}{2}, z)$$

and $a_k(m)$ are the coefficients defined in (2.1.2).

From (C.2) and the duplication formula for the gamma function $(2k)! = 2^{2k}\pi^{-1/2}k!\,\Gamma(k+\frac{1}{2})$, we then obtain the 2-stage expansion

$$I_m(z) = \frac{e^z}{\sqrt{2\pi z}} \left\{ \sum_{k=0}^{\infty} \frac{a_k(m)}{(2z)^k}\, P(k+\tfrac{1}{2}, z) + \right.$$
$$\left. (-)^m e^{-2z} \sum_{k=0}^{\infty} \frac{a_k(m)}{(2z)^k}\, P^*(k+\tfrac{1}{2}, -z) \right\} \tag{C.3}$$

valid when $\mathrm{Re}(z) > 0$ and $m = 0, 1, 2, \ldots$, where $P^*(a, -z)$ is defined in (A.3).

The above expansion has clearly resulted from forward-reverse expansion of the amplitude function in the first integral in (C.1) about the points $t = 0$ and $t = \pi$. From (A.6), the decay of the terms in both series is easily seen to be controlled by the geometric factor 2^{-k}. When v is a positive integer m, the 2-stage Hadamard expansion obtained in (2.2.24) agrees with (C.3), except for the presence of the term v in the first argument of the incomplete gamma functions. It is shown in Paris (2009) that when $v = m$ the expansions (2.2.24) and (C.3) are, in fact, equivalent.

When v is non-integer there is an additional contribution from the first integral [1] in (C.1) together with that from the second integral over $[0, \infty)$. Borwein, Borwein and Chan (2008) dealt with this integral by using an integration by parts followed by the change of variable $s = \cosh t$ to find

$$\int_0^\infty e^{-z \cosh t - vt} dt = v^{-1} e^{-z} - \frac{z}{v} \int_1^\infty e^{-zs} e^{-v \operatorname{arccosh} s} ds$$

when $\operatorname{Re}(z) > 0$. Then by appropriate expansion of the factor $e^{-v \operatorname{arccosh} s}$ as another exp-arc series, they obtained the representation of $I_v(z)$ in $\operatorname{Re}(z) > 0$. This mode of expansion is shown in Paris (2009) to involve an infinite sequence of Hadamard series.

---

[1] An alternative way of expanding the first integral in (C.1) for arbitrary v is given in Example 2.5.

References

Abramowitz, M. and Stegun, I. (eds.) 1965. *Handbook of Mathematical Functions*, New York: Dover.

Airey, J. R. 1937. The 'converging factor' in asymptotic series and the calculation of Bessel, Laguerre and other functions, *Philos. Mag.* **24**, 521–552.

Bakhoom, N. G. 1933. Asymptotic expansions of the function $F_k(x) = \int_0^\infty e^{-u^k + xu} du$, *Proc. London Math. Soc.* **35**(2), 83–100.

Barnes, E. W. 1906. The asymptotic expansion of integral functions defined by Taylor's series, *Philos. Trans. R. Soc. London* **A206**, 249–297.

Bender, C. M. and Orszag, S. A. 1978. *Advanced Mathematical Methods for Scientists and Engineers*, New York: McGraw-Hill.

Berry, M. V. 1989. Uniform asymptotic smoothing of Stokes's discontinuities, *Proc. R. Soc. London* **A412**, 7–21.

Berry, M. V. 1991a. Asymptotics, superasymptotics, hyperasymptotics..., in *Asymptotics Beyond All Orders* (ed. H. Segur, H. Tanveer and H. Levine), pp. 1–14, New York: Plenum Press.

Berry, M. V. 1991b. Infinitely many Stokes smoothings in the gamma function, *Proc. R. Soc. London* **A434**, 465–472.

Berry, M. V. 1991c. Stokes's phenomenon for superfactorial asymptotic series, *Proc. R. Soc. London* **A435**, 437–444.

Berry, M. V. and Howls, C. J. 1990. Hyperasymptotics, *Proc. R. Soc. London* **A430**, 653–667.

Berry, M. V. and Howls, C. J. 1991. Hyperasymptotics for integrals with saddles, *Proc. R. Soc. London* **A434**, 657–675.

Berry, M. V. and Howls, C. J. 1993. Infinity interpreted, *Phys. World* **6**, 35–39.

Bleistein, N. 1966. Uniform asymptotic expansions of integrals with stationary point near an algebraic singularity, *Commun. Pure Appl. Math.* **19**, 353–370.

Bleistein, N. and Handelsman, R. A. 1975. *Asymptotic Expansions of Integrals*, New York: Holt, Rinehart and Winston.

Borwein, D., Borwein, J. M. and Chan, O.-Y. 2008. The evaluation of the Bessel functions via exp-arc integrals, *J. Math. Anal. Appl.* **341**, 478–500.

Borwein, D., Borwein, J. M. and Crandall, R. 2008. Effective Laguerre asymptotics, *SIAM J. Num. Anal.* **341**, 478–500.

Boyd, J. P. 1999. The Devil's invention: asymptotic, superasymptotic and hyperasymptotic series, *Acta Appl. Math.* **56**, 1–98.

Boyd, W. G. C. 1990. Stieltjes transforms and the Stokes phenomenon. *Proc. R. Soc. London* **A429**, 227–246.

236   *References*

Boyd, W. G. C. 1993. Error bounds for the method of steepest descents, *Proc. R. Soc. London* **A440**, 493–518.

Brillouin, L. 1916. Sur une méthode de calcul approchée de certaines intégrales, dite méthode de col, *Ann. Sci. Ec. Norm. Super.* **33**, 17–69.

Burwell, W. R. 1924. Asymptotic expansions of generalised hypergeometric functions, *Proc. London Math. Soc.* **53**, 599–611.

Byatt-Smith, J. G. B. 1998. Formulation and summation of hyperasymptotic expansions obtained from integrals, *Eur. J. Appl. Math.* **9**, 159–185.

Cauchy, A. L. 1829. Mémoire sur divers points d'analyse, *Mem. Acad. Fr.* **8**, 130–138; reprinted in *Oeuvres Complètes d'Augustin Cauchy*, série I, vol. 2, pp. 52–58, 1908, Paris: Gauthier-Villars.

Chester, C., Friedman, B. and Ursell, F. J. 1957. An extension of the method of steepest descents, *Proc. Cambridge Philos. Soc.* **53**, 599–611.

Connor, J. N. L. and Curtis, P. R. 1982. A method for the numerical evaluation of the oscillatory integrals associated with the cuspoid catastrophes: application to Pearcey's integral and its derivatives, *J. Phys. A* **15**, 1179–1190.

Copson, E. T. 1935. *Theory of Functions of a Complex Variable*, Oxford: Oxford University Press.

Copson, E. T. 1965. *Asymptotic Expansions*, Cambridge: Cambridge University Press.

De Bruijn, N. G. 1958. *Asymptotic Methods in Analysis*, New York: Dover.

Debye, P. 1909. Näherungsformeln für die Zylinderfunktionen für große Werte des Arguments und unbeschränkt veränderliche Werte des Index, *Math. Ann.* **67**, 535–558.

Dingle, R. B. 1973. *Asymptotic Expansions: Their Derivation and Interpretation*, London: Academic Press.

Écalle, J. 1981. *Les Fonctions Résurgentes* (3 volumes), Orsay: Publ. Math. Université Paris-Sud.

Edwards, H. M. 1974. *Riemann's Zeta Function*, New York: Academic Press.

Erdélyi, A. (ed.) 1953. *Higher Transcendental Functions*, vol. 1, New York: McGraw-Hill.

Erdélyi, A. 1956. *Asymptotic Expansions*, New York: Dover.

Faxén, H. 1921. Expansion in series of the integral $\int_y^\infty \exp\{-x(t \pm t^{-\mu})\}t^\nu dt$, *Ark. Mat. Astron. Fys.* **15**, 1–57.

Ferreira, C., López, J. L., Pagola, P. and Pérez Sinusía, E. 2007. The Laplace's and steepest descents methods revisited, *Int. Math. Forum* **2**, 297–314.

Franklin, J. and Friedman, B. 1957. A convergent asymptotic representation for integrals, *Proc. Cambridge Philos. Soc.* **53**, 612–619.

Gautschi, W., Harris, F. E. and Temme, N. M. 2003. Expansions of the exponential integral in incomplete gamma functions, *Appl. Math. Lett.* **16**, 1095–1099.

Gillespie, C. C. 1997. *Pierre-Simon Laplace, 1749–1827: A Life in Exact Science*, Princeton, NJ: Princeton University Press.

Gradshteyn, I. S. and Ryzhik, I. M. 1980. *Table of Integrals, Series and Products*, New York: Academic Press.

Greene, D. H. and Knuth, D. E. 1982. *Mathematics for the Analysis of Algorithms*, Boston, MA: Birkhauser.

Hadamard, J. 1908. Sur l'expression asymptotique de la fonction de Bessel, *Bull. Soc. Math. Fr.* **36**, 77–85. Reprinted in *Oeuvres de Jacques Hadamard*, vol. 1, pp. 365–373, 1968, Paris: CNRS.

Hadamard, J. 1912. Sur la série de Stirling, Fifth Int. Cong. Math., Cambridge, 1912. Reprinted in *Oeuvres de Jacques Hadamard*, vol. 1, pp. 375–377, 1968, Paris: CNRS.

Hardy, G. H. and Littlewood, J. E. 1920. Some problems of 'Partitio Numerorum' I: a new solution of Waring's Problem, *Göttinger Nachr. (Math. Phys. Kl.)* 13–17.

Hobson, E. W. 1925. *A Treatise on Plane Trigonometry*, Cambridge: Cambridge University Press.

Jeffreys, H. 1958. The remainder in Watson's lemma, *Proc. R. Soc. London* **A248**, 88–92.

Jeffreys, H. and Jeffreys, B. S. 1972. *Methods of Mathematical Physics*, Cambridge: Cambridge University Press.

Jones, D. S. 1972. Asymptotic behavior of integrals, *SIAM Rev.* **14**, 286–317.

Jones, D. S. 1997. *Introduction to Asymptotics: A Treatment using Nonstandard Analysis*, Singapore: World Scientific.

Kaminski, D. 1989. Asymptotic expansion of the Pearcey integral near the caustic, *SIAM J. Math. Anal.* **20**, 987–1005.

Kaminski, D. and Paris, R. B. 1999. On the zeroes of the Pearcey integral. *J. Comput. Appl. Math.* **107**, 31–52.

Kaminski, D. and Paris, R. B. 2003. Hadamard expansions for a Laplace integral with two nearly coincident poles, *Technical Report MS 03:03*, University of Abertay Dundee.

Kaminski, D. and Paris, R. B. 2004. On the use of Hadamard expansions in hyperasymptotic evaluation: differential equations of hypergeometric type, *Proc. R. Soc. Edinburgh* **134A**, 159–178.

Kelvin, Lord, 1903. The scientific work of Sir George Stokes, *Nature* **67**, 337–338.

Kibble, W. F. 1939. A Bessel function in terms of incomplete gamma functions, *J. Indian Math. Soc.* **3**, 271–294.

Kowalenko, V. 2001. Towards a theory of divergent series and its importance to asymptotics, *Recent Res. Dev. Phys.* **2**, 17–67.

Kowalenko, V. 2002. Exactification of the asymptotics for Bessel and Hankel functions, *Appl. Math. Comput.* **133**, 487–518.

Kruskal, M. D. and Segur, H. 1991. Asymptotics beyond all orders in a model of crystal growth, *Stud. Appl. Math.* **85**, 129–181.

Laplace, P. S. 1820. *Théorie Analytique des Probabilités*, 3rd edn., Paris: Courcier, reprinted in *Complete Works*, vol. 7, 1886, Paris: Gauthier-Villars.

Lauwerier, H. A. 1966. *Asymptotic Expansions*, Mathematical Centre Tracts 13, Amsterdam: Mathematisch Centrum.

López, J. L. 2007. The Stokes phenomenon as a boundary-value problem, *J. Phys. A: Math. Theor.* **40**, 10807–10812.

López, J. L., Pagola, P. and Pérez Sinusía, E. 2009. A simplification of Laplace's method: applications to the gamma function and Gauss hypergeometric function, *J. Approx. Theory* **161**, 280–291.

Muller, K. E. 2001. Computing the confluent hypergeometric function, $M(a, b, x)$, *Numer. Math.* **90**, 179–196.

Murphy, B. T. M. and Wood, A. D. 1997. Hyperasymptotic solutions of second-order differential equations with a singularity of arbitrary integer rank, *Methods Appl. Anal.* **3**, 250–260.

Nishikov, A. I. and Ritus, V. I. 1992. Stokes line width, *Teor. Mat. Fiz.* **92**, 24–40.

Olde Daalhuis, A. B. 1996. Hyperterminants I, *J. Comput. Appl. Math.* **76**, 255–264.

Olde Daalhuis, A. B. 1997. Hyperterminants II, *J. Comput. Appl. Math.* **89**, 87–95.

Olde Daalhuis, A. B. and Olver, F. W. J. 1995. Hyperasymptotic solutions of second-order linear differential equations I, *Methods Appl. Anal.* **2**, 173–197.

Olde Daalhuis, A. B., Chapman, S. J., King, J. R., Ockendon, J. R. and Tew, R. H. 1995. Stokes phenomenon and matched asymptotic expansions, *SIAM J. Appl. Math.* **55**, 1469–1483.

Olver, F. W. J. 1954. The asymptotic expansion of Bessel functions of large order, *Philos. Trans. R. Soc. London* **A247**, 328–368.

Olver, F. W. J. 1968. Error bounds for the Laplace approximation for definite integrals, *J. Approx. Theory* **1**, 293–313.

Olver, F. W. J. 1970. Why steepest descents?, *SIAM Rev.* **12**, 228–247.

Olver, F. W. J. 1990. On Stokes' phenomenon and converging factors, *Proc. Int. Conf. on Asymptotic and Computational Analysis, Winnipeg, Canada, 5–7 June 1989* (ed. R. Wong), pp. 329–355, New York: Marcel Dekker.

Olver, F. W. J. 1991a. Uniform, exponentially improved, asymptotic expansions for the confluent hypergeometric function and other integral transforms, *SIAM J. Math. Anal.* **22**, 1475–1489.

Olver, F. W. J. 1991b. Uniform, exponentially improved, asymptotic expansions for the generalized exponential integral, *SIAM J. Math. Anal.* **22**, 1460–1474.

Olver, F. W. J. 1992. Converging factors, in *Wave Asymptotics* (ed. P. A. Martin and G. R. Wickham), pp. 54–68, Cambridge: Cambridge University Press.

Olver, F. W. J. 1995. On an asymptotic expansion of a ratio of gamma functions, *Proc. R. Irish Acad.* **95A**, 5–9.

Olver, F. W. J. 1997. *Asymptotics and Special Functions*, New York: Academic Press, 1974. Reprinted Wellesley, MA: A. K. Peters.

Olver, F. W. J., Lozier, D. W., Boisvert, R. F. and Clark, C. W. 2010. *NIST Handbook of Mathematical Functions*, Cambridge: Cambridge University Press.

Paris, R. B. 1991. The asymptotic behaviour of Pearcey's integral for complex variables, *Proc. R. Soc. London* **A 432**, 391–426.

Paris, R. B. 1996. The mathematical work of G. G. Stokes, *Math. Today* **32**, 43–46.

Paris, R. B. 1997. The asymptotic expansion of Gordeyev's integral, *Z. Angew. Math. Phys.* **49**, 322–338.

Paris, R. B. 2000 The Hadamard expansion for $\log \Gamma(z)$, *Technical Report MS 00:03*, Division of Mathematical Sciences, University of Abertay Dundee.

Paris, R. B. 2001a On the use of Hadamard expansions in hyperasymptotic evaluation. I. Real variables, *Proc. R. Soc. London* **A457**, 2835–2853.

Paris, R. B. 2001b On the use of Hadamard expansions in hyperasymptotic evaluation. II. Complex variables, *Proc. R. Soc. London* **A457**, 2855–2869.

Paris, R. B. 2004a On the use of Hadamard expansions in hyperasymptotic evaluation of Laplace-type integrals. I. Real variable, *J. Comput. Appl. Math.* **167**, 293–319.

Paris, R. B. 2004b On the use of Hadamard expansions in hyperasymptotic evaluation of Laplace-type integrals. II. Complex variable, *J. Comput. Appl. Math.* **167**, 321–343.

Paris, R. B. 2004c Exactification of the method of steepest descents: the Bessel functions of large order and argument, *Proc. R. Soc. London* **460A**, 2737–2759.

Paris, R. B. 2005. The Stokes phenomenon associated with the Hurwitz zeta function $\zeta(s, a)$, *Proc. R. Soc. London* **A461**, 297–304.

Paris, R. B. 2007a. On the use of Hadamard expansions in hyperasymptotic evaluation of Laplace-type integrals. III. Clusters of saddle points, *J. Comput. Appl. Math.* **207**, 273–290.

Paris, R. B. 2007b. On the use of Hadamard expansions in hyperasymptotic evaluation of Laplace-type integrals. IV. Poles, *J. Comput. Appl. Math.* **206**, 454–472.

Paris, R. B. 2009. High-precision evaluation of the Bessel functions via Hadamard series, *J. Comput. Appl. Math.* **224**, 84–100.

Paris, R. B. 2010. Asymptotic expansion of n-dimensional Faxén-type integrals, *Eur. J. Pure Appl. Math.* **3**(6), 1006–1031.

Paris, R. B. and Kaminski, D. 2001. *Asymptotics and Mellin–Barnes Integrals*, Cambridge: Cambridge University Press.

Paris, R. B. and Kaminski, D. 2005. Hadamard expansions for integrals with saddles coalescing with an endpoint, *Appl. Math. Lett.* **18**, 1389–1395.

Paris, R. B. and Kaminski, D. 2006. Hyperasymptotic evaluation of the Pearcey integral via Hadamard expansions, *J. Comput. Appl. Math.* **190**, 437–452.

Paris, R. B. and Liakhovetski, G. V. 2000. Asymptotics of the multidimensional Faxén integral, *Frac. Calc. Appl. Anal.*, **3**, 63–73.

Paris, R. B. and Wood, A. D. 1986. *Asymptotics of High Order Differential Equations*, Pitman Research Notes in Mathematics, **129**, Harlow: Longman Scientific & Technical.

Paris, R. B. and Wood, A. D. 1992. Exponentially-improved asymptotics for the gamma function, *J. Comput. Appl. Math.* **41**, 135–143.

Paris, R. B. and Wood, A. D. 1995. Stokes phenomenon demystified, *IMA Bull.* **31**, 21–28.

Petrova, S. S. and Solov'ev, A. D. 1997. The origin of the method of steepest descent, *Hist. Math.* **24**, 361–375.

Poincaré, H. 1886. Sur les intégrales irrégulières des équations linéaires, *Acta Math.* **8**, 295–344.

Riemann, B. 1863. Sullo svolgimento del quoziente di due serie ipergeometriche in frazione continua infinita. In *Collected Works of Bernhard Riemann* (ed. H. Weber), pp. 424–430, 1953, New York: Dover.

Rosser, J. B. 1951. Transformations to speed the convergence of series, *J. Res. Nat. Bur. Stand.* **46**, 56–64.

Rosser, J. B. 1955. Explicit remainder terms for some asymptotic series, *J. Rational Mech. Anal.* **4**, 595–626.

Schmitt, J. P. M. 1974. The magnetoplasma dispersion function: some mathematical properties, *J. Plasma Phys.* **12**, 51–59.

Senouf, D. 1996. Asymptotic and numerical approximations of the zeros of Fourier integrals, *SIAM J. Math. Anal.* **27**, 1102–1128.

Shi, W. and Wong, R. 2009. Hyperasymptotic expansions of the modified Bessel function of the third kind of purely imaginary order, *Asymptotic Anal.* **63**, 101–123.

Stieltjes, T. J. 1886. Recherches sur quelques séries semi-convergentes, *Ann. Sci. Ec. Norm. Super.* **3**, 201–258. Reprinted in *Complete Works*, vol. 2, pp. 2–58, 1918, Groningen: Noordhoff.

Stix, T. H. 1962. *The Theory of Plasma Waves*, New York: McGraw-Hill.

Stokes, G. G. 1850. On the numerical calculation of a class of definite integrals and infinite series. In *Collected Mathematical and Physical Papers*, vol. 2, pp. 329–357, Cambridge: Cambridge University Press.

Stokes, G. G. 1857. On the discontinuity of arbitrary constants which appear in divergent developments. In *Collected Mathematical and Physical Papers*, vol. 4, pp. 79–109, Cambridge: Cambridge University Press.

Temme, N. M. 1994. Steepest descent paths for integrals defining the modified Bessel functions of imaginary order, *Methods Appl. Anal.* **1**, 14–24.

Temme, N. M. 1996. *Special Functions: An Introduction to the Classical Functions of Mathematical Physics*, New York: Wiley.

Tricomi, F. G. 1950. Sulla funzione gamma incompleta, *Ann. Mat.* **33**, 263–279.

Ursell, F. 1991. Integrals with a large parameter. A strong form of Watson's lemma, in *Elasticity, Mathematical Methods and Applications* (ed. G. Eason and R. W. Ogden), pp. 391–395, Chichester: Ellis-Horwood.

Van der Waerden, B. L. 1951. On the method of saddle points, *Appl. Sci. Res. Ser. B* **2**, 33–45.

Watson, G. N. 1918. Harmonic functions associated with the parabolic cylinder, *Proc. London Math. Soc.* **17**(2), 116–148.

Watson, G. N. 1952. *Treatise on the Theory of Bessel Functions*, Cambridge: Cambridge University Press.

Whittaker, E. T. and Watson, G. N. 1952. *A Course of Modern Analysis*, Cambridge: Cambridge University Press.

Wojdylo, J. 2006. On the coefficients that arise from Laplace's method, *J. Comput. Appl. Math.* **196**, 241–266.

Wong, R. 1980. Error bounds for asymptotic expansions of integrals, *SIAM Rev.* **22**, 401–435.

Wong, R. 1989. *Asymptotic Approximations of Integrals*, London: Academic Press.

Wrench, J. W. 1968. Concerning two series for the gamma function, *SIAM J. Math. Anal.* **22**, 617–626.

Wright, F. J. 1980. The Stokes set of the cusp diffraction catastrophe, *J. Phys. A* **13**, 2913–2928.

Wyman, M. and Wong, R. 1969. The asymptotic behaviour of $\mu(z, \beta, \alpha)$, *Can. J. Math.* **21**, 1013–1023.

Yang, S. and Srivastava, H. M. 2004. Some generalizations of the Hadamard expansion for the modified Bessel function, *Appl. Math. Lett.* **17**, 591–596.

Index